Methods in Cell Biology

VOLUME 91

Cilia: Structure and Motility

Series Editors

Leslie Wilson
Department of Molecular, Cellular and Developmental Biology
University of California
Santa Barbara, California

Paul Matsudaira
Department of Biological Sciences
National University of Singapore
Singapore

Methods in Cell Biology

VOLUME 91

Cilia: Structure and Motility

Edited by

Stephen M. King

Department of Molecular, Microbial and Structural Biology
University of Connecticut Health Center
Farmington, Connecticut

Gregory J. Pazour

Program in Molecular Medicine
University of Massachusetts Medical School
Worcester, Massachusetts

AMSTERDAM • BOSTON • HEIDELBERG • LONDON
NEW YORK • OXFORD • PARIS • SAN DIEGO
SAN FRANCISCO • SINGAPORE • SYDNEY • TOKYO
Academic Press is an imprint of Elsevier

Academic Press is an imprint of Elsevier
30 Corporate Drive, Suite 400, Burlington, MA 01803, USA
525 B Street, Suite 1900, San Diego, CA 92101-4495, USA
32, Jamestown Road, LondonNW1 7BY, UK
LinacreHouse, JordanHill, OxfordOX2 8DP, UK

First edition 2009

ISBN–13: 978-0-12-374973-4
ISSN: 0091-679X

For information on all Academic Press publications
visit our website at elsevierdirect.com

Printed and bound in United States of America
Transferred to Digital Printing, 2012

CONTENTS

CONTRIBUTORS

Numbers in parentheses indicate the pages on which the authors' contributions begin.

Karsten Boldt (143), Department of Protein Science, Helmholtz Zentrum München, 85764 Neuherberg, Germany

Stan A. Burgess (41), Astbury Centre for Structural Molecular Biology, Faculty of Biological Sciences, Institute of Molecular and Cellular Biology, University of Leeds, Leeds LS2 9JT, United Kingdom

Jacqueline S. Domire (111), Department of Pharmacology, Department of Internal Medicine Division of Human Genetics, College of Medicine, The Ohio State University, Columbus, Ohio 43210

John A. Follit (81), Program in Molecular Medicine, University of Massachusetts Medical School, Worcester, Massachusetts 01605

Kenneth W. Foster (173), Department of Physics, Syracuse University, Syracuse, New York 13244-1130

Stefan Geimer (63), Zellbiologie/Elektronenmikroskopie, Universität Bayreuth, 95440 Bayreuth, Germany

Christian Johannes Gloeckner (143), Department of Protein Science, Helmholtz Zentrum München, 85764 Neuherberg, Germany

Gregory Hendricks (81), Department of Cell Biology, University of Massachusetts Medical School, Worcester, Massachusetts 01655

Nobutaka Hirokawa (265), Department of Cell Biology and Anatomy, University of Tokyo, Graduate School of Medicine, 7-3-1 Hongo, Bunkyo-ku, Tokyo 113-0033, Japan

Hiroyuki Iwamoto (89), Research and Utilization Division, SPring-8, Japan Synchrotron Radiation Research Institute, Hyogo 679-5198, Japan

Shinji Kamimura (89), Department of Biological Sciences, Faculty of Science and Engineering, Chuo University, Kasuga 1-13-27, Bunkyo, Tokyo 112-8551, Japan

Ritsu Kamiya (241), Department of Biological Sciences, Graduate School of Science, University of Tokyo, Bunkyo-ku, Tokyo 113-0033, Japan

Karl-Ferdinand Lechtreck (255), Department of Cell Biology, University of Massachusetts Medical School, Worcester, Massachusetts 01655

Niki T. Loges (123), Department of Pediatrics and Adolescent Medicine, University Hospital Freiburg, Mathildenstrasse 1, 79106 Freiburg, and Klinik und Poliklinik für Kinder- und Jugendmedizin - Allgemeine Pädiatrie - Universitätsklinikum Münster, Albert-Schweitzer-Strasse 33, 48149 Münster, Germany

Kirk Mykytyn (111), Department of Pharmacology, Department of Internal Medicine Division of Human Genetics, College of Medicine, The Ohio State University, Columbus, Ohio 43210

Daniela Nicastro (1), Biology Department, Rosenstiel Center, MS029, Brandeis University, Waltham, Massachusetts 02454-9110

Shigenori Nonaka (287), Laboratory for Spatiotemporal Regulations, National Institute for Basic Biology, Nishigonaka 38, Myodaiji, Okazaki 444-8585 Aichi, Japan

Kazuhiro Oiwa (89), Kobe Advanced ICT Research Center, National Institute of Information and Communications Technology, 588-2 Iwaoka, Nishi-ku, Kobe 651-2492, Japan, and Graduate School of Life Science, University of Hyogo, Harima Science Park City, Hyogo 678-1297, Japan

Yasushi Okada (265), Department of Cell Biology and Anatomy, University of Tokyo, Graduate School of Medicine, 7-3-1 Hongo, Bunkyo-ku, Tokyo 113-0033, Japan

Heymut Omran (123), Department of Pediatrics and Adolescent Medicine, University Hospital Freiburg, Mathildenstrasse 1, 79106 Freiburg, and Klinik und Poliklinik für Kinder- und Jugendmedizin - Allgemeine Pädiatrie - Universitätsklinikum Münster, Albert-Schweitzer-Strasse 33, 48149 Münster, Germany

Gregory J. Pazour (81), Program in Molecular Medicine, University of Massachusetts Medical School, Worcester, Massachusetts 01605

Helle A. Praetorius (299), Department of Physiology and Biophysics, Aarhus University, 8000 Aarhus C, Denmark

Anthony J. Roberts* (41), Astbury Centre for Structural Molecular Biology, Faculty of Biological Sciences, Institute of Molecular and Cellular Biology, University of Leeds, Leeds LS2 9JT, United Kingdom

Ronald Roepman (143), Department of Human Genetics, Radboud University Nijmegen Medical Centre, 6500 HB Nijmegen, The Netherlands, and Nijmegen Centre for Molecular Life Sciences, Radboud University Nijmegen Medical Centre, 6500 HB Nijmegen, The Netherlands

Miho Sakato (161), Department of Molecular Biology and Biochemistry, Wesleyan University, Middletown, Connecticut 06459

Jovenal T. SanAgustin (81), Program in Molecular Medicine, University of Massachusetts Medical School, Worcester, Massachusetts 01605

Michael J. Sanderson (255), Department of Physiology, University of Massachusetts Medical School, Worcester, Massachusetts 01655

Marius Ueffing (143), Department of Protein Science, Helmholtz Zentrum München, 85764 Neuherberg, Germany, and Institute of Human Genetics, Klinikum rechts der Isar, Technical University of Munich, Munich 81675, Germany

Jeroen van Reeuwijk (143), Department of Human Genetics, Radboud University Nijmegen Medical Centre, 6500 HB Nijmegen, The Netherlands, and Nijmegen Centre for Molecular Life Sciences, Radboud University Nijmegen Medical Centre, 6500 HB Nijmegen, The Netherlands

George B. Witman (255), Department of Cell Biology, University of Massachusetts Medical School, Worcester, Massachusetts 01655

* Present address of Anthony J Robers: Department of Cell Biology, Harvard Medical School, Boston, Massachusetts 02115.

PREFACE

Cilia and flagella have long been the subject of intense study and a previous volume of *Methods in Cell Biology* dedicated to this organelle was published in 1995. However, in the 15 years since that publication, interest in the organelle has dramatically increased as it has come to be appreciated that these tiny structures play fundamental roles in the development and health of mammals and are vital for vertebrates to perceive their environment and respond to it. In humans the list of ciliary diseases, or ciliopathies, has grown tremendously since the publication of the previous volume. In 1995 the field recognized that cilia and flagella played critical roles in male fertility and respiratory disease and were recognized as being important in the determination of left–right asymmetry of vertebrates but the mechanism was not known. In addition, it was known that the senses of vision and smell depended on receptors localized to modified cilia. It is now appreciated that ciliary defects underlie a wide range of human diseases. These include polycystic kidney disease (PKD), nephronophthisis, Bardet–Biedl syndrome (BBS), Meckel–Gruber syndrome, Joubert syndrome, Jeune syndrome, and short rib-polydactyly syndrome that are thought to result from defects in primary cilia. Other diseases such as male infertility, hydrocephaly, juvenile myoclonic epilepsy, primary ciliary dyskinesia, Kartagener's syndrome, and left-right asymmetry defects of the heart are thought to result from defects in motile cilia. In addition, anosmia and blindness can derive from dysfunction of the highly specialized sensory cilia of the olfactory epithelium and retina. It is clear from studies in mouse that this collection of diseases is just the tip of the iceberg for ciliary disorders of man.

Eukaryotic cilia and flagella are complex organelles composed of hundreds of different proteins. This complexity likely reflects the diverse motility and sensory roles played by these organelles. The motility functions of cilia have long been recognized and in mammals these are important for moving mucus in the lungs, moving cerebrospinal fluid in the brain, and propelling the male gametes. The sensory functions are less well known but include roles in olfaction in the nose and light detection in the eye. In addition, nearly every cell type in vertebrate organisms is ciliated by nonmotile primary cilia that are thought to sense the extracellular environment. The proteins of the cilium are organized around a microtubule-based cytoskeleton termed the axoneme and a specialized domain of the plasma membrane that covers the axoneme. The ciliary membrane is contiguous with the plasma membrane of the cell but is a separate domain containing a unique set of proteins, many of which play roles in sensory perception. The axonemes of motile cilia typically have a $9+2$ arrangement of microtubules while nonmotile sensory and primary cilia typically have a $9 + 0$ arrangement. These microtubules serve as scaffolding to bind and organize the multitude of proteins needed to carry out the motility and sensory functions of cilia. The microtubules of the axoneme are templated from a centriole at the center of the centrosome. When the cell is ciliated, the centriole (which is now called a basal body) and centrosome remain at the base of the cilium. The centrosome is best known for its role in organizing the cytoskeleton and also is postulated to be an important control center of the cell, integrating signals that regulate morphology, migration, and proliferation.

With the explosion of interest in cilia, the model organisms available to study cilia and flagella have grown much more diverse, and the techniques available for assessing cilia structure and function have become more sophisticated. In these three volumes, we have asked top researchers in the field to provide methods used in their laboratories to study cilia and flagella. *Cilia: Structure and Motility,* Volume 91, focuses on general methods to study these organelles covering microscopic techniques for both structural analysis and detailing motility parameters, as well as biochemical approaches to define protein–protein associations and complexes. *Cilia: Motors and Regulation*, Volume 92, focuses on techniques for studying dynein structure and function and the varied mechanisms by which these motor complexes are regulated. *Cilia: Model Organisms and Intraflagellar Transport*, Volume 93, focuses on the methods for studying intraflagellar transport which is required for assembly of the organelle and provides general approaches for studying this and other cilia-related phenomena in all of the major model organisms that are currently being used to study cilia and flagella.

CHAPTER 1

Cryo-Electron Microscope Tomography to Study Axonemal Organization

Daniela Nicastro

Biology Department, Rosenstiel Center, MS029, Brandeis University, Waltham, Massachusetts 02454-9110

Abstract

Cilia and flagella are important organelles that perform both motile and sensory functions. For more than half a century, electron microscopy has provided crucial insights into the fundamental architecture and function of these organelles, such as the characteristic [9 + 2] microtubule arrangement of the axoneme or the dynein-driven microtubule sliding as the basis of motility. However, we are just starting to explore the

978-0-12-374973-4
DOI: 10.1016/S0091-679X(08)91001-3

molecular organization and mechanisms that drive and regulate axonemal bending. Recently, electron tomography (ET) of rapidly frozen, that is, life-like preserved specimen, has emerged as a cutting-edge technique that provides three-dimensional (3D) views of cellular structures. Cryo-ET and subtomogram averaging has provided high-resolution 3D images of intact flagella and axonemes, allowing us to discover new structures and gain a better understanding of their molecular organization. This chapter provides an overview of the principles of cryo-preservation, ET, and tomographic averaging, and it highlights both strengths and limitations of combining these methods to study axonemal organization. The chapter gives a comprehensive overview of the major technical steps involved in cryo-ET and 3D averaging, and explains successful strategies to generate structural data of the axoneme with 3 to 4 nm resolution. Basic equipment requirements, available software packages and how to use them, as well as common problems, artifacts and future challenges are discussed. The chapter is addressed to both scientists who already use or consider using cryo-tomography of cilia and flagella, as well as researchers who would like to learn more about the process and how to "read" these new 3D images.

I. Introduction and Rational

A. Introduction to Cilia and Flagella, the Axoneme, and Dynein

Cilia and flagella are highly conserved and important eukaryotic organelles that perform both motile and sensory functions in a wide variety of species and cell types. In humans, the normal function of several organs requires the activity of cilia (Snell *et al.*, 2004), and various diseases are associated with ciliary malfunction. Defects in ciliary motility, or their assembly and sensory functions, have been implicated in human genetic diseases, including polycystic kidney disease, Bardet-Biedl syndrome, and primary ciliary dyskinesia (Fliegauf *et al.*, 2007; Gerdes *et al.*, 2009).

The motion of cilia and flagella has fascinated cell biologists for more than 150 years, and thus it is not surprising that these organelles were among the earliest biological samples studied when electron microscopy became available (Faucett, 1981). The typical [9 + 2] arrangement of "fibrils" in the axoneme, the microtubule-based core of cilia and flagella, was described in the early 1950s (Fawcett and Porter, 1954; Manton *et al.*, 1952), and in the 1960s the nature of the nine peripheral "fibrils" was established as microtubule doublets (Afzelius, 1959; Gibbons and Grimstone, 1960; Pease, 1963). The importance of ATP for ciliary and flagellar movement was known early on (Gibbons, 1963), and biochemical dissection of axonemes led to the discovery of proteins with ATPase activity, called dyneins, which were directly linked to the presence of "arms" attached to the doublets and the capability of axonemes to move (Gibbons and Rowe, 1965).

Originally a "contractile mechanism" for the motility of cilia and flagella was favored, until electron micrographs of the tips of cilia in different phases of their beat cycle contradicted a shortening of the doublets on the concave side of the bend, laying the foundation for the sliding microtubule theory (Satir, 1968). This theory was then supported with more direct evidence by sliding disintegration experiments, in which after brief trypsination of isolated axonemes and upon addition of ATP the microtubule doublets slide by each other so that the axoneme extended on both ends like a telescope (Summers and Gibbons, 1971). The interdoublet sliding motions caused by dyneins are thought to be converted into bending by constraints on this sliding that are imposed by the interdoublet nexin links and the radial spokes [for reviews see Mitchell (1994); Porter (1996)]. In sliding disintegration experiments the trypsin treatment seems to digest these restrictions, uncoupling the sliding of the doublets from axonemal bending. For an axoneme to generate the complex motions typical of beating cilia and flagella, dynein's action must be controlled both around the circumference and along the length of the axoneme. Key complexes that are thought to be involved in dynein regulation and coordination are the central pair complex, the radial spokes, the dynein regulatory complex (DRC), the nexin link and the I1 inner-arm dynein intermediate-light chain complex (King and Kamiya, 2009; Mitchell, 2009; Wirschell *et al.*, 2009; Yang and Smith, 2009) (Fig. 1).

The axonemal dyneins are organized in two rows along each microtubule doublet: the outer dynein arms repeat every 24 nm along the doublets, whereas the inner arms are arranged in complex groups within the 96 nm axonemal repeat. Dyneins are large minus-end directed microtubule motors that convert chemical energy derived from ATP hydrolysis into mechanical force (Gibbons, 1981; Gibbons and Rowe, 1965; Sale and Satir, 1977; Satir, 1984). In addition to their crucial role in the motility of cilia and flagella, cytoplasmic dynein has a major impact on cell behavior, including cell division, signaling, retrograde transport, cell shape, and polarized cell growth (Vallee *et al.*, 2004). Dyneins are strikingly different from the other cytoskeletal motors, kinesin and myosin (Hackney, 1996). All dyneins are complexes of multiple proteins referred to as heavy, intermediate, and light chains (Pfister *et al.*, 2006; Porter, 1996; Vallee *et al.*, 2004). They are built around 1–3 heavy chains, which consist of three main domains: the ring-shaped head domain containing 6 AAA- and a C-terminal domain, the cargo-binding N-terminal tail, and a long stalk that emerges between AAA-domains 4 and 5, and binds to the "track"-microtubule in an ATP-sensitive manner via a small globular microtubule-binding domain at its tip (Mizuno *et al.*, 2004). In contrast to kinesin and myosin, the MT-binding domain is well separated (~20 nm) from the site of ATP hydrolysis in AAA-domain 1 (Burgess *et al.*, 2003). It has been shown that the cargo-binding tail is important for dynein motility (Shima *et al.*, 2006), and mechanistic models about dynein's action and mechano-chemical cycle have been proposed (Burgess *et al.*, 2003; Mallik *et al.*, 2004; Reck-Peterson *et al.*, 2006; Roberts *et al.*, 2009; Ross *et al.*, 2006; Sakato and King, 2004; Samso and Koonce, 2004; Toba *et al.*, 2006); however, we need detailed and comprehensive structural information of dynein under native conditions and in different nucleotide states to be able to test these models.

B. Introduction to Electron Microscopy

Electron microscopy (EM) has been an essential technique in characterizing the structure of cilia and flagella, the axoneme, and dynein. In an electron microscope,

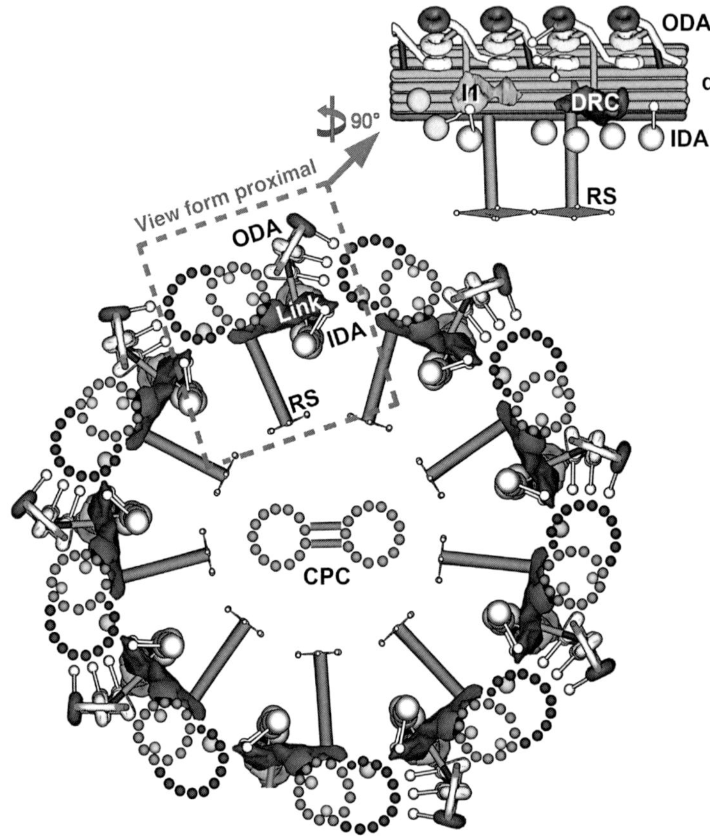

Fig. 1 Schematic models showing the general organization of cilia and flagella. The simplified scheme in the *center* shows a cross-sectional view of an axoneme with nine outer microtubule doublets surrounding the central pair complex (*CPC*); the viewing direction is from the flagellar base (*proximal*). Three structures connect neighboring doublets: the outer and inner dynein arms (*ODA*, *IDA*), and the nexin *link* (also called circumferential link). One doublet is boxed and shown 90° rotated at the *top*; here, one 96 nm repeat unit from one microtubule doublet is shown in longitudinal view as seen from a neighboring doublet. Two central complexes have been identified that regulate dynein activity: the intermediate/light-chain complex of the *I1* dynein close to the proximal radial spoke (*RS*), and the dynein regulatory complex (*DRC*) near the distal (*d*) radial spoke. (Images are modified from Nicastro *et al.*, 2006).

images are generated when the electron beam interacts with the specimen. Electrons have a much smaller wavelength than visible light and, therefore, the resolution that can be achieved in EM images is significantly better than with light microscopy. However, electron microscopes need to be operated under vacuum to increase the mean free path of the electrons, that is, to allow the electrons to travel through the microscope only scattered by the specimen that should be imaged. For biological material to withstand both the high vacuum and the aggressive electron beam they have to be fixed, that is, EM images usually represent only static snapshots of cellular life. The preparation of tissues, cellular and molecular specimens for EM is critically important to preserve the cellular structures as close to the native state as possible, and to avoid distortions or artifacts, which would irreversibly misrepresent the structure in the final micrographs. Therefore, the development and refinement of EM specimen preparation techniques have been the focus of many studies (Dubochet *et al.*, 1988; Moor *et al.*, 1980; Sabatini *et al.*, 1964) (see also Fig. 2A).

In "conventional" EM the specimen is chemically fixed (e.g., with aldehydes), dehydrated and stained with heavy metals (e.g., osmium tetroxide and uranyl acetate), resin-embedded, sectioned, and poststained before observing the sections in the EM at room temperature. The metal stains strongly scatter electrons and generate well-contrasted EM images that have been invaluable for visualizing the basic [9 + 2] arrangement of microtubules and associated components of the axoneme (Fig. 1) (Gibbons, 1981; Mitchell, 1994, 2000; Porter, 1996; Porter and Sale, 2000; Satir, 1968; Tyler, 1949; Warner, 1976). This method has also successfully been used to compare wild-type and mutant axonemes from *Chlamydomonas* in 2D difference maps (Gardner *et al.*, 1994; Mastronarde *et al.*, 1992; Perrone *et al.*, 2000). Although EM of metal replicas of freeze-fractured and deep-etched cilia and flagella can only reveal "surfaces," this technique has provided some of the most informative images of axoneme architecture in the early days of structural studies (e.g., Burgess *et al.*, 1991; Goodenough and Heuser, 1985; Lupetti *et al.*, 2005). The quick-frozen specimens are fractured, and some of the water is allowed to freeze-sublimate in the vacuum before a metal replica is deposited either from a rotating source or by unidirectional shadowing. The metal atoms again strongly scatter electrons, providing high contrast in the micrographs.

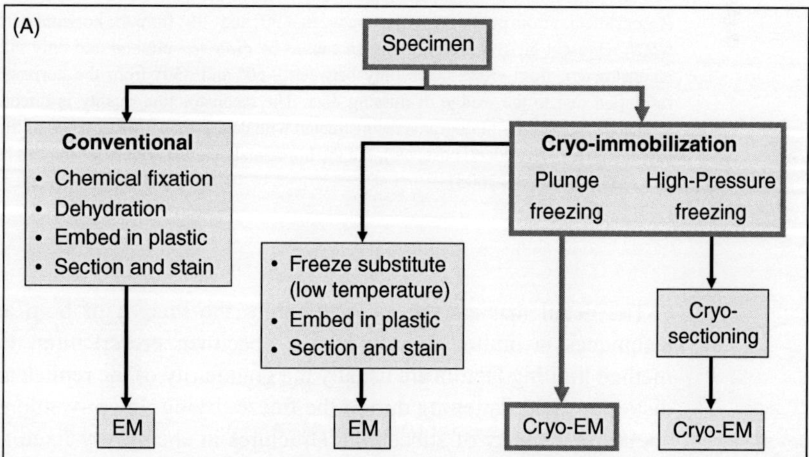

Fig. 2 Flowchart of specimen preparation in EM and principle of electron tomography. (A) Flowchart showing key steps for sample preparation in EM (Refer parts B–G on next page).

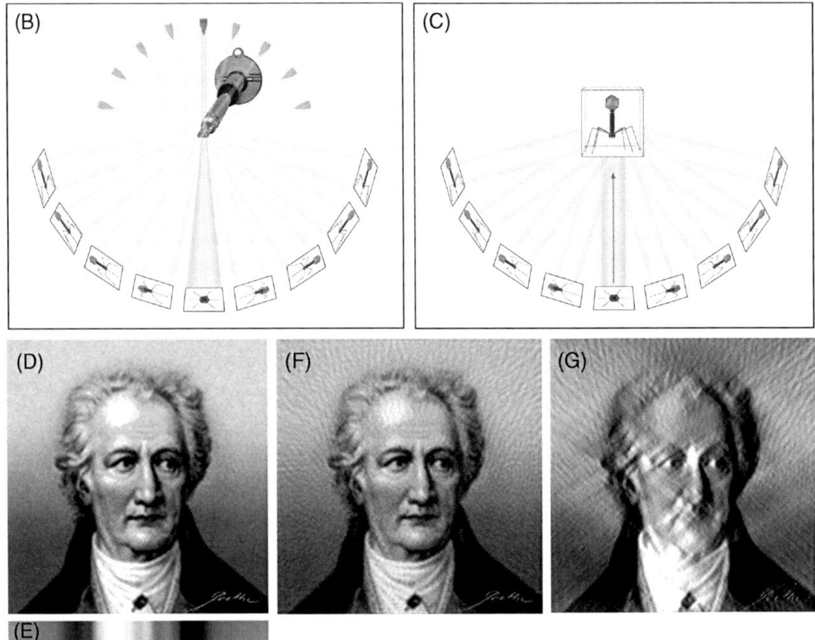

Fig. 2 (*Continued*) (B–G) Principle of electron tomography. (B) A set of projection images with different viewing directions is recorded from a biological specimen (here a bacteriophage) that is mounted in an EM sample holder by tilting the holder in the microscope. (C) To compute a 3D reconstruction of the original structure each tilted view is projected back into a common volume at the same angle that it was recorded. (D) A 2D image of the face of Goethe. (E) 1D projection of the 2D object in (D) generated by the summation of all of the brightness in the 2D picture along a set of vertical lines. (F–G) Two reconstructions of Goethe's face achieved by back-projecting 1D projection images (as in E) that were generated by tilting with 2° angular increments. (F) Reconstruction from projections taken between +90° and –90° from the horizontal. Note the ripples in the image which represent the resolution limitation caused by discrete sampling and only 90 projections images. (G) A reconstruction from views taken only between +60° and –60° from the horizontal shows a further loss in resolution due to the wedge of missing data. The reconstruction quality is directionally degraded, which is typical for single-axis tomograms reconstructed with data from a limited range of tilt. Note that vertical detail is still sharp (e.g., shoulders, nose, ear), but the horizontal detail is poor (e.g., mouth) [Images (B–G) are modified from McIntosh *et al.*, 2005].

The detail that can be extracted from the images of both above-mentioned EM techniques is limited mainly by the specimen preservation. For the metal replica method limiting factors are usually the granularity of the replica and possible structural distortions and flattening during the freeze-drying. In conventional EM concerns exist about the integrity of subcellular structures in chemically fixed samples, the selective stain deposition, the fact that the images are not collected of the biological structures themselves but of the stain, and about possible aggregation of "floating" cellular components upon dehydration (Kistler and Kellenberger, 1977), which together is believed to limit the achievable resolution to about 5 nm (reviewed in McIntosh *et al.* 2005). The aggregation effect becomes evident by comparing chemical fixed with high-pressure-frozen/freeze-substituted or cryo-specimen of loose or network-like structures that compact due to dehydration and chemical fixation, such as the kinetochore (McEwen *et al.*, 1998), chromatin, intermediate filaments, desmosomes (Al-Amoudi *et al.*, 2004), and the outer dynein arms in axonemes (Nicastro *et al.*, 2005, 2006). Alternative preservation methods based on cryo-immobilization by rapid freezing (Fig. 2A) were developed in the 1980s (Adrian *et al.*, 1984; Dubochet and McDowall, 1981; Dubochet *et al.*, 1988; Gilkey and Staehelin, 1986; Moor *et al.*, 1980; Roos *et al.*, 1990; Taylor and Glaeser, 1974, 1976) and used in the cilia and flagella field for the above-mentioned freeze-fracture metal replica method and isolated attempts at cryo-EM, for example, of plunge frozen sea urchin sperm flagella (Murray,1986).

The combination of live cell imaging, FRET, biochemical fractionation, *in vitro* motility assays, structural studies, and computer-based simulations has refined our ideas about the organization and function of axonemes and dynein. Proteomic studies of the flagella of the green alga and model organism *Chlamydomonas* have shown that these complex organelles comprise over 650 different proteins (Pazour *et al.*, 2005; see also http://labs.umassmed.edu/chlamyfp); however, to date we know little to nothing about the majority of these proteins, for example, where they locate or how they function. Despite all the excellent work that has already been done, the complexity of axonemes and the large size of the dynein motor proteins (a single dynein heavy chain is >500 kDa) have made it difficult to elucidate the details of the molecular mechanisms that underlie ciliary and flagellar beating, and dynein's motion. To fully understand the functional and regulatory interactions of all the players in axoneme motility, one will need the three-dimensional (3D) structure of the organelle in its native state and at a resolution that is sufficient both to identify and localize its macromolecular components and to characterize the structural changes they undergo during their functional cycles.

Over the past decade, electron microscope tomography (ET) of rapidly frozen specimens in combination with image processing has emerged as a cutting-edge imaging technique of macromolecular complexes, organelles, viruses, and intact cells (Medalia *et al.*, 2002; Nicastro *et al.*, 2000; for review see also Lucic *et al.* 2005). Cryo-ET has also successfully been used for studying isolated microtubule doublets (Sui and Downing, 2006), isolated axonemes from *Chlamydomonas* (Bui *et al.*, 2008; Heuser *et al.*, in press; Ishikawa *et al.*, 2007; Nicastro *et al.*, 2006), and intact sea urchin sperm flagella (Nicastro *et al.*, 2005, 2006). Therefore in this chapter, we will focus on describing the strengths and limitations of cryo-preservation and ET, and provide a comprehensive overview of the technical steps involved in cryo-ET of axonemes to generate structural data with better than 4 nm resolution.

C. Introduction to Cryo-Preservation

In the life sciences, EM of samples prepared by rapid freezing has been particularly successful due to the outstanding structural preservation in a near-to-native state, and the good time resolution of dynamic cellular processes (Adrian *et al.*, 1984; Dubochet *et al.*, 1988; Echlin, 1991; Fernandez-Moran, 1960; Gilkey and Staehelin, 1986; Roos *et al.*, 1990; Taylor and Glaeser, 1974, 1976). Biological samples inherently contain high amounts of water, therefore, obtaining well-frozen specimen that allows a reliable description of their ultrastructure, requires very high cooling rates to cryo-immobilize the sample in vitreous (amorphous) ice within milliseconds and without ice crystal formation that would disrupt the structure (Bruggeller and Mayer, 1980; Dubochet and McDowall, 1981; Dubochet *et al.*, 1988; Stewart and Vigers, 1986). The requirement for fast cooling rates limits the suitable specimen thickness, which in practice varies for different types of samples, for example, depending on the water content or use of cryoprotectants; in general, plunge freezing into liquid ethane or propane (Dubochet *et al.*, 1988; Templeton *et al.*, 1997) will yield good preservation of samples that are a few micrometers thick (5–10 μm), while high-pressure freezing (Moor *et al.*, 1980) extends this size by up to two orders of magnitude (~300–600 μm) (Gilkey and Staehelin, 1986; McIntosh *et al.*, 2005).

In the absence of any staining or dehydration, that is, with all the water present as immobilized, amorphous ice, cryo-EM of frozen-hydrated samples provides a faithful, high-time resolution "snapshot" of physiological conditions, which is essential for structural work that will improve our understanding, for example, of the interactions among molecules in a cell. Once vitrified, the sample must be kept below the devitrification temperature of about −130 to −140°C at all times. In practice, the observation of frozen-hydrated material in the EM is performed at liquid nitrogen temperature. At this temperature, water practically does not evaporate in the vacuum and radiation damage of the sample is reduced.

The pristine structural preservation of frozen-hydrated specimen, however, comes at a price: (1) the radiation sensitivity and (2) the relatively low electron contrast of unfixed and unstained samples. Radiation sensitivity is an important issue, as this means that the vitreous samples must be imaged under low electron dose conditions, which in turn results in images with a low signal-to-noise ratio (SNR), limiting both the resolution due to poor image statistics and the suitable sample thickness (McIntosh *et al.*, 2005). Cryo-EM images arise from the direct interaction of electrons with the specimen, that is, the real distribution of the biological material within the thickness of the specimen is visualized. However, in contrast to conventional EM where electrons scattered by the stain are blocked outside the objective aperture forming strong amplitude contrast, the scattering of electrons by native biological material is much lower and would not reveal many details in the sample. Therefore, in cryo-EM we use appropriate defocusing to cause a small phase shift between the scattered and the nonscattered electron waves, generating phase contrast in the resulting images (Dubochet *et al.*, 1988). Even though cryo-EM has become more widely available and a standard tool in structural cell biology, there are some special equipment requirements that are not inexpensive: special cryo-sample holders to maintain the specimen at a low temperature during EM observation, a stable specimen stage to avoid sample drift, coherent electron beam, the capability to image the radiation-sensitive samples under low-dose conditions, high vacuum quality, and anti-contamination blades (or box).

D. Electron Tomography and Volume Averaging

The images recorded by EM are two-dimensional (2D) projections of the sample along the beam axis, because the depth of focus of the imaging lens is large compared to the sample thickness. Biological structures, however, are intrinsically 3D and the superimposition of the 3D density of the specimen into 2D images discards valuable information and restricts resolution. This is not just an isolated problem in EM and one of the most flexible and successful techniques to solve the issue is tomography, which is based on Johann Radon's theory of projection (Frank, 1992). In biology this method was implemented by Aaron Klug for single particle reconstructions of the 3D structure of viruses (Derosier and Klug, 1968), and by Godfrey Hounsfield (1973) and Allan Cormack (Shampo and Kyle, 1996) for the medical applications of computed axial tomography (CAT-scan). The radiation sources might be different in these examples, but in all methods projection images recorded from different viewing angles are combined to generate an accurate 3D image (Fig. 2B and C). In ET the specimen is tilted inside the microscope around the specimen holder axis with angular increments between 1 and 4°, a 2D image is recorded at each angle, and from this tilt series the tomographic reconstruction is calculated using different algorithms (Fig. 2B–G), such as weighted back-projection (Radermacher *et al.*, 1986), or iterative methods like SIRT (Penczek *et al.*, 1992) and ART (Marabini *et al.*, 1998). The virtue of ET is that it allows the reconstruction and analysis of the 3D information of unique and polymorphic biological structures in a noninvasive manner, that is, one can rotate or dissect the resulting 3D information voxel-by-voxel in the computer at high resolution without having to slice the object physically (McIntosh *et al.*, 2005).

ET is by no means a new technique, but its practical application to frozen-hydrated specimen, to take advantage of the near-to-native structure preservation, was not trivial as one has to record many images (usually ~100) of a radiation-sensitive sample. Theoretically the cumulative electron dose that a specimen can tolerate before radiation damage can be detected at a certain resolution, can be fractionated out over any number of images, and the resulting resolution of the 3D reconstruction should approach the resolution of a single 2D image if this dose had been applied all at once (Hoppe *et al.*, 1974; McEwen *et al.*, 1995). In practice, however, the dose per image needs to be high enough so that the SNR still allows an accurate tilt-series alignment before the reconstruction. Several technical advances were key for making cryo-ET a successful imaging technique: increased computational capacity and software developments, for example, for automated low-dose tilt-series acquisition, electron guns with improved coherence, sensitive CCD cameras, and zero-loss energy-filtering to improve image quality by removing inelastically scattered electrons (Koster *et al.*, 1997; Lucic *et al.*, 2005; McIntosh *et al.*, 2005).

Over the last decade, cryo-ET in combination with image processing has emerged as the imaging method of choice for many biological specimens as it can provide valuable 3D views of the molecular structure of native complexes, organelles, and cells, with resolutions ranging from ~3 to 12 nm (e.g., Borgnia *et al.*, 2008; Bostina *et al.*, 2007; Briegel *et al.*, 2008; Butan *et al.*, 2008; Cardone *et al.*, 2007; Cheng *et al.*, 2007; Cyrklaff *et al.*, 2007; Grunewald *et al.*, 2003; Henderson *et al.*, 2007; Iancu *et al.*, 2007; Izard *et al.*, 2008; Kurner *et al.*, 2005; Medalia *et al.*, 2002; Murphy *et al.*, 2006; Nicastro *et al.*, 2000; Rouiller *et al.*, 2008; Wright *et al.*, 2007). This cutting-edge technique has now also provided unprecedented detail of isolated microtubule doublets (Sui and Downing, 2006), isolated axonemes from *Chlamydomonas* (Bui *et al.*, 2008; Heuser *et al.*, in press; Ishikawa *et al.*, 2007; Nicastro *et al.*, 2006) (Fig. 3), and intact sea urchin sperm flagella

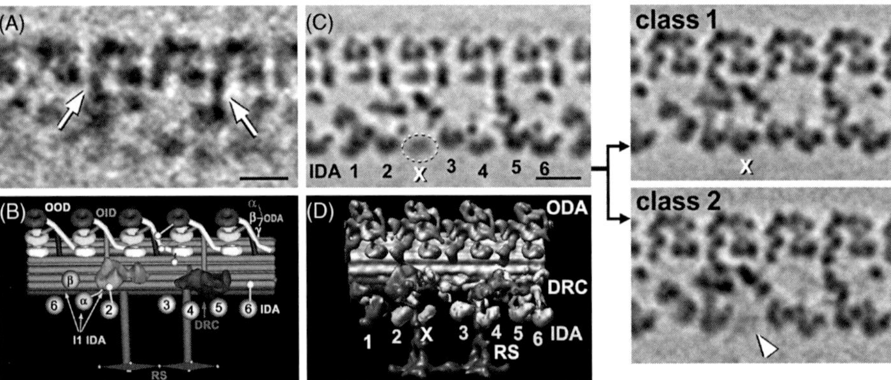

Fig. 3 Cryo-electron tomography and averaging of subtomographic volumes provide new insights into the organization of the 96 nm axonemal repeat of *Chlamydomonas* flagella. (A, B) Shown are a tomographic slice (A) and a graphical model (B) in longitudinal orientation of the first published cryo-tomographic average of the 96 nm repeat unit of vitrified, intact axonemes [images modified from Nicastro *et al.* (2006)]. The improved resolution (~4.3 nm) allowed the discovery of several novel structures, including the two outer–inner dynein (OID) links (*orange arrows*) and the outer–outer dynein (*OOD*) links that could both be important for fast signal transduction in beating flagella (Nicastro *et al.*, 2006). (C, D) Shown are—also in longitudinal orientation—a tomographic slice (**C**) and a surface-rendering representation (D) of our improved averages of the 96 nm axonemal repeat with ~3.2 nm resolution. Over the past 3 years we have made advancements in almost every step of the procedures involved in cryo-ET and 3D correlation averaging, which has led to this "leap" in resolution (see also Heuser *et al.*, in press). The new averages have a greatly increased signal-to-noise ratio compared to the 2006 data, and showed an additional, much weaker density located within the inner dynein arm row between IDA 2 and IDA 3 [*red dotted circle* in (C) and marked with an *X* in (C) and (D)]. We separated the total pool of subtomographic volumes (720 particles) into two classes based on the presence or absence of density at the position of the weak density (*X*); we then computed class-averages, whereby *class 1* clearly contains an additional dynein at position IDA-*X*, while this dynein arm is missing in *class 2* (*red arrow*). Particles from the two classes come from different microtubule doublets, indicating that this dynein is a doublet-specific feature. This example also highlights one of the future challenges: heterogeneity, and the need for larger data sets and sophisticated classification tools, so that we do not loose potentially important variances during the averaging process (see *Discussion*). Other labels: *DRC*, dynein regulatory complex; *IDA*, inner dynein arms (*1–6*); *ODA*, outer dynein arms; *RS*, radial spokes. Scale bars: 20 nm. (See Plate no. 1 in the Color Plate Section.)

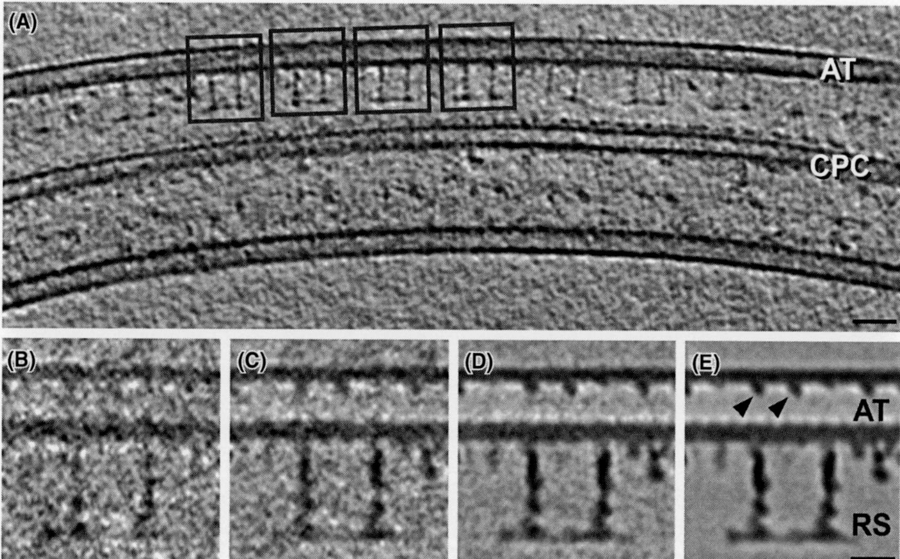

Fig. 4 Quality improvements by averaging subtomographic volumes. (A) An ~10 nm thick, longitudinal slice through a tomogram of a frozen hydrated axoneme of *Chlamydomonas* shows parts of three microtubules; the microtubule in the center of the image is part of the central pair complex (*CPC*), whereas the microtubules at the top and bottom of the image are part of two microtubule doublets on opposite sides of the axoneme. The top microtubule can clearly be identified as the A-tubule (*AT*) of the doublet by the attached radial spokes (*RS*). Four 96 nm repeat units have been highlighted by *boxes*. (B–E) Each of the four images shows a 1 nm thick tomographic slice of a tomographic average of the 96 nm axonemal repeat units. The slice shows a longitudinally oriented A-tubule (*AT*) with attached radial spokes (*RS*). The images show the improvement in SNR and resolution due to increasing numbers of particles included in the tomographic averages, that is, averaged are 2 (B), 10 (C), 80 (D), and 720 repeat units (E), respectively. Note the increasing level of axonemal details, including the repeating Microtubule Inner Proteins (MIP2) that are attached to the inner side of the microtubule wall (*arrowheads*).

(McEwen *et al.*, 2002; Nicastro *et al.*, 2005, 2006) (Fig. 6A and B). Using cryo-ET we discovered several novel structures in axonemes, such as the outer–inner dynein (OID) linker, the outer–outer dynein (OOD) linkers (Fig. 3), and the microtubule inner proteins (MIPs) inside doublet microtubules (arrowheads in Fig. 4E) (Nicastro *et al.*, 2006).

Cryo-ET has also limitations: those that apply to cryo-EM in general include: (1) the intrinsically low image contrast and (2) the radiation sensitivity of frozen-hydrated specimen; the latter means the requirement for relatively thin specimen (up to ~500-m thickness) and low dose imaging, which leads to low SNR in the resulting tomograms; issues that are specific to ET include: (1) the tilt-series alignment accuracy and (2) a restricted tilt-angle range, causing missing information in the shape of a "missing wedge" in the Fourier space (Fig. 2G). Approaches for reducing the missing information are dual-axis ET (Mastronarde, 1997; Penczek *et al.*, 1995), that is, the acquisition and combination of data from two perpendicular tilt axes, which is routinely used in plastic section ET, but is less useful in cryo-ET due to the dose limitation, or conical tomography (Lanzavecchia *et al.*, 2005).

Overall the poor image statistics, that is, the low SNR, is a key issue, as the noise obscures the intrinsic resolution and thus valuable structural information in the cryo-tomograms. If, however, the structure of interest is repetitive and present in multiple copies in one or many tomograms (Fig. 4A), the SNR—and thus the resolution—can in principle be improved dramatically by averaging (Fig. 4B–E). This improvement is only achieved by averaging features that are supposedly alike, because then the random noise gets suppressed while the signal contributions are maintained (Heymann *et al.*, 2008; Nicastro *et al.*, 2006). Three-dimensional correlation averaging of tomographic sub-volumes builds on techniques used for single-particle reconstruction (Frank, 1975), with the primary difference that alignment and voxel estimation occur over the 3D tomographic volume instead of 2D projection images (Bartesaghi *et al.*, 2008; Forster *et al.*, 2008; Nicastro *et al.*, 2006; Walz *et al.*, 1997). The advantage of combining cryo-ET with volume averaging over conventional single-particle reconstruction, which usually achieves higher resolutions (0.5–3 nm), is that the structure of interest can be studied *in situ* instead of having to be removed from the cellular context. This approach also allows overcoming the "missing wedge" problem, if 3D images with different orientations to the tilt axis can be combined in the average. Axonemes are excellent specimens for cryo-ET combined with tomographic averaging, thanks to their relatively small diameter (~220 nm), the highly ordered arrangement of microtubules, and repetitive distribution of associated protein complexes in the 96 nm repeats (Fig. 4A). Combining the 96 nm axonemal repeats from all microtubule doublets in a tomogram allows for automatic compensation of the missing wedge (Nicastro *et al.*, 2006).

The combination of these technologies has defined the field of modern EM and structural research of intact axonemes, which is now providing high-resolution 3D views of these remarkable organelles and contributing significantly to our understanding of their molecular organization (Fig. 3) (Bui *et al.*, 2008; Nicastro *et al.*, 2005, 2006). Therefore, this chapter provides an overview of the strengths and limitations of cryo-ET of cilia, flagella, and isolated axonemes, as well as the major technical steps involved; as such the chapter is addressed to both scientists who already use or consider using this method as well as researchers who are confronted with cryo-ET data, and would like to learn more about the process and how to "read" these 3D images.

II. Methods and Materials

The first cryo-electron tomograms of sea urchin sperm axonemes were published by McEwen and colleagues (McEwen *et al.*, 2002), but the axonemes were severely flattened within the ice layer and the resolution of 8–10 nm was not sufficient to reveal new details. A few years later we imaged intact sea urchin sperm flagella (Nicastro *et al.*, 2005, 2006) and *Chlamydomonas* axonemes from wild-type and an inner dynein arm mutant (Nicastro *et al.*, 2006) embedded in thicker ice to avoid any significant distortions, and applied first linear averaging along the flagellum axis (Nicastro *et al.*, 2005) and finally 3D correlation averaging of all 96 nm repeat units in the tomograms (Fig. 3A and B) (Nicastro *et al.*, 2006). The latter study pushed the resolution to ~4.3 nm, providing new details previously not seen in conventional EM studies, and therefore it opened a new window into axonemal structure. Since then, the Ishikawa group (Bui *et al.*, 2008) has published tomographic averages of axonemes from several *Chlamydomonas* mutants with 3.8 to 4.1 nm resolution, and our group has now reached up to 3.2 nm resolution studying the dynein regulatory complex (DRC) in wild-type and mutant axonemes, also of *Chlamydomonas* (Heuser *et al.*, in press) (Fig. 3C and D). Various protocols for ET have been published in previous volumes of *Methods in Cell Biology* and *Methods in Molecular Biology* (Hoenger and Nicastro, 2007; Marko and Hsieh, 2007; McEwen *et al.*, 2008; O'Toole *et al.*, 2007). Here we focus specifically on cryo-ET methods used to image isolated axonemes and flagella by providing an overview of the procedures, hard- and software used for data acquisition, reconstruction, averaging and visualization in our studies; we provide a summary of how to evaluate these data, what artifacts to be aware of, and an outlook into future developments and challenges.

A. Cryo-Preparation

The specimen preparation is critically important for the quality of the cryo-ET results; we are extra careful during all steps to avoid, for example, mechanical stress as much as possible. We use well-established protocols; however, other than for biochemical preparations we need only very small amounts of specimen and so during the preparation we follow these guidelines: quality not quantity, specimen should not have been frozen before plunge freezing (i.e., shipments on wet not dry ice), and they should be free from cryo-protectants such as glycerol and sucrose that interfere with imaging.

1. Specimen Preparation

For a sea urchin sperm flagella preparation we order ripe sea urchins, for example, *Strongylocentrotus purpuratus* (Marinus, Long Beach, CA, USA), and sometimes cultivated them in the laboratory for a few days. Spawning is induced by intracoelomic injection of 1–2 ml 0.5 M potassium chloride (Tyler, 1949). The sperm are then collected directly with a glass pipette (without dilution in sea water, which would activate the flagella), kept on ice, and processed within 1 h (Nicastro *et al.*, 2005).

Preparations of axonemes are performed as previously described (Nicastro *et al.*, 2006; Rupp *et al.*, 1996): strains of *Chlamydomonas reinhardtii* are grown on solid tris-acetate-phosphate medium (Gorman and Levine, 1965). After 5–7 days, cells are resuspended in liquid minimal medium for at least 1 h to induce flagellar growth. Cells are collected by centrifugation, washed two times to remove the cell wall and other debris, and resuspended in 10 mM HEPES buffer (pH 7.4, 1 mM $SrCl_2$, 4% sucrose, 1 mM DTT). Flagella are then detached from the cells using the pH-shock method (Witman *et al.*, 1972) and added to 5 mM $MgSO_4$, 1 mM EGTA, 0.1 mM EDTA, and 5 μg/mL aprotinin, leupeptin, and pepstatin; other laboratories also use dibucaine to induce deflagellation

(Witman, 1986). After centrifugation (1000 × *g* at 4°C for 5–10 min) the flagella-containing supernatant is purified from remaining cell bodies and debris by two additional centrifugation steps over a 20% sucrose cushion. Flagella are demembranated with 0.1% IGEPAL CA-630 (Sigma Aidrich, St. Louis, MO) and axonemes collected by centrifugation at 35,000 × *g* for 60 min at 4°C. The pellet is washed and finally resuspended in HMEEN buffer (30 mM HEPES, pH 7.4, 5 mM $MgSO_4$, 1 mM EGTA, 0.1 mM EDTA, 25 mM NaCl, 0.1 μg/mL aprotinin, leupeptin, and pepstatin). Axonemes (at ~2 mg/mL concentration) are stored at 4°C and usually plunge-frozen within 24 h.

2. Grid Preparation

Quantifoil grids (type Cu 200 mesh R2/2; Quantifoil Micro Tools GmbH, Jena, Germany) or C-flat grids (Protochips, Inc., Raleigh, NC, USA) coated with holey carbon support film are glow discharged for 30 s at −40 mA, which makes the surface hydrophilic and allows a better distribution of the sample across the grid. To apply gold clusters that can later be used as fiducial markers in the tilt-series alignment process, 5 μl of 10 nm colloidal gold in aqueous suspension (Sigma, St. Louis, MO, USA) is air-dried onto the grid, which is then briefly dipped into distilled water to wash crystallized salts off and air-dried again.

3. Plunge Freezing

For the rapid freezing we have used four different types of plunge freezers: home-made guillotine-like devices with and without automated blotting (Dubochet *et al.*, 1988; Templeton *et al.*, 1997), a Vitrobot (FEI Company, Hillsboro, OR, USA) and a cryo-plunge 3 (Gatan Inc., Pleasanton, CA, USA), and have achieved similar results with all instruments.

Typically, we apply 3–4 μl of the sea urchin sperm or axoneme preparation to the grid and mix 1 μl 10× spin-concentrated 10 nm colloidal gold (table-top centrifuge, 14.000 × *g* for 15 min) into the drop with specimen. After allowing the specimen to adsorb to the grid for a few seconds, excess fluid is blotted for a few seconds either from the front or in automated plunge freezers from both sides using Whatman #1 filter; the blotting time depends on many factors (e.g., humidity), but the goal is to embed the axonemes or flagella in an ~200 to 250 nm thick layer of ice. For rapid freezing the blotted grid is immediately plunged into liquid ethane cooled by a surrounding bath of liquid nitrogen. The vitrified samples are stored in liquid nitrogen until examination by cryo-EM.

In cryo-ET it can be difficult to achieve a nice distribution of gold clusters around the specimen for later fiducial alignment of the tilt series; we therefore apply gold twice, dried to the grid, as it is common to find that some of the markers detach from the carbon film and appear in the frozen ice layer, and directly into the specimen solution briefly before blotting.

B. Cryo-Electron Microscopy and Data Acquisition

1. Cryo-Electron Microscope

As mentioned in the introduction, specific equipment and software is required or recommended for cryo-EM and cryo-ET; we expand on a few of these items in this section.

- Choosing the right accelerating voltage depends on the type and thickness of samples that will be imaged. In cryo-EM the dominant contrast mode is phase contrast, which requires a sufficient amount of elastically scattered electrons to reach the imaging plane; thus the sample thickness should not greatly exceed the inelastic mean free path of the electrons. The mean free path of electrons in vitreous ice at 120 kV is ~200 nm (Grimm *et al.*, 1996) and increases with higher accelerating voltages, for example, at 300 kV to ~350 nm. Therefore, intermediate voltage instruments 200–400 kV are preferred for vitrified cellular specimen, especially in cryo-ET, where the tilting increases the effective specimen thickness that is penetrated by the beam by the inverse cosine of the tilt angle (e.g., doubled at 60°). Higher voltages were also used for ET, but at some point the reduced contrast and efficiency of the CCD cameras for these fast electrons outweigh their advantage in sample penetration.
- Our Tecnai F30 microscope (FEI, Eindhoven, the Netherlands) is equipped with a postcolumn energy filter (GIF, Gatan, Pleasanton, CA, USA); however, similar principles apply also for in-column omega-type filters (Egerton, 1996). In the energy filter, a magnetic prism spectrometer, the electrons are dispersed according to their kinetic energy or velocity. A slit in the dispersion plane (we use a slit width of 20 eV) is used for selecting electrons of a specific energy range; in our case for zero-loss electrons that have not lost energy due to inelastic collisions with the specimen. Inelastically scattered electrons are not useless, but without correction of chromatic aberrations of the objective lens these electrons contribute blurring or noise to the EM image, and so for relatively thick frozen-hydrated specimen (e.g., intact axonemes) zero-loss energy filtering improves the image SNR (Grimm *et al.*, 1997).
- For cryo-ET of relatively thick, frozen-hydrated specimen a sensitive CCD camera is recommended, that is, for intermediate voltage microscopes a relatively thick phosphor scintillator is needed to achieve a high yield of photons captured by the CCD per incident electron (10–20 photons per incident electron); one has to find the best compromise between sensitivity and resolution. For low-dose imaging the electron dose that the specimen is exposed to has to be determined; this can be done by calibrating the CCD camera with respect to the incident electron dose using a Faraday cup. Then at the beginning of every EM session the microscope is aligned and the incident dose measured without specimen in the electron beam. The array size of the camera—together with the chosen magnification—will determine the field of view and thus the size of the tomogram; typical are 2k × 2k or 4k × 4k CCD cameras.

- Some new-generation cryo-electron microscopes are equipped with specific cryo-transfer mechanisms, automatically refilling nitrogen-cooling system, storage of multiple grids inside the microscope, and new holder design; however, we have the widely used system, where the vitrified specimen is transferred into a high-tilt cryo-holder (e.g., a Gatan 626, Pleasanton, CA, USA) using a specific transfer station that allows the grid to be kept under liquid nitrogen during the transfer, before insertion into the EM.

2. Tilt-Series Acquisition Software

For cryo-ET, computer control of the microscope including the tilting stage is essential, as this allows the application of automated tilt-series acquisition to minimize the cumulative electron dose that the beam-sensitive specimen is exposed to (Koster *et al.*, 1997). There are different software packages available for cryo-ET data acquisition, such as "precalibration" software (Ziese *et al.*, 2002), UCSF tomography (Zheng *et al.*, 2007), SerialEM (Mastronarde, 2005), TOM (Nickell *et al.*, 2005), and Leginon (Suloway *et al.*, 2009), which are all free for academic users, and the commercial Xplore3D package (FEI Company, Hillsboro, OR, USA).

In our laboratory we use SerialEM (Mastronarde, 2005), because it was the first automated tomography package with robust prediction of specimen movements and highly flexible user interface. The robust predictions make this an ideal package for cryo-ET. It uses changes in specimen position at previous tilt angles to predict the position at the current tilt angle. This is similar to previous implementations (Zheng *et al.*, 2007; Ziese *et al.*, 2002), but it decides automatically—based on statistical errors—between rapid data acquisition when conditions are good, and more controlled data collection by taking more frequently focusing and tracking images when conditions are not as good (Mastronarde, 2005). Other notable features of SerialEM that we use are: (1) low-dose imaging mode, in which tracking and focusing occur away from the area of interest, (2) control of the energy filter, including alignment of the slit-position, (3) automatic acquisition of montages and the navigator tool, which allows one to generate low magnification maps of the entire grid and medium magnification maps of specific areas for efficient screening for good specimens under low-dose conditions, and (4) flexible interface for adjustment of imaging conditions and user intervention.

3. Strategies for Tilt-Series Acquisition

Automated digital image acquisition greatly facilitates both time-efficient and very low-dose acquisition schemes, minimizing the radiation damage accumulated by the sample.

- Part of our recent improvement in resolution of cryo-ET of axonemes (Fig. 3C and D) has been rigorous screening of the grid for well-preserved axonemes, that is, without compression (Fig. 6A and B), flaring or twisted doublets, uniform vitreous ice of 200 to 250 nm thickness, and good imaging conditions, including a good distribution of gold fiducial markers and axonemes in a good orientation, that is, close to parallel to the tilt axis. This screening process using low and medium magnification maps is time consuming, but necessary for the best possible results.

Choosing the parameters for a cryo-tilt series is usually a compromise between electron dose and resolution, but in practice many factors need to be considered and selected for optimal results:

- The maximum dose that can be tolerated by the specimen is determined by the nature of the radiation-sensitive specimen and the desired resolution of specimen detail; for axonemes we usually keep the cumulative dose below 100 electrons/Å^2. One of the antagonistic issues in cryo-ET is that in order to improve resolution the number of electrons would have to be increased with the fourth power of the targeted resolution improvement (McEwen *et al.*, 2002; Saxberg and Saxton, 1981), yet to conserve more specimen detail the dose would have to be reduced.
- In principle, the illumination should remain constant during the tilt series, but in practice images at the high tilts (and increased effective specimen thickness due to the slab geometry) become too noisy at constant illumination; therefore, the mean image pixel value is kept close to a set target value that incrementally increases the exposure as the tilt angle increases.
- The tilting scheme determines the data coverage in the Fourier space and the theoretical resolution of the tomogram (Crowther *et al.*, 1970), though in practice the determination of tomogram resolution is very complicated. For intact axonemes we usually record single-axis tilt series over a tilt-angle range of ±60–70° with constant 1–1.5° tilt increment. Graduated tilt increments, that is finer increments at higher tilt angles, would theoretically correct for the oversampling at low tilt angles and undersampling at high tilt angles (Saxton *et al.*, 1984); however, in practice, we have not noticed improvements using the Saxton scheme, probably because any advantages are offset by the disadvantage of shifting more dose toward the noisier high tilt images. We tilt continuously from one extreme angle to the other (i.e., from −65° to +65°) instead of starting at 0°, then tilting to one extreme angle, and returning to 0° before tilting to the other extreme; we found that the tilt-series alignment and reconstruction is easier with the continuous scheme, as any changes during the series acquisition occur gradually, avoiding a large jump in the middle of the tilt series.
- Underfocusing is used to increase phase contrast of frozen-hydrated samples. The defocus of the objective lens is set relatively high for thick cellular samples (10–15 μm), but for high-resolution cryo-ET of intact axonemes we use 6–8 μm underfocus. Lower defocus is better in terms of the phase-contrast transfer function (CTF), where spatial frequencies are not transferred with a uniform intensity, but at the same time lower defocus will result in lower contrast and noisier images. While images of specimen in thinner ice (<200 nm) have inherently better SNR and therefore allow smaller defocus (e.g., 2–4 μm), too thin ice can introduce artifacts (see Section II.E.1 *compression*). Focusing is performed automatically by collecting an image pair at small

beam tilts, and the resulting shift between the images is proportional to the defocus by an amount that has been previously calibrated, thus allowing focus to be corrected to the desired value.

- The magnification is selected so that the resulting pixel size on the CCD does not become the resolution-limiting factor, that is, the pixel size should be $\sim 3\times$ smaller than the targeted resolution; we record tilt series typically with ~ 1 nm pixel size, but smaller sizes can be appropriate for thinner specimens.
- The basic sequence of steps for recording a tilt series are *tilting* the goniometer to the next angle, *tracking* to keep the region of interest within the field of view, *focusing*, *recording* a full-resolution image of the region of interest, *tilting* to the next angle, and so on. In low-dose mode tracking and focusing images are acquired from a region that is shifted away (along the tilt axis) from the region of interest. Tracking during a cryo-tilt-series can be challenging and is one of the most common errors during data acquisition. A good tracking area has enough features to allow accurate correlation between the current tracking image and the tracking image from the previous tilt angle that has been automatically stored as reference. Especially at high tilts specimen drift can be high, so that tracking before and after focusing, as well as short exposure times are recommended. Using the steps outlined here a cryo-ET tilt series can be automatically recorded in 30–50 min (depending on the size and readout time of different CCD cameras).

C. Building and Visualizing the Tomogram

1. Software

Once a tilt series is recorded, subsequent processing requires the use of specialized software for the computation and analysis of 3D reconstructions. The following is a list of software packages commonly used for ET in the biological community: the tomogram reconstruction and modeling software package IMOD (Kremer *et al.*, 1996; Mastronarde, 2008), SPIDER (Frank *et al.*, 1996), EM3D (Ress *et al.*, 1999), Protomo (Winkler, 2007), UCSF tomography (Zheng *et al.*, 2007), TOM (Nickell *et al.*, 2005), Bsoft (Heymann *et al.*, 2008), the commercial Xplore3D (FEI Company, Hillsboro, OR, USA), and for the alignment process only RAPTOR (Amat *et al.*, 2008). The 3D reconstruction process is—similar to data acquisition—also relatively automated and most software packages include graphical user interfaces (GUIs) for ease of use; for example, Etomo is the reconstruction GUI of IMOD, one of the early and most popular tomography software packages (Kremer *et al.*, 1996; Mastronarde, 2008); it is offered by the Boulder laboratory for 3D electron microscopy, and their webpage provides many useful resources including tutorials for new users (bio3d.colorado.edu). Although the following three sections describe the tomogram reconstruction process as it is implemented in IMOD, including preprocessing, alignment, computation of the 3D volume, and visualization of the tomographic data, these basic steps and principles are more or less also part of the remaining software packages.

2. Preprocessing and Alignment of the Tilt Series

Before calculating the tomogram, all tilt-series images must be aligned to a common origin; otherwise densities in different images that correspond to the same feature would not be projected back to the same location in the 3D reconstruction. Therefore, the accuracy of the alignment, which includes the determination of the image rotation, translation, and various corrections, is critically important for the reconstruction quality, and even small misalignments will lead to artifacts in the tomogram. Tilt-series images can be aligned by cross-correlation methods (Winkler and Taylor, 2006), but currently better reliability is achieved if colloidal gold particles are used as fiducial markers (Lawrence, 1992; Mastronarde, 1997, 2006; Penczek *et al.*, 1995). Especially in cryo-ET images with low SNR these electron-dense markers are often the only discernable features that will facilitate a robust alignment. We use 10 nm colloidal gold markers, which have a diameter of about 10 pixels in our tilt-series images; this is a good compromise between being visible even in the noisier high tilt images, and also being small enough so that their center can be determined accurately.

- *Setup*: In this step several parameters that will be needed during the reconstruction process are set, such as the pixel size, size of the fiducial markers, the nominal tilt angles, and tilt-axis angle (as it was calibrated at the electron microscope).
- *Preprocessing*: Here the tilt-series images are processed (and sometimes normalized) to ensure that their means and standard deviations are in a similar range, for example, by removing X-rays from the images, which are typically a few pixels large, but have counts that are 1–2 orders of magnitude higher or lower than the image mean.
- ***Coarse alignment***: Using cross-correlation of successive tilt-series images, the images are brought into an initial, coarse alignment, mainly to ease the following step of generating a fiducial marker model throughout the tilt series.
- ***Fiducial model generation***: The user picks a set of gold fiducial markers on a single image of the tilt series, usually the 0° projection; ideally many evenly distributed markers are selected, but in practice, often only a few fiducials might be available in the field of view. The program then automatically tracks these gold particles on each projection image of the tilt series and determines their 2D coordinates. This is usually a multi-iteration process until all gaps are filled and the fiducials are tracked on every image that they are visible on.
- ***Fine alignment***: The marker coordinates are used to calculate a 3D model of the fiducial locations and the alignment algorithm computes a global alignment solution with translation and rotation for each image, and various corrections, for example, for changes in magnification, the tilt angles and rotation of the tilt-axis angle. By highlighting gold markers with high errors in relation to the global solution, the program aids in inspecting and manually correcting errant localization of markers; after corrections are made, the 3D marker model is updated and the refinement iterated until the overall residual error stops decreasing and the best possible solution is found.

Images of cryo-tilt series are relatively noisy and grouping of sequential images (5–8) before computing certain corrections, such as rotation of the tilt-axis angle and refinement of the tilt angles, will provide less noisy and more reliable solutions. The low SNR could in part also explain why the residual error is typically higher for cryo-tilt series (~0.7–1.1× pixel size) than for plastic section tilt series. We have also found that the linear distortion correction and local alignment options, which were developed in IMOD to correct for the changes that plastic sections undergo during beam exposure, usually do not seem to improve the alignment of the biological material embedded in the ice layer.

3. Tomogram Reconstruction

- ***Aligned tilt series***: Once the user settles on the best alignment, the rotations, transformations and potential distortion corrections are applied in all to the original tilt-series images to generate a stack of the original images but with each image aligned to a common coordinate system and the tilt axis aligned to the *y*-axis. At this step IMOD offers several additional options, such as 2D filtering (for noise reduction), CTF correction (Xiong *et al.*, 2009) and removal of nonspecimen features with extreme pixel values like the gold markers; by replacing the dark pixel values of the gold markers with the image mean, streaking artifacts radiating from the edges of the fiducials can be avoided; these ray artifacts are mainly visible in *xz*-slices through the tomogram (with the *z*-axis being parallel to the electron beam in the 0° tilt image), but in our case this procedure is only useful if gold markers are close to the ice embedded axonemes.
- ***Tomogram positioning***: This is a useful step that allows the user to quickly calculate the rotation and shift parameters required for positioning the ice layer with the specimen optimally within the to-be-reconstructed volume. Orienting the specimen layer flat in the 3D reconstruction minimizes the size of the tomogram (which can be 1–2 GB large) and eases later data visualization. Some of the parameters are actually applied to the original images during the previous step that generates the aligned tilt series, others are fed into the following tomogram generation process.
- ***Tomogram generation***: Based on the aligned tilt-series images the tomogram is reconstructed in real space using weighted backprojection. IMOD also offers the option to use simultaneous iterative reconstruction technique (SIRT) as the reconstruction algorithm.

4. Visualizing Tomographic Data

- ***Histogram processing of the generated tomogram***: Ideally, the features of interest in the tomogram should be displayed by a visualization program utilizing the full dynamic range of gray values. However, for example, the voxel values of the gold fiducials and white halos around them are much higher/lower than the biological material in the tomogram. Therefore, it is recommended to correct for this by saturating undesired features and stretch the tomogram histogram in the postprocessing step.

Tomographic reconstructions are volumetric data and a large arsenal of both general 3D visualization software and specialized tools, for example, for denoising (filtering) or segmentation, are available (many of them free for academic use). Here we focus on a brief overview of programs and tools typically used for visualizing and analyzing cryo-ET data in our laboratory, but in addition we provide references for further reading.

- ***Tomographic slices***: Most software packages for ET have integrated programs for displaying the 3D reconstructions as 2D slices through the tomographic volume, for example, 3dmod is the model editing and image display program in IMOD (Kremer *et al.*, 1996; Mastronarde, 2008); in this section the terminology of display windows refers to that used by the 3dmod program. The first step in analyzing tomographic reconstructions is viewing 2D *xy*-slices through the tomogram and then moving along the *z*-axis through the volume (ZAP-window) (Figs. 3A and C; 4; 6A–D). Tomographic *xy*-slices (which are normal to the electron beam) have the best resolution and are least influenced by the missing wedge that is typical for single-axis tomograms. Instead of displaying *xy*-slices from different *z*-heights sequentially, they can be displayed at once as gallery of 2D slices (Multi-Z window). One of the great advantages of tomographic data is that they can be inspected voxel-by-voxel and rotated arbitrarily to view 2D slices in a favorable orientation of the feature of interest (Slicer window). When displaying noisy cryo-tomograms it can be useful to increase the thickness of the displayed slice by adding (projecting through) consecutive slices.

  Tomographic slices are typically displayed as gray-value images. Conventionally, EM images are shown without inverting the original image contrast, so that the images represent the contrast generated in the microscope; that is, in classical EM images stain appears black (e.g., a stained membrane), because stain strongly scatters electrons that then do not contribute to the image, generating amplitude contrast by leaving areas with stain dark in the image; negatively stained molecules, on the other hand, appear white surrounded by dark stain. The same convention applies for frozen-hydrated specimen; the biological material (and gold fiducials) scatters electrons and appears dark in cryo-EM images. Changing this convention by displaying tomographic slices with inverted contrast (i.e., white structure on black background) can lead to confusion of the viewer.
- ***Graphical modeling, surface and volume rendering***: To fully take advantage of the three dimensionality of tomographic data, for example, to display the complex spatial arrangements of microtubule-associated structures in axonemes, volume and surface-rendering methods are important. Among the most popular software packages for graphical modeling and 3D rendering of ET data are IMOD (Kremer *et al.*, 1996; Mastronarde, 2008), UCSF Chimera

(Goddard and Ferrin, 2007), and the commercial AMIRA (TGS, Mercury Computer Systems, San Diego, CA, USA). Additional information about free and commercial software packages for 3D visualization of biological structures can be found at the World Index of Molecular Visualization Resources (http://www.molvisindex.org).

Graphic modeling is often used where ET data are too crowded or noisy, for example, because of low SNR or uneven stain distribution. Features of interest in the tomogram, such as membranes or microtubules, can be hand- or automatically traced by placing model points and contours on the tomographic slices. From these contours 3D surfaces can be computed (skinning), which are then represented by meshes of simple primitive structures, ranging from triangles to polygons (Kremer *et al.*, 1996; McEwen and Marko, 1999). Once the mesh is calculated, different properties, like color, projection mode (orthogonal or perspective), lighting, reflectivity, and transparency can be assigned, before displaying the 3D surface (Fig. 3B).

For surface rendering a skin is computed in the same manner as above; however, rather than basing the calculations on traced contours, boundaries between objects and the background are extracted and represented by meshes by setting a certain voxel value as threshold (Fig. 3D). The voxels are subdivided to facilitate smooth surfaces. This display form, in which light is reflected from surfaces, is popular as it simulates our vision. In both cases of surface calculations, however, information about the inner densities of the volume is omitted. Skinning by threshold is also not particularly suited for noisy data, because random noise in the background with voxel values above the threshold will also be displayed, obscuring the view of features of interest. Surface renderings can easily be transformed into file formats suitable for 3D printers (Gillet *et al.*, 2005; Goddard and Ferrin, 2007).

In volume rendering the voxel intensities are converted to values of transparency and the transmission of illumination through the volume is computed, that is, the entire density within the tomogram is visualized and not just borders (surfaces) between features. Overlapping transparent features can make the analysis difficult, but under appropriate conditions volume rendering can be a valuable visualization technique, especially for noisy 3D data. Volume rendering is computationally intensive, making the visualization of large tomographic data in real time difficult. However, with easy availability of graphics cards with large amounts of memory and powerful graphical processing units, fast volume rendering is now more achievable for moderately sized volumes.

- ***Filtering (denoising) and segmentation***: Many tools have been developed for denoising tomograms and segmenting information that belong to specified objects within the 3D reconstruction. We use only the most basic procedures, such as smoothing tomograms (a type of denoising filter in IMOD that computes weighted averages of each pixel and its neighboring pixels) (Kremer *et al.*, 1996) and coloring axonemal subcomplexes in surface renderings using the placement of markers in UCSF Chimera (Goddard and

Ferrin, 2007), for visualizing high-resolution cryo-ET data of axonemes (Nicastro *et al.*, 2006); the major tool for improving the SNR of our raw cryo-tomographic data is 3D correlation averaging as described in Section II.D. The following is therefore just a list of further reading about median, smoothing, wavelet, nonlinear anisotropic diffusion, and bilateral filtering (Frangakis and Hegerl, 2001; Jiang *et al.*, 2003; Narasimha *et al.*, 2008; Stoschek and Hegerl, 1997; van der Heide *et al.*, 2007). To learn more about different approaches of data mining and segmentation consult further reading on automatic modeling, masking, and template matching, Eigenvector and watershed segmentation, and more (Bazan *et al.*, 2009; Bohm *et al.*, 2000; Frangakis and Forster, 2004; Frangakis and Hegerl, 2002; Frangakis *et al.*, 2002; Garduno *et al.*, 2008; Jiang *et al.*, 2004; Lebbink *et al.*, 2007; Martin *et al.*, 2005; Noske *et al.*, 2008; Salvi *et al.*, 2008; Sandberg and Brega, 2007; Volkmann, 2002).

D. Volume Averaging of the 96 nm Axonemal Repeat and Resolution Measurement

Averaging repetitive particles is an ideal method for improving the image SNR and thus resolution, assuming the combined structures are indeed alike and can be accurately aligned before summation (see Section I.D). Tomographic averaging was first used a decade ago (Walz *et al.*, 1997), but user-friendly software was not available until recently and additional tools are still under development. For 3D correlation averaging of subtomographic volumes of intact axonemes we use the Particle Estimation for ET (PEET) software of the Boulder software group (Nicastro *et al.*, 2006); the Ishikawa group has used a combination of Bsoft (Heymann *et al.*, 2008), SPIDER (Frank *et al.*, 1996) and TOM (Nickell *et al.*, 2005), and other tomography groups have developed their own programs (Schmid *et al.*, 2006; Winkler *et al.*, 2009). The basic steps (Fig. 5) are common among these software packages:

- ***Particle picking***: So far, the volume averaging programs rely on manually selecting repetitive particles (subtomographic volumes) in the tomogram, that is, we mark the center of the axonemal 96 nm repeat units in our 3D reconstructions (Fig. 4A). For averaging with PEET (Nicastro *et al.*, 2006) the particle model can be generated using 3dmod in IMOD (Kremer *et al.*, 1996; Mastronarde, 2008).
- ***3D correlation alignment***: The selected subvolumes are extracted from the tomogram in 3D and aligned with respect to a reference; initially this reference can be one of the selected particle or a previous average. Without prior knowledge about the orientation of the particles a computationally intensive, brute force angular search is required to determine the three rotation angles and three shifts from 3D cross-correlations between the particle and the reference volume. In the case of averaging 96 nm axonemal repeats we can limit the search range considerably (see *Strategies for averaging axonemes* below). The alignment is a multi-iteration process, typically starting with a large search range and angular increment, which is then reduced in later iterations. Before the following iteration the reference is updated with an average

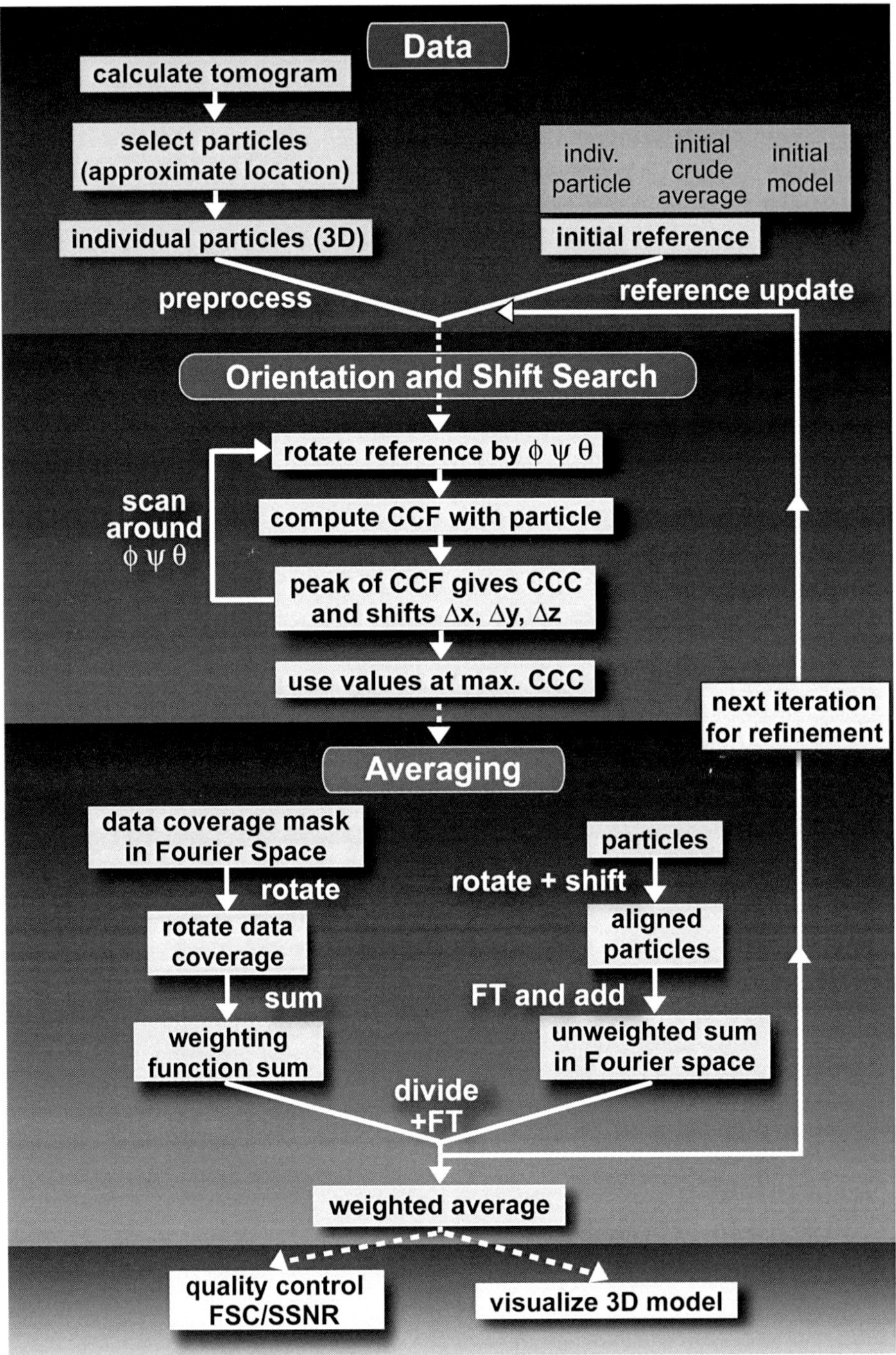

Fig. 5 (*continued*)

using only the best of all particles (i.e., with the highest cross-correlation coefficient) and using the newly determined particle alignment parameters.

- ***Computation of a weighted average***: Once all particles are oriented to the same reference volume a weighted average with missing wedge compensation and improved SNR is computed by adding the aligned subtomograms in the Fourier space and dividing it by the weight from the contributions of all particles to avoid uneven sampling (Figs. 4 and 5).
- ***Strategies for averaging axonemes***: Two different strategies have been employed: (1) alignment and 3D averaging of axonemal repeats first along each microtubule doublet, and then the subaverages of all doublets are aligned to a chosen subaverage, which is used as reference (Bui *et al.*, 2008; Ishikawa *et al.*, 2007). (2) In our laboratory, we use the geometry of axonemes to estimate the coarse alignment angles between all selected 96 nm repeats of one intact axoneme; we then align the particles from all doublets in a common run to a reference starting the search at the previously calculated angles (Nicastro *et al.*, 2006). This has the advantages that the initial search range can be kept small and by choosing the right search parameters the missing wedge bias can be reduced during the alignment procedure (Frangakis *et al.*, 2002). By making use of particles from all nine microtubule doublets, which have different orientations within the tomogram, the missing wedge can be filled in, resulting in isotropic resolution in the final average (Heuser *et al.*, in press).
- ***Resolution measurement***: The quality of image data is often assessed by their resolution. The overall resolution of tomograms, however, can be limited by different factors (e.g., specimen preservation, missing wedge, and low SNR), which makes it difficult to derive quantitative estimations. Universal criteria have not been found, but suggestions range from using features in the tomogram as indicator (e.g., can the bilayer of membranes be identified?) (McIntosh *et al.*, 2005), to a simple formula using undersampling in the 3D Fourier transform as

Fig. 5 Flowchart of the basic steps of tomographic averaging using the PEET program. Particle Estimation for Electron Tomography (PEET, Nicastro *et al.*, 2006) is a software program for 3D correlation averaging specifically designed for electron tomography. The computational work can be divided into three major parts that repeat iteratively: (1) input *Data* extraction, (2) *Orientation and Shift Search*, and (3) *Averaging*. (1) Input Data extraction: The position of repetitive structures (3D particles) is selected in the raw tomogram, for example, the 96 nm repeat unit of axonemes; the particles are then extracted and aligned to a reference volume in the following step. (2) Orientation and Shift Search: The position and orientation of a 3D particle is uniquely described by their Cartesian coordinates (x, y, z) and Euler angles (ϕ, θ, φ—their rotation within the 3D space). To determine the six parameters (shifts and rotations) necessary to align the subtomographic volumes to the same reference (in the simplest case: one specific particle) an angular search is performed in 3D. At each angle the particle is cross-correlated with the reference and the value of the cross-correlation coefficient (*CCC*) is stored. At the end the best orientation is selected based on the highest *CCC*. This brute force search is a computationally intensive step; therefore a wide search range with large search increments is used in the beginning and iteratively refined to smaller search ranges and increments. (3) Averaging: All particles are shifted and rotated according to their best reference match and then combined in Fourier space (*FT*, Fourier transform). A weighting algorithm is applied to account for data coverage in the Fourier space. Particles are ranked according to their *CCC* and weighted averages with different amounts of included particles can be computed.

limiting factor (Crowther *et al.*, 1970), to recent methods, such as spectral SNR (SSNR) (Penczek, 2002; Unser *et al.*, 1987) or noise-compensated leave-one-out (NLOO) (Cardone *et al.*, 2005).

The Fourier shell correlation (FSC) is a well-established resolution criterion for 3D reconstructions in the single-particle EM community, and has now also become the accepted method for assessing data quality and resolution of tomographic averages. Most widely used are the 0.5 and sigma-factor curves as FSC threshold (Bottcher *et al.*, 1997; Saxton and Baumeister, 1982; van Heel and Schatz, 2005). As important as it is to have quantitative resolution estimates to guide data interpretation and technical development, the absolute resolution value estimated by the FSC criterion might not be as meaningful as it seems; it does not provide information if the resolution is isotropic or varies between regions and structures within the tomographic average, and the value can greatly vary depending on the voxel number used for calculating the FSC (van Heel and Schatz, 2005). Therefore, a single number for the resolution should not substitute a thorough inspection of the actual data: How well are the microtubule protofilaments resolved? Are features, like the dyneins, well defined or do they appear "blobby"? Also the quality of tomographic slices is a more direct criterion for quality than renderings, which incorporate more subjective decisions, like threshold setting or segmentation.

E. Limitations, Data Quality, and Artifacts

The previous sections gave a comprehensive description of the major steps involved in providing stunning new views of an old organelle; cryo-ET and tomographic averaging is now generating 3D reconstructions of intact axonemes in a near-to-native state, and molecular complexes in the 96 nm repeats can be resolved with 3 to 4 nm resolution. In practice, however, the process does not always go so smoothly. In fact, at each step of this pipeline, factors can influence the final data quality or introduce undesirable artifacts that are not always easy to spot. As mentioned above, there is a variety of factors that limit the resolution of cryo-tomograms, such as low SNR, the specimen thickness, alignment errors, the missing wedge, or heterogeneity among the averaged particles. Overall, care must be exercised when attempting to characterize structures smaller than the estimated resolution limit. Here we summarize some of the most common problems seen with cryo-ET both in general and of intact axonemes.

1. Factors that Limit the Specimen Quality

- ***Specimen Preparation***: Cryo-ET has the potential to show great detail of a specimen, but only if it is preserved during the specimen preparation. Established protocols that might have worked well for chemically fixed EM specimen in the past can turn out to be too harsh when imaged using ET, especially of frozen-hydrated specimen. Therefore, the specimen should be prepared fresh, not frozen before plunge freezing, and mechanical stress avoided. Signs of problems with the preservation, for example, of isolated and vitrified axonemes, would be single microtubule doublets, flared or twisted axonemes, or "sticky" axonemes that are densely covered with gold clusters (possibly due to proteolysis or denaturation) in the vitrified sample.

- ***Compression***: When force is applied to soft specimens they deform, which can introduce distortions of the native structure. While insufficient blotting before plunge freezing produces ice that is too thick (low SNR in the images), excessive blotting causes compression of the sample due to surface tension (McEwen *et al.*, 2008), that is, the usually cylindrical-shaped, intact axonemes are flattened as illustrated in Figure 6. During the screening process of the grid, deformed axonemes that are embedded in too thin ice are easy to identify even in low-dose images, because the compression along the dimension of the ice layer is compensated by a proportional increase of the axoneme width (Fig. 6C and D). On the other hand, cryo-EM images of thinner specimen have a better SNR and specimen thickness is a resolution-limiting factor in cryo-ET. Therefore, it is not surprising that flattened axonemes appear attractive, and their tomographic reconstructions often exhibit a very good SNR. However, tomographic averages of compressed axonemes clearly show that most axonemal structures, with the exception of the microtubule doublets, suffer to some degree from distortion artifacts (Fig. 6G and G') and can lead to misinterpretations (Ishikawa *et al.*, 2007). It seems that blotting-induced deformations ultimately limit the achievable resolution, while data from noncompressed axonemes provide high-resolution averages (Fig. 3C and D) (Heuser *et al.*, in press).

2. Factors that Limit the Ice and Image Quality

- ***Ice quality***: Similar to specimen preparation, good vitrification is a crucial prerequisite for high-resolution cryo-ET data without artifacts. Bad freezing rates during the plunge freezing or warming up of the grids to higher than about –135°C during grid transfers or storage lead to ice crystal formation and ice damage of the sample, because the crystal growth displaces and aggregates normal cellular structures (Dubochet *et al.*, 1988). Hexagonal crystals have sharp borders and are easily recognized in cryo-EM images, while changes due to cubic ice can be identified in the electron diffraction pattern (Dubochet *et al.*, 1988) or by contrast inversions during specimen tilting (Bragg diffraction); the latter can be seen as "blinking patches" in the specimen when moving through the stack of tilt-series images.
- ***Cumulative electron dose***: Vitreous material is radiation-sensitive; obvious signs of damage caused by too much electron dose are gross distortions by "bubbling," which is the appearance of first small, then growing bubbles in the specimen due to ice radiolysis while the produced gas remains entrapped in the ice layer. Radiation damage results in loss of structural details before

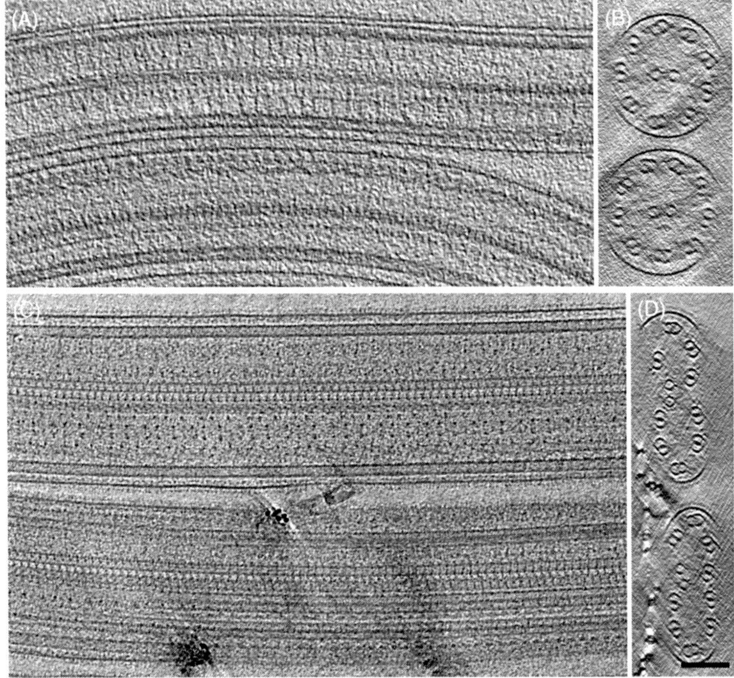

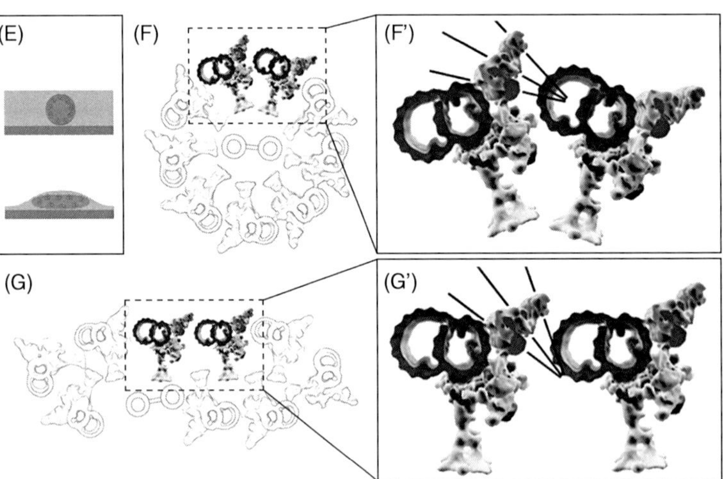

Fig. 6 (*continued*)

bubbling is visible in the specimen (McEwen *et al.*, 2002). Even if the bubbling is not within the material of interest and only in the surrounding area (e.g., over the carbon film), heterogeneous movements in the ice can make accurate tilt-series alignment difficult or impossible. In an electron dose test series (without tilting) of the base of a sea urchin sperm with emerging flagellum, we found that bubbling appeared first in "crowded" and thick areas of the cell, like the mitochondrion at the sperm head base, but was not visible in the flagellum till ~250 electrons/$Å^2$ (D. Nicastro, unpublished data); therefore, cryo-tilt series of axonemes are usually recorded with ≤100 electrons/$Å^2$. Vitreous biological samples are also bad electrical conductors, and the applied dose can lead to charging effects that decrease image quality. A too low electron dose can of course be just as problematic, if the center of the fiducial markers cannot be identified accurately due to the low SNR in the projection images.

3. Factors that Limit the Reconstruction Quality

- ***Alignment problems***: Only few or no gold markers in a tilt series, or markers that are located only in a certain region of the field of view, can create alignment artifacts in the tomogram, because the information is insufficient, or too noisy, or biased to accurately align the tilt-series images. Alignment errors can lead to severe distortions of the features in the 3D reconstruction, which can make interpretation of the data difficult or impossible. In case of missing fiducials in a tilt series, marker-free alignment based on cross-correlation can be attempted, but this method tends to produce systematic shifts in the direction perpendicular to the tilt axis, especially when the SNR is low (data points are smeared out to curves in *xz*-slices of the reconstruction). Vitreous material is a

Fig. 6 Blotting-induced distortions of the axoneme structure are an easily overlooked artifact of cryo-ET of cilia and flagella. (A–D) Comparison between 3D reconstructions of vitrified sea urchin sperm flagella that were (A, B) embedded in an ice layer of suitable thickness (~250 nm) and (C, D) in too thin ice, respectively. The image pairs (A, B) and (C, D) show longitudinal (left) and cross-sectional (right) tomographic slices of two different tomograms, each of which contains two parallel oriented flagella. The cross-sectional views (B, D) clearly show that the flagella in thick ice (B) are nicely cylindrical-shape with a diameter of ~250 nm, while the organelles in thin ice (D) are severely flattened to only ~150 nm thickness (direction of the beam, here in the horizontal direction), which is compensated by a increase in their width to ~350 nm (parallel to the ice layer, here in the vertical direction). The images, however, also show that the SNR is greatly improved in tomograms with compressed flagella (C). (E) The most likely explanation for these distortion artifacts is surface tension exerted during excessive blotting before plunge freezing; the cartoons show cross-sectional views of flagella (*red*) in a layer of solution (*blue*) on top of the carbon support film (*gray*); the bottom drawing shows how a too thin liquid layer can compress soft specimens. (F-G') The comparison between simplified cross-sections of noncompressed (F, F') and compressed axonemes (G, G') demonstrates that the blotting-induced deformations (G, G') change the relative positions between microtubule doublets and cause distortion artifacts, which could lead to increased structural heterogeneity and misinterpretation of data. In well-preserved axonemes all outer dynein arms project toward the center of the neighboring doublet (*angled lines* in F') [see also Nicastro *et al.* (2006)]; a common sign of compressed data is the misalignment of the outer dynein arms, now projecting toward the axoneme center (*angled lines* in G') [see also Ishikawa *et al.* (2007)]. Scale bars: 100 nm. (See Plate no. 2 in the Color Plate Section.)

high-viscosity liquid, that is, it slowly flows; it also seems that its apparent viscosity decreases under the effect of an electron beam (Dubochet *et al.*, 1988). Whereas conventional plastic sections change slightly at the beginning of the irradiation and then remain remarkably stable (McIntosh *et al.*, 2005), vitreous specimens undergo small changes throughout the tilt-series acquisition. Beam-induced movements of the specimen can be different from those of the fiducial markers ("swimming" gold fiducials), leading to erroneous alignment solutions. This could explain why sometimes seemingly good cryo-tilt series cannot be aligned without large residual error and alignment artifacts that do not seem to improve with the distortion correction and local alignment tools provided in IMOD (Kremer *et al.*, 1996).

- ***Missing information in the 3D Fourier transform***: The missing wedge in single-axis ET, arising from the limited tilt range, is a prominent, directional artifact in 3D reconstructions. In practice, this means that some features are distorted (elongated) in the direction that corresponds to the missing angular information (mainly along the *z*-direction), while other structures with an arbitrary density distribution close to perpendicular to the tilt axis are not displayed at all (McIntosh *et al.*, 2005). Averaging subtomographic volumes of particles with different orientation to the tilt axis allows for compensating for the missing information to achieve isotropic resolution in the reconstruction. However, this is not possible if the repetitive particles show preferred orientation on the grid, like isolated microtubule doublets (Sui and Downing, 2006).

III. Discussion

ET has made particularly impressive progress over the last 5–10 years, and its application to frozen-hydrated specimens, including cilia and flagella, will undoubtedly continue to grow. However, further advances (e.g., in hardware and software) and innovative hybrid approaches will be necessary to tackle new challenges in the near future, such as heterogeneity and the localization of specific proteins *in situ*.

A. Hardware Developments

As described in this chapter, in cryo-ET it is important to find the best compromise between different imaging parameters, like electron dose, defocus, and contrast. However, if contrast could be improved without having to increase the electron dose, then we should be able to increase the achievable resolution. Phase plates, which are under development by several laboratories and EM manufacturers, introduce a phase shift between the unscattered and the elastically scattered electrons, producing enhanced phase contrast; this would omit the need for underfocusing the objective lens and thus result in a nearly flat CTF in the spatial frequency range that is critical for cellular cryo-ET (Cambie *et al.*, 2007; Danev and Nagayama, 2001; Majorovits *et al.*, 2007; Nagayama and Danev, 2008). Therefore, close-to-focus imaging with a phase plate in the back focal plane of the objective lens holds currently probably the greatest promise as the single technical advance that would allow for a significant improvement in resolution for tomographic applications, which could bring 2 to 2.5 nm resolution into reach for studies of intact axonemes, cilia, and flagella. Other promising developments that can drive progress in the cryo-EM field in the near future include highly sensitive direct electron detectors (Deptuch *et al.*, 2007; Milazzo *et al.*, 2005), refined high-contrast lens technology, and aberration correctors (Haider *et al.*, 2008).

B. Structural Heterogeneity

The method of subtomogram averaging to improve the SNR is based on the assumption that the averaged particles have identical structure. In practice, however, cellular structures are likely to have some degree of heterogeneity, for example, different conformational states, regions with structural flexibility, or certain components that are present or absent in some but not all of the repetitive particles. This variability could be biologically important; for example, the nucleotide-dependent conformational changes of dynein motors, the recently shown localization of some low-abundance dynein heavy chains to the proximal region of *Chlamydomonas* flagella (Yagi *et al.*, 2009), the doublet specificity of axonemal structures like the "beak"-structure in the B-tubule of some doublets (Hoops and Witman, 1983) or doublet-specific dyneins in *Chlamydomonas* flagella (Fig. 3C-class 1 vs 2). The problem is that features with considerable variability will be lost in the average. Therefore, to understand the biologically significant variation and to be able to average repetitive particles without losing the meaningful differences that will make this analysis informative, it will be crucial to use classification methods that can separate the many 3D maps of individual particles in a tomogram into homogenous subgroups that can then be safely averaged. Three-dimensional classification of tomographic data is under development (Bartesaghi *et al.*, 2008; Walz *et al.*, 1997), but is not trivial due to the low SNR of cryo-tomograms, the anisotropy caused by the missing wedge, and the relatively large size of subtomographic volumes, which makes extensive analytical calculations time-consuming.

C. Structural Proteomics

Identifying and localizing specific proteins *in situ* and understanding how they interact is one of the timely challenges in cell biology. Using cryo-ET and tomographic averaging in combination with difference mapping between wild-type and mutant axonemes has allowed direct visualization of the affected gene products and their interactions with other components in flagella (Bui *et al.*, 2008; Nicastro *et al.*, 2006; Heuser *et al.*, in press). For a comprehensive exploration of the structural proteome of cilia and flagella, however, just comparing mutants will probably not be sufficient. For a more direct approach the development of improved labeling techniques with high-density labels for tagging proteins within frozen-hydrated cells, will be important (McIntosh *et al.*, 2005).

This chapter has described the application and future potential of cryo-ET and 3D averaging for studying the organization of axonemes. This technique has already become an invaluable tool for exploring cellular architecture with sufficiently high resolution to characterize the structural and functional organization of molecular complexes *in situ*, including axonemal dyneins and regulatory complexes (Bui *et al.*, 2008; Heuser *et al.*, in preparation; Nicastro *et al.*, 2006).

Acknowledgments

I thank David Mastronarde, Steve King, and Greg Pazour for critical comments on the manuscript, my own group at Brandeis University for their help, and Mary Porter for our great collaboration. I am also grateful to the National Institutes of Health (GM083122) and the Pew Foundation for their support.

References

Adrian, M., Dubochet, J., Lepault, J., and McDowall, A.W. (1984). Cryo-electron microscopy of viruses. *Nature* **308**, 32–36.

Afzelius, B. (1959). Electron microscopy of the sperm tail – Results obtained with a new fixative. *J. Biophys. Biochem. Cytol.* **5**, 269–278.

Al-Amoudi, A., Chang, J.J., Leforestier, A., McDowall, A.W., Salamin, L.M., Norlen, L.P., Richter, K., Blanc, N.S., Studer, D., and Dubochet, J. (2004). Cryo-electron microscopy of vitreous sections. *EMBO J.* **23**, 3583–3588.

Amat, F., Moussavi, F., Comolli, L.R., Elidan, G., Downing, K.H., and Horowitz, M. (2008). Markov random field based automatic image alignment for electron tomography. *J. Struct. Biol.* **161**, 260–275.

Bartesaghi, A., Sprechmann, P., Liu, J., Randall, G., Sapiro, G., and Subramaniam, S. (2008). Classification and 3D averaging with missing wedge correction in biological electron tomography. *J. Struct. Biol.* **162**, 436–450.

Bazan, C., Miller, M., and Blomgren, P. (2009). Structure enhancement diffusion and contour extraction for electron tomography of mitochondria. *J. Struct. Biol.* **166**, 144–155.

Bohm, J., Frangakis, A.S., Hegerl, R., Nickell, S., Typke, D., and Baumeister, W. (2000). Toward detecting and identifying macromolecules in a cellular context: Template matching applied to electron tomograms. *Proc. Natl. Acad. Sci. USA* **97**, 14245–14250.

Borgnia, M.J., Subramaniam, S., and Milne, J.L.S. (2008). Three-dimensional imaging of the highly bent architecture of Bdellovibrio bacteriovorus by using cryo-electron tomography. *J. Bacteriol.* **190**, 2588–2596.

Bostina, M., Bubeck, D., Schwartz, C., Nicastro, D., Filman, D.J., and Hogle, J.M. (2007). Single particle cryoelectron tomography characterization of the structure and structural variability of poliovirus-receptor-membrane complex at 30 A resolution. *J. Struct. Biol.* **160**, 200–210.

Bottcher, B., Wynne, S.A., and Crowther, R.A. (1997). Determination of the fold of the core protein of hepatitis B virus by electron cryomicroscopy. *Nature* **386**, 88–91.

Briegel, A., Ding, H.J., Li, Z., Werner, J., Gitai, Z., Dias, D.P., Jensen, R.B., and Jensen, G.J. (2008). Location and architecture of the Caulobacter crescentus chemoreceptor array. *Mol. Microbiol.* **69**, 30–41.

Bruggeller, P., and Mayer, E. (1980). Complete vitrification in pure liquid water and dilute aqueous-solutions. *Nature* **288**, 569–571.

Bui, K.H., Sakakibara, H., Movassagh, T., Oiwa, K., and Ishikawa, T. (2008). Molecular architecture of inner dynein arms in situ in Chlamydomonas reinhardtii flagella. *J. Cell Biol.* **183**, 923–932.

Burgess, S.A., Dover, S.D., and Woolley, D.M. (1991). Architecture of the outer arm dynein ATPase in an avian sperm flagellum, with further evidence for the B-link. *J. Cell Sci.* **98**, 17–26.

Burgess, S.A., Walker, M.L., Sakakibara, H., Knight, P.J., and Oiwa, K. (2003). Dynein structure and power stroke. *Nature* **421**, 715–718.

Butan, C., Winkler, D.C., Heymann, J.B., Craven, R.C., and Steven, A.C. (2008). RSV capsid polymorphism correlates with polymerization efficiency and envelope glycoprotein content: Implications that nucleation controls morphogenesis. *J. Mol. Biol.* **376**, 1168–1181.

Cambie, R., Downing, K.H., Typke, D., Glaeser, R.M., and Jin, J. (2007). Design of a microfabricated, two-electrode phase-contrast element suitable for electron microscopy. *Ultramicroscopy* **107**, 329–339.

Cardone, G., Grunewald, K., and Steven, A.C. (2005). A resolution criterion for electron tomography based on cross-validation. *J. Struct. Biol.* **151**, 117–129.

Cardone, G., Winkler, D.C., Trus, B.L., Cheng, N., Heuser, J.E., Newcomb, W.W., Brown, J.C., and Steven, A.C. (2007). Visualization of the herpes simplex virus portal in situ by cryo-electron tomography. *Virology* **361**, 426–434.

Cheng, Y., Boll, W., Kirchhausen, T., Harrison, S.C., and Walz, T. (2007). Cryo-electron tomography of clathrin-coated vesicles: Structural implications for coat assembly. *J. Mol. Biol.* **365**, 892–899.

Crowther, R.A., Derosier, D.J., and Klug, A. (1970). Reconstruction of 3 Dimensional Structure from projections and its application to electron microscopy. *Proc. R. Soc. Lond. A-Math. Phys. Sci.* **317**, 319–340.

Cyrklaff, M., Linaroudis, A., Boicu, M., Chlanda, P., Baumeister, W., Griffiths, G., and Krijnse-Locker, J. (2007). Whole cell cryo-electron tomography reveals distinct disassembly intermediates of vaccinia virus. *PLoS ONE* **2**, e420.

Danev, R., and Nagayama, K. (2001). Transmission electron microscopy with Zernike phase plate. *Ultramicroscopy* **88**, 243–252.

Deptuch, G., Besson, A., Rehak, P., Szelezniak, M., Wall, J., Winter, M., and Zhu, Y. (2007). Direct electron imaging in electron microscopy with monolithic active pixel sensors. *Ultramicroscopy* **107**, 674–684.

Derosier, D.J., and Klug, A. (1968). Reconstruction of 3 dimensional structures from electron micrographs. *Nature* **217**, 130–134.

Dubochet, J., Adrian, M., Chang, J.J., Homo, J.C., Lepault, J., McDowall, A.W., and Schultz, P. (1988). Cryo-electron microscopy of vitrified specimens. *Q. Rev. Biophys.* **21**, 129–228.

Dubochet, J., and McDowall, A.W. (1981). Vitrification of pure water for electron-microscopy. *J. Microsc. Oxford* **124**, Rp3–Rp4.

Echlin, P. (1991). Ice crystal damage and radiation effects in relation to microscopy and analysis at low-temperatures. *J. Microsc. Oxford* **161**, 159–170.

Egerton, R.F. (1996). "Electron Energy-Loss Spectroscopy in the Electron Microscope." Springer, New York.

Faucett, D.W. (1981). Cilia and flagella. *In* "The Cell," pp. 575–603. W. B. Saunders Company, Philadelphia, PA.

Fawcett, D.W., and Porter, K.R. (1954). A study of the fine structure of ciliated epithelia. *J. Morphol.* **94**, 221–282.

Fernandez-Moran, H. (1960). Low-temperature preparation techniques for electron microscopy of biological specimens based on rapid freezing with liquid helium II. *Ann. NY Acad. Sci.* **85**, 689–713.

Fliegauf, M., Benzing, T., and Omran, H. (2007). When cilia go bad: Cilia defects and ciliopathies. *Nat. Rev. Mol. Cell Biol.* **8**, 880–893.

Forster, F., Pruggnaller, S., Seybert, A., and Frangakis, A.S. (2008). Classification of cryo-electron sub-tomograms using constrained correlation. *J. Struct. Biol.* **161**, 276–286.

Frangakis, A.S., Bohm, J., Forster, F., Nickell, S., Nicastro, D., Typke, D., Hegerl, R., and Baumeister, W. (2002). Identification of macromolecular complexes in cryoelectron tomograms of phantom cells. *Proc. Natl. Acad. Sci. USA* **99**, 14153–14158.

Frangakis, A.S., and Forster, F. (2004). Computational exploration of structural information from cryo-electron tomograms. *Curr. Opin. Struct. Biol.* **14**, 325–331.

Frangakis, A.S., and Hegerl, R. (2001). Noise reduction in electron tomographic reconstructions using nonlinear anisotropic diffusion. *J. Struct. Biol.* **135**, 239–250.

Frangakis, A.S., and Hegerl, R. (2002). Segmentation of two- and three-dimensional data from electron microscopy using eigenvector analysis. *J. Struct. Biol.* **138**, 105–113.

Frank, J. (1975). Averaging of low exposure electron micrographs of non-periodic objects. *Ultramicroscopy* **1**, 159–162.

Frank, J. (ed.) (1992). "Electron Tomography: Three-Dimensional Imaging with the Transmission Electron Microscope." Plenum Press, New York.

Frank, J., Radermacher, M., Penczek, P., Zhu, J., Li, Y., Ladjadj, M., and Leith, A. (1996). SPIDER and WEB: Processing and visualization of images in 3D electron microscopy and related fields. *J. Struct. Biol.* **116**, 190–199.

Gardner, L.C., Otoole, E., Perrone, C.A., Giddings, T., and Porter, M.E. (1994). Components of a "dynein-regulatory-complex" are located at the junction between the radial spokes and the dynein arms in Chlamydomonas-flagella. *J. Cell Biol.* **127**, 1311–1325.

Garduno, E., Wong-Barnum, M., Volkmann, N., and Ellisman, M.H. (2008). Segmentation of electron tomographic data sets using fuzzy set theory principles. *J. Struct. Biol.* **162**, 368–379.

Gerdes, J.M., Davis, E.E., and Katsanis, N. (2009). The vertebrate primary cilium in development, homeostasis, and disease. *Cell* **137**, 32–45.

Gibbons, I.R. (1963). Studies on the protein components of cilia from Tetrahymena pyriformis. *Proc. Natl. Acad. Sci. USA* **50**, 1002–1010.

Gibbons, I.R. (1981). Transient flagellar waveforms during intermittent swimming in sea urchin sperm. II. Analysis of tubule sliding. *J. Muscle Res. Cell Motil.* **2**, 83–130.

Gibbons, I.R., and Grimstone, A.V. (1960). On flagellar structure in certain flagellates. *J. Biophys. Biochem. Cytol.* **7**, 697–715.

Gibbons, I.R., and Rowe, A.J. (1965). Dynein—A protein with adenosine triphosphatase activity from cilia. *Science* **149**, 424–426.

Gilkey, J.C., and Staehelin, L.A. (1986). Advances in ultra-rapid freezing for the preservation of cellular ultrastructure. *J. Electron Microsc. Tech.* **3**, 177–210.

Gillet, A., Sanner, M., Stoffler, D., and Olson, A. (2005). Tangible interfaces for structural molecular biology. *Structure* **13**, 483–491.

Goddard, T.D., and Ferrin, T.E. (2007). Visualization software for molecular assemblies. *Curr. Opin. Struct. Biol.* **17**, 587–595.

Goodenough, U.W., and Heuser, J.E. (1985). Substructure of inner dynein arms, radial spokes, and the central pair/projection complex of cilia and flagella. *J. Cell Biol.* **100**, 2008–2018.

Gorman, D.S., and Levine, R.P. (1965). Cytochrome f and plastocyanin: Their sequence in the photosynthetic electron transport chain of Chlamydomonas reinhardi. *Proc. Natl. Acad. Sci. USA* **54**, 1665–1669.

Grimm, R., Barmann, M., Hackl, W., Typke, D., Sackmann, E., and Baumeister, W. (1997). Energy filtered electron tomography of ice-embedded actin and vesicles. *Biophys J.* **72**, 482–489.

Grimm, R., Typke, D., Barmann, M., and Baumeister, W. (1996). Determination of the inelastic mean free path in ice by examination of tilted vesicles and automated most probable loss imaging. *Ultramicroscopy* **63**, 169–179.

Grunewald, K., Desai, P., Winkler, D.C., Heymann, J.B., Belnap, D.M., Baumeister, W., and Steven, A.C. (2003). Three-dimensional structure of herpes simplex virus from cryo-electron tomography. *Science* **302**, 1396–1398.

Hackney, D.D. (1996). The kinetic cycles of myosin, kinesin, and dynein. *Annu. Rev. Physiol.* **58**, 731–750.

Haider, M., Muller, H., Uhlemann, S., Zach, J., Loebau, U., and Hoeschen, R. (2008). Prerequisites for a Cc/Cs-corrected ultrahigh-resolution TEM. *Ultramicroscopy* **108**, 167–178.

Henderson, G.P., Gan, L., and Jensen, G.J. (2007). 3-D ultrastructure of O. tauri: Electron cryotomography of an entire eukaryotic cell. *PLoS ONE* **2**, e749.

Heuser, T., Raytchev, M., Krell, J., Porter, M.E., and Nicastro, D., The dynein regulatory complex is the nexin link and a major regulatory node in cilia and flagella (in press in *J. Cell Biol.*)

Heymann, J.B., Cardone, G., Winkler, D.C., and Steven, A.C. (2008). Computational resources for cryo-electron tomography in Bsoft. *J. Struct. Biol.* **161**, 232–242.

Hoenger, A., and Nicastro, D. (2007). Electron microscopy of microtubule-based cytoskeletal machinery. *Methods Cell Biol.* **79**, 437–462.

Hoops, H.J., and Witman, G.B. (1983). Outer doublet heterogeneity reveals structural polarity related to beat direction in Chlamydomonas flagella. *J. Cell Biol.* **97**, 902–908.

Hoppe, W., Gassmann, J., Hunsmann, N., Schramm, H.J., and Sturm, M. (1974). Three-dimensional reconstruction of individual negatively stained yeast fatty-acid synthetase molecules from tilt series in the electron microscope. *Hoppe-Seyler's Z. Physiol. Chem.* **355**, 1483–1487.

Hounsfield, G.N. (1973). Computerized transverse axial scanning (tomography). 1. Description of system. *Br. J. Radiol.* **46**, 1016–1022.

Iancu, C.V., Ding, H.J., Morris, D.M., Dias, D.P., Gonzales, A.D., Martino, A., and Jensen, G.J. (2007). The structure of isolated Synechococcus strain WH8102 carboxysomes as revealed by electron cryotomography. *J. Mol. Biol.* **372**, 764–773.

Ishikawa, T., Sakakibara, H., and Oiwa, K. (2007). The architecture of outer dynein arms in situ. *J. Mol. Biol.* **368**, 1249–1258.

Izard, J., Hsieh, C.E., Limberger, R.J., Mannella, C.A., and Marko, M. (2008). Native cellular architecture of Treponema denticola revealed by cryo-electron tomography. *J. Struct. Biol.* **163**, 10–17.

Jiang, M., Ji, Q., and McEwen, B. (2004). Model-based automated segmentation of kinetochore microtubule from electron tomography. *Conf. Proc. IEEE Eng. Med. Biol. Soc.* **3**, 1656–1659.

Jiang, W., Baker, M.L., Wu, Q., Bajaj, C., and Chiu, W. (2003). Applications of a bilateral denoising filter in biological electron microscopy. *J. Struct. Biol.* **144**, 114–122.

King, S.M., and Kamiya, R. (2009). Axonemal dyneins: Assembly, structure, and force generation. *In* "The *Chlamydomonas* Sourcebook" (E. Harris, D. Stern, and G. Witman, eds.), Vol. 3, pp. 131–208. Academic Press, Oxford, GB.

Kistler, J., and Kellenberger, E. (1977). Collapse phenomena in freeze-drying. *J. Ultrastruct. Res.* **59**, 70–75.

Koster, A.J., Grimm, R., Typke, D., Hegerl, R., Stoschek, A., Walz, J., and Baumeister, W. (1997). Perspectives of molecular and cellular electron tomography. *J. Struct. Biol.* **120**, 276–308.

Kremer, J.R., Mastronarde, D.N., and McIntosh, J.R. (1996). Computer visualization of three-dimensional image data using IMOD. *J. Struct. Biol.* **116**, 71–76.

Kurner, J., Frangakis, A.S., and Baumeister, W. (2005). Cryo-electron tomography reveals the cytoskeletal structure of Spiroplasma melliferum. *Science* **307**, 436–438.

Lanzavecchia, S., Cantele, F., Bellon, P.L., Zampighi, L., Kreman, M., Wright, E., Zampighi, G.A., (2005). Conical tomography of freeze-fracture replicas: a method for the study of integral membrane proteins inserted in phospholipid bilayers. *J. Struct. Biol.* **149**, 87–98.

Lawrence, M.C. (1992). Least-squares method of alignment using markers. *In* "Electron Tomography: Three-Dimensional Imaging with the Transmission Electron Microscope" (J. Frank, ed.), pp. 197–204. Plenum Press, New York.

Lebbink, M.N., Geerts, W.J.C., van der Krift, T.P., Bouwhuis, M., Hertzberger, L.O., Verkleij, A.J., and Koster, A.J. (2007). Template matching as a tool for annotation of tomograms of stained biological structures. *J. Struct. Biol.* **158**, 327–335.

Lucic, V., Forster, F., and Baumeister, W. (2005). Structural studies by electron tomography: From cells to molecules. *Annu. Rev. Biochem.* **74**, 833–865.

Lupetti, P., Lanzavecchia, S., Mercati, D., Cantele, F., Dallai, R., and Mencarelli, C. (2005). Three-dimensional reconstruction of axonemal outer dynein arms in situ by electron tomography. *Cell Motil. Cytoskeleton* **62**, 69–83.

Majorovits, E., Barton, B., Schultheiss, K., Perez-Willard, F., Gerthsen, D., and Schroder, R.R. (2007). Optimizing phase contrast in transmission electron microscopy with an electrostatic (Boersch) phase plate. *Ultramicroscopy* **107**, 213–226.

Mallik, R., Carter, B.C., Lex, S.A., King, S.J., and Gross, S.P. (2004). Cytoplasmic dynein functions as a gear in response to load. *Nature* **427**, 649–652.

Manton, I., Clarke, B., Greenwood, A.D., and Flint, E.A. (1952). Further observations on the structure of plant cilia, by a combination of visual and electron microscopy. *J. Exp. Bot.* **3**, 204–215.

Marabini, R., Herman, G.T., and Carazo, J.M. (1998). 3D reconstruction in electron microscopy using ART with smooth spherically symmetric volume elements (blobs). *Ultramicroscopy* **72**, 53–65.

Marko, M., and Hsieh, C.E. (2007). Three-dimensional cryotransmission electron microscopy of cells and organelles. *Methods Mol. Biol.* **369**, 407–429.

Martin, K., Ibanez, L., Avila, L., Barre, S., and Kaspersen, J.H. (2005). Integrating segmentation methods from the Insight Toolkit into a visualization application. *Med. Image Anal.* **9**, 579–593.

Mastronarde, D.N. (1997). Dual-axis tomography: An approach with alignment methods that preserve resolution. *J. Struct. Biol.* **120**, 343–352.

Mastronarde, D.N. (2005). Automated electron microscope tomography using robust prediction of specimen movements. *J. Struct. Biol.* **152**, 36–51.

Mastronarde, D.N. (2006). Fiducial marker and hybrid alignment methods for single- and double-axis tomography. *In* "Electron Tomography: Methods for Three-Dimensional Visualization of Structures in the Cell" (J. Frank, ed.), pp. 163–185. Springer, Berlin.

Mastronarde, D.N. (2008). Correction for non-perpendicularity of beam and tilt axis in tomographic reconstructions with the IMOD package. *J. Microsc.Oxford* **230**, 212–217.

Mastronarde, D.N., O'Toole, E.T., McDonald, K.L., McIntosh, J.R., and Porter, M.E. (1992). Arrangement of inner dynein arms in wild-type and mutant flagella of Chlamydomonas. *J. Cell Biol.* **118**, 1145–1162.

McEwen, B.F., Downing, K.H., and Glaeser, R.M. (1995). The relevance of dose-fractionation in tomography of radiation-sensitive specimens. *Ultramicroscopy* **60**, 357–373.

McEwen, B.F., Hsieh, C.E., Mattheyses, A.L., and Rieder, C.L. (1998). A new look at kinetochore structure in vertebrate somatic cells using high-pressure freezing and freeze substitution. *Chromosoma* **107**, 366–375.

McEwen, B.F., and Marko, M. (1999). Three-dimensional transmission electron microscopy and its application to mitosis research. *Methods Cell Biol.* **61**, 81–111.

McEwen, B.F., Marko, M., Hsieh, C.E., and Mannella, C. (2002). Use of frozen-hydrated axonemes to assess imaging parameters and resolution limits in cryoelectron tomography. *J. Struct. Biol.* **138**, 47–57.

McEwen, B.F., Renken, C., Marko, M., and Mannella, C. (2008). Chapter 6: Principles and practice in electron tomography. *Methods Cell Biol.* **89**, 129–168.

McIntosh, R., Nicastro, D., and Mastronarde, D. (2005). New views of cells in 3D: An introduction to electron tomography. *Trends Cell Biol.* **15**, 43–51.

Medalia, O., Weber, I., Frangakis, A.S., Nicastro, D., Gerisch, G., and Baumeister, W. (2002). Macromolecular architecture in eukaryotic cells visualized by cryoelectron tomography. *Science* **298**, 1209–1213.

Milazzo, A.C., Leblanc, P., Duttweiler, F., Jin, L., Bouwer, J.C., Peltier, S., Ellisman, M., Bieser, F., Matis, H.S., Wieman, H., Denes, P., Kleinfelder, S., and Xuong, N.H. (2005). Active pixel sensor array as a detector for electron microscopy. *Ultramicroscopy* **104**, 152–159.

Mitchell, D.R. (1994). Cell and molecular biology of flagellar dyneins. *Int. Rev. Cytol.* **155**, 141–180.

Mitchell, D.R. (2000). Chlamydomonas flagella. *J. Phycol.* **36**, 261–273.

Mitchell, D.R. (2009). The flagellar central pair apparatus. In "The Chlamydomonas Sourcebook" (E. Harris, D. Stern, and G. Witman, eds.), Vol. 3, pp. 235–252. Academic Press, Oxford, GB.

Mizuno, N., Toba, S., Edamatsu, M., Watai-Nishii, J., Hirokawa, N., Toyoshima, Y.Y., and Kikkawa, M. (2004). Dynein and kinesin share an overlapping microtubule-binding site. *EMBO J.* **23**, 2459–2467.

Moor, H., Bellin, G., Sandri, C., and Akert, K. (1980). The influence of high pressure freezing on mammalian nerve tissue. *Cell Tissue Res.* **209**, 201–216.

Murphy, G.E., Leadbetter, J.R., and Jensen, G.J. (2006). In situ structure of the complete Treponema primitia flagellar motor. *Nature* **442**, 1062–1064.

Murray, J.M. (1986). Electron microscopy of frozen hydrated eukaryotic flagella. *J. Ultrastruct. Mol. Struct. Res.* **95**, 196–209.

Nagayama, K., and Danev, R. (2008). Phase contrast electron microscopy: Development of thin-film phase plates and biological applications. *Philos Trans. R. Soc. B* **363**, 2153–2162.

Narasimha, R., Aganj, I., Bennett, A.E., Borgnia, M.J., Zabransky, D., Sapiro, G., McLaughlin, S.W., Milne, J.L.S., and Subramaniam, S. (2008). Evaluation of denoising algorithms for biological electron tomography. *J. Struct. Biol.* **164**, 7–17.

Nicastro, D., Frangakis, A.S., Typke, D., and Baumeister, W. (2000). Cryo-electron tomography of Neurospora mitochondria. *J. Struct. Biol.* **129**, 48–56.

Nicastro, D., McIntosh, J.R., and Baumeister, W. (2005). 3D structure of eukaryotic flagella in a quiescent state revealed by cryo-electron tomography. *Proc. Natl. Acad. Sci. USA* **102**, 15889–15894.

Nicastro, D., Schwartz, C., Pierson, J., Gaudette, R., Porter, M.E., and McIntosh, J.R. (2006). The molecular architecture of axonemes revealed by cryoelectron tomography. *Science* **313**, 944–948.

Nickell, S., Forster, F., Linaroudis, A., Del Net, W., Beek, F., Hegerl, R., Baumeister, W., and Plitzko, J.M. (2005). TOM software toolbox: Acquisition and analysis for electron tomography. *J. Struct. Biol.* **149**, 227–234.

Noske, A.B., Costin, A.J., Morgan, G.P., and Marsh, B.J. (2008). Expedited approaches to whole cell electron tomography and organelle mark-up in situ in high-pressure frozen pancreatic islets. *J. Struct. Biol.* **161**, 298–313.

O'Toole, E.T., Giddings, T.H., Jr., and Dutcher, S.K. (2007). Understanding microtubule organizing centers by comparing mutant and wild-type structures with electron tomography. *Methods Cell Biol.* **79**, 125–143.

Pazour, G.J., Agrin, N., Leszyk, J., and Witman, G.B. (2005). Proteomic analysis of a eukaryotic cilium. *J. Cell Biol.* **170**, 103–113.

Pease, D.C. (1963). Ultrastructure of flagellar fibrils. *J. Cell Biol.* **18**, 313–326.

Penczek, P.A. (2002). Three-dimensional spectral signal-to-noise ratio for a class of reconstruction algorithms. *J. Struct. Biol.* **138**, 34–46.

Penczek, P.A., Marko, M., Buttle, K., and Frank, J. (1995). Double-tilt electron tomography. *Ultramicroscopy* **60**, 393–410.

Penczek, P.A., Radermacher, M., and Frank, J. (1992). Three-dimensional reconstruction of single particles embedded in ice. *Ultramicroscopy* **40**, 33–53.

Perrone, C.A., Myster, S.H., Bower, R., O'Toole, E.T., and Porter, M.E. (2000). Insights into the structural organization of the I1 inner arm dynein from a domain analysis of the 1 beta dynein heavy chain. *Mol. Biol. Cell* **11**, 2297–2313.

Pfister, K.K., Shah, P.R., Hummerich, H., Russ, A., Cotton, J., Annuar, A.A., King, S.M., and Fisher, E.M. (2006). Genetic analysis of the cytoplasmic dynein subunit families. *PLoS Genet.* **2**, e1.

Porter, M.E. (1996). Axonemal dyneins: Assembly, organization, and regulation. *Curr. Opin. Cell Biol.* **8**, 10–17.

Porter, M.E., and Sale, W.S. (2000). The 9 + 2 axoneme anchors multiple inner arm dyneins and a network of kinases and phosphatases that control motility. *J. Cell Biol.* **151**, F37–F42.

Radermacher, M., Wagenknecht, T., Verschoor, A., and Frank, J. (1986). A new 3-D reconstruction scheme applied to the 50S ribosomal subunit of *E. coli*. *J. Microsc.* **141**, RP1–RP2.

Reck-Peterson, S.L., Yildiz, A., Carter, A.P., Gennerich, A., Zhang, N., and Vale, R.D. (2006). Single-molecule analysis of dynein processivity and stepping behavior. *Cell* **126**, 335–348.

Ress, D., Harlow, M.L., Schwarz, M., Marshall, R.M., and McMahan, U.J. (1999). Automatic acquisition of fiducial markers and alignment of images in tilt series for electron tomography. *J. Electron Microsc.* **48**, 277–287.

Roberts, A.J., Numata, N., Walker, M.L., Kato, Y.S., Malkova, B., Kon, T., Ohkura, R., Arisaka, F., Knight, P.J., Sutoh, K., and Burgess, S.A. (2009). AAA+ Ring and linker swing mechanism in the dynein motor. *Cell* **136**, 485–495.

Roos, N., Kinde, U., and Morgan, J.A. (1990). Morphology of rat exocrine pancreas prepared by anhydrous cryo-procedures. *J. Electron Microsc. Tech.* **14**, 39–45.

Ross, J.L., Wallace, K., Shuman, H., Goldman, Y.E., and Holzbaur, E.L. (2006). Processive bidirectional motion of dynein-dynactin complexes in vitro. *Nat. Cell Biol.* **8**, 562–570.

Rouiller, I., Xu, X.P., Amann, K.J., Egile, C., Nickell, S., Nicastro, D., Li, R., Pollard, T.D., Volkmann, N., and Hanein, D. (2008). The structural basis of actin filament branching by the Arp2/3 complex. *J. Cell Biol.* **180**, 887–895.

Rupp, G., O'Toole, E., Gardner, L.C., Mitchell, B.F., and Porter, M.E. (1996). The sup-pf-2 mutations of Chlamydomonas alter the activity of the outer dynein arms by modification of the gamma-dynein heavy chain. *J. Cell Biol.* **135**, 1853–1865.

Sabatini, D.D., Miller, F., and Barrnett, R.J. (1964). Aldehyde fixation for morphological and enzyme histochemical studies with the electron microscope. *J. Histochem. Cytochem.* **12**, 57–71.

Sakato, M., and King, S.M. (2004). Design and regulation of the AAA+ microtubule motor dynein. *J. Struct. Biol.* **146**, 58–71.

Sale, W.S., and Satir, P. (1977). Direction of active sliding of microtubules in Tetrahymena cilia. *Proc. Natl. Acad. Sci. USA* **74**, 2045–2049.

Salvi, E., Cantele, F., Zampighi, L., Fain, N., Pigino, G., Zampighi, G., and Lanzavecchia, S. (2008). JUST (Java User Segmentation Tool) for semi-automatic segmentation of tomographic maps. *J. Struct. Biol.* **161**, 287–297.

Samso, M., and Koonce, M.P. (2004). 25 Angstrom resolution structure of a cytoplasmic dynein motor reveals a seven-member planar ring. *J. Mol. Biol.* **340**, 1059–1072.

Sandberg, K., and Brega, M. (2007). Segmentation of thin structures in electron micrographs using orientation fields. *J. Struct. Biol.* **157**, 403–415.

Satir, P. (1968). Studies on Cilia. 3. Further studies on cilium tip and a sliding filament model of ciliary motility. *J. Cell Biol.* **39**, 77–94.

Satir, P. (1984). The generation of ciliary motion. *J. Protozool.* **31**, 8–12.

Saxberg, B.E.H., and Saxton, W.O. (1981). Quantum noise in 2d projections and 3d reconstructions. *Ultramicroscopy* **6**, 85–90.

Saxton, W.O., and Baumeister, W. (1982). The correlation averaging of a regularly arranged bacterial cell envelope protein. *J. Microsc.* **127**, 127–138.

Saxton, W.O., Baumeister, W., and Hahn, M. (1984). 3-Dimensional reconstruction of imperfect two-dimensional crystals. *Ultramicroscopy* **13**, 57–70.

Schmid, M.F., Paredes, A.M., Khant, H.A., Soyer, F., Aldrich, H.C., Chiu, W., and Shively, J.M. (2006). Structure of Halothiobacillus neapolitanus carboxysomes by cryo-electron tomography. *J. Mol. Biol.* **364**, 526–535.

Shampo, M.A., and Kyle, R.A. (1996). Allan Cormack—codeveloper of computed tomographic scanner. *Mayo Clin Proc.* **71**, 288.

Shima, T., Imamula, K., Kon, T., Ohkura, R., and Sutoh, K. (2006). Head-head coordination is required for the processive motion of cytoplasmic dynein, an AAA+ molecular motor. *J. Struct. Biol.* **156**, 182–189.

Snell, W.J., Pan, J., and Wang, Q. (2004). Cilia and flagella revealed: From flagellar assembly in Chlamydomonas to human obesity disorders. *Cell* **117**, 693–697.

Stewart, M., and Vigers, G. (1986). Electron-microscopy of frozen-hydrated biological-material. *Nature* **319**, 631–636.

Stoschek, A., and Hegerl, R. (1997). Denoising of electron tomographic reconstructions using multiscale transformations. *J. Struct. Biol.* **120**, 257–265.

Sui, H., and Downing, K.H. (2006). Molecular architecture of axonemal microtubule doublets revealed by cryo-electron tomography. *Nature* **442**, 475–478.

Suloway, C., Shi, J., Cheng, A., Pulokas, J., Carragher, B., Potter, C.S., Zheng, S.Q., Agard, D.A., and Jensen, G.J. (2009). Fully automated, sequential tilt-series acquisition with Leginon. *J. Struct. Biol.* **167**, 11–18.

Summers, K.E., and Gibbons, I.R. (1971). Adenosine triphosphate-induced sliding of tubules in trypsin-treated flagella of sea-urchin sperm. *Proc. Natl. Acad. Sci. USA* **68**, 3092–3096.

Taylor, K.A., and Glaeser, R.M. (1974). Electron diffraction of frozen, hydrated protein crystals. *Science* **186**, 1036–1037.

Taylor, K.A., and Glaeser, R.M. (1976). Electron microscopy of frozen hydrated biological specimens. *J. Ultrastruct. Res.* **55**, 448–456.

Templeton, N.S., Lasic, D.D., Frederik, P.M., Strey, H.H., Roberts, D.D., and Pavlakis, G.N. (1997). Improved DNA: Liposome complexes for increased systemic delivery and gene expression. *Nat. Biotechnol.* **15**, 647–652.

Toba, S., Watanabe, T.M., Yamaguchi-Okimoto, L., Toyoshima, Y.Y., and Higuchi, H. (2006). Overlapping hand-over-hand mechanism of single molecular motility of cytoplasmic dynein. *Proc. Natl. Acad. Sci. USA* **103**, 5741–5745.

Tyler, A. (1949). A simple non-injurious method for inducing repeated spawning of sea urchins and sand-dollars. *Collect. Net.* **19**, 19–20.

Unser, M., Trus, B.L., and Steven, A.C. (1987). A new resolution criterion based on spectral signal-to-noise ratios. *Ultramicroscopy* **23**, 39–51.

Vallee, R.B., Williams, J.C., Varma, D., and Barnhart, L.E. (2004). Dynein: An ancient motor protein involved in multiple modes of transport. *J. Neurobiol.* **58**, 189–200.

van der Heide, P., Xu, X.P., Marsh, B.J., Hanein, D., and Volkmann, N. (2007). Efficient automatic noise reduction of electron tomographic reconstructions based on iterative median filtering. *J. Struct. Biol.* **158**, 196–204.

van Heel, M., and Schatz, M. (2005). Fourier shell correlation threshold criteria. *J. Struct. Biol.* **151**, 250–262.

Volkmann, N. (2002). A novel three-dimensional variant of the watershed transform for segmentation of electron density maps. *J. Struct. Biol.* **138**, 123–129.

Walz, J., Typke, D., Nitsch, M., Koster, A.J., Hegerl, R., and Baumeister, W. (1997). Electron tomography of single ice-embedded macromolecules: Three-dimensional alignment and classification. *J. Struct. Biol.* **120**, 387–395.

Warner, F.D. (1976). Ciliary inter-microtubule bridges. *J. Cell Sci.* **20**, 101–114.

Winkler, H. (2007). 3D reconstruction and processing of volumetric data in cryo-electron tomography. *J. Struct. Biol.* **157**, 126–137.

Winkler, H., and Taylor, K.A. (2006). Accurate marker-free alignment with simultaneous geometry determination and reconstruction of tilt series in electron tomography. *Ultramicroscopy* **106**, 240–254.

Winkler, H., Zhu, P., Liu, J., Ye, F., Roux, K.H., and Taylor, K.A. (2009). Tomographic subvolume alignment and subvolume classification applied to myosin V and SIV envelope spikes. J. Struct. Biol. 165, 64–77.

Wirschell, M., Nicastro, D., Porter, M.E., and Sale, W.S. (2009). The regulation of axonemal bending. In "The Chlamydomonas Sourcebook" (E. Harris, D. Stern, and G. Witman, eds.), Vol. 3, pp. 253–282. Academic Press, Oxford, GB.

Witman, G.B. (1986). Isolation of Chlamydomonas flagella and flagellar axonemes. *Meth. Enzymol.* **134**, 280–290.

Witman, G.B., Carlson, K., Berliner, J., and Rosenbaum, J.L. (1972). Chlamydomonas flagella. I. Isolation and electrophoretic analysis of microtubules, matrix, membranes, and mastigonemes. *J. Cell Biol.* **54**, 507–539.

Wright, E.R., Schooler, J.B., Ding, H.J., Kieffer, C., Fillmore, C., Sundquist, W.I., and Jensen, G.J. (2007). Electron cryotomography of immature HIV-1 virions reveals the structure of the CA and SP1 Gag shells. *EMBO J.* **26**, 2218–2226.

Xiong, Q., Morphew, M.K., Schwartz, C.L., Hoenger, A.H., Mastronarde, D.N. (2009). CTF determination and correction for low dose tomographic tilt series. *J. Struct. Biol.*

Yagi, T., Uematsu, K., Liu, Z., and Kamiya, R. (2009). Identification of dyneins that localize exclusively to the proximal portion of Chlamydomonas flagella. *J. Cell Sci.* **122**, 1306–1314.

Yang, P., and Smith, E.F. (2009). The flagellar radial spokes. In "The Chlamydomonas Sourcebook" (E. Harris, D. Stern, G. Witman, eds.), Vol. 3, pp. 209–234. Academic Press, Oxford, GB.

Zheng, S.Q., Keszthelyi, B., Branlund, E., Lyle, J.M., Braunfeld, M.B., Sedat, J.W., and Agard, D.A. (2007). UCSF tomography: An integrated software suite for real-time electron microscopic tomographic data collection, alignment, and reconstruction. *J. Struct. Biol.* **157**, 138–147.

Ziese, U., Janssen, A.H., Murk, J.L., Geerts, W.J.C., Van der Krift, T., Verkleij, A.J., and Koster, A.J. (2002). Automated high-throughput electron tomography by pre-calibration of image shifts. *J. Microsc. Oxford* **205**, 187–200.

CHAPTER 2

Electron Microscopic Imaging and Analysis of Isolated Dynein Particles

Anthony J. Roberts *and* Stan A. Burgess

Astbury Centre for Structural Molecular Biology, Institute of Molecular and Cellular Biology, Faculty of Biological Sciences, University of Leeds, Leeds LS2 9JT, United Kingdom

Abstract

Despite more than 40 years of investigation since the discovery of dynein [Gibbons, I. R. and Rowe, A. J. (1965). *Science* **149**, 424–426] our understanding of how this microtubule-based motor generates force and movement remains frustratingly incomplete at the atomic level. Electron microscopy (EM) has played a major role in establishing dynein's complex architecture and its nucleotide-dependent conformational changes. In this chapter we review recent structural studies and describe in

978-0-12-374973-4
DOI: 10.1016/S0091-679X(08)91002-5

detail negative stain EM and computational single-particle image processing techniques that have been used to investigate dynein. We describe studies of both *Chlamydomonas* flagellar inner arm dynein-c and recombinant cytoplasmic dynein from *Dictyostelium*. We also detail methods for locating green fluorescent protein (GFP) and blue fluorescent protein (BFP) tags inserted at specific locations within the dynein motor, which can be used to map subdomains and conformational changes.

I. Introduction

The dynein motor is large. Its heavy chain, which contains the motor domain, is typically ~500–540 kDa with the motor domain itself ~380 kDa (Koonce and Samsó, 1996). This is about ten times greater than the motor domain of kinesin, the other class of microtubule motor. Despite being discovered about 20 years after dynein (Vale *et al.*, 1985), our understanding of kinesin's structure and mechanism far exceeds that of dynein. Nevertheless, recent progress with dynein has been made, and in this chapter we review the electron microscopy (EM) techniques that have shed new light on dynein's mechanism of action.

Dynein's motor domain is not only larger than kinesin's, but is also much more extended in structure. Whereas kinesin's motor is contained within a globular structure ~6 nm in diameter (Vale and Milligan, 2000), dynein's motor comprises a ringlike head domain ~13 nm in diameter, together with an ~15 nm-long coiled-coil stalk that protrudes from the head (Fig. 1A). The head contains dynein's ATP hydrolysis sites, and the stalk carries at its distal end the critical ATP-sensitive microtubule-binding site within a small ~4 nm-diameter globular domain (Fig. 1). This organization places the

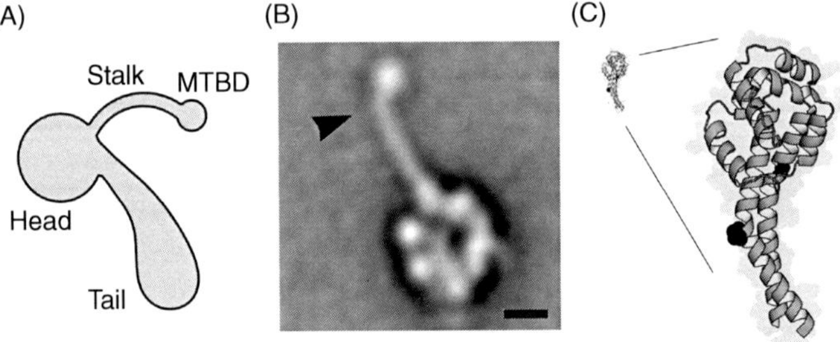

Fig. 1 Overview of dynein structure. (A) Cartoon showing the domain organization of dynein, including the microtubule-binding domain (MTBD) at the distal end of the coiled-coil stalk. (B) Cytoplasmic dynein motor domain (i.e., head and stalk) imaged by negative-stain EM followed by single-particle image processing. The stalk has a kink (arrowhead) at its distal end close to the globular MTBD and corresponding closely to the kink seen in the atomic model (C). Scale bar 5 nm. (C) Atomic model of the distal stalk and MTBD (PDB accession code 3ERR; Carter *et al.*, 2008). Small cartoon (left) is shown at approximately the same scale as in (B). Conserved prolines responsible for the kink are shown in black space fill (P3285 and P3409: mouse cytoplasmic dynein 1 sequence). The figure was made using PyMol (DeLano, W.L., 2002).

sites of ATP hydrolysis and microtubule binding in dynein ~28 nm apart: about ten times the equivalent distance of their counterparts in kinesin.

These features of dynein present significant challenges to structural studies of its intact motor domain. Dynein's large size rules out structure determination by nuclear magnetic resonance (NMR). Crystallographic studies are problematic because the motor is extended and the stalk flexible (Burgess *et al.*, 2003). Thus, while numerous atomic resolution structures have been obtained of kinesin (and also of the actin-based motor myosin) in various nucleotide states, no atomic resolution structures exist for the entire motor domain of dynein. Truncating the motor domain of dynein to isolate subdomains renders the motor nonfunctional (Gee *et al.*, 1997), making any resulting structures potentially difficult to interpret (or even to assay for functionality and correct folding). Nevertheless, this approach has been applied to dynein's microtubule-binding domain (MTBD), producing the first atomic resolution structure of dynein's heavy chain (Carter *et al.*, 2008). This significant breakthrough was achieved by creating a chimeric protein in which the MTBD was fused to the coiled coil of a small compact protein, seryl tRNA synthetase (Gibbons *et al.*, 2005). This required a considerable amount of work to characterize the behavior of the resulting fusion proteins (assayed for MT-binding affinity). This approach seems less suitable for the rest of the motor, making EM as the most feasible technique for structural studies of the intact dynein motor.

Structures of kinesin and myosin bound to their cytoskeletal tracks have been obtained by exploiting the ability of these motors to saturate fully the lattices of their tracks. Because their tracks have helical symmetry, it is possible to obtain 3D reconstructions from electron micrographs of frozen-hydrated specimens. Indeed, this is the major technique used to obtain structural information about motors bound to their tracks, and recent advances have improved the resolution of the kinesin–microtubule complex to ~9 Å—sufficient to resolve secondary structure elements (Sindelar and Downing, 2007). Unfortunately, this approach is unfeasible for intact dynein as its motor domain is too large to fully saturate all available binding sites on the microtubule lattice. Conversely, it has been used successfully to obtain the structures of various MTBD fragments of dynein bound to the microtubule (Carter *et al.*, 2008; Mizuno *et al.*, 2004). To date, the resolution achieved in these studies (~20–30 Å) has not yet approached the best obtained for kinesin-microtubule reconstructions, but the MTBD-microtubule density maps do suggest the geometry of attachment of the distal stalk and the site on tubulin where binding occurs.

The earliest and arguably most striking dynein EM images are those from freeze-etch replicas of cilia and flagella, showing the outer dynein arms *in situ* in unprecedented detail for their time (Goodenough and Heuser, 1982). Indeed, these remain some of the best images of the motor bound to its microtubule track—not only because they show the motors *in situ*, but also because the coiled-coil stalks are visible directly in micrographs without any image processing. The advantage of this technique is that the shadowing material (in this case platinum/carbon) favorably deposits on the fine structure of the stalk, thereby revealing its presence. The disadvantage is that only the outermost features of the shadowed surface are visible, so heavy chains closer to the interior of the axoneme are obscured.

Recent improvements in cryo-electron tomography (see Chapter 1 by Nicastro, this volume) have produced remarkable structural insights into the entire intact axoneme, including the multiple dynein heavy chains of the inner and outer dynein arms (Bui *et al.*, 2008; Ishikawa *et al.*, 2007; Nicastro *et al.*, 2005, 2006). The chief advantage of this approach is that the entire 3D structure is visualized rather than just the surfaces of favorable fracture planes (as in the freeze-etch technique). However, to date, the resolution of this technique (30–40 Å) does not reveal the stalks, leaving the motor-track interaction elusive. There is optimism, however, that technical improvements may overcome this limitation.

Other cryo-EM techniques have also been applied to *in vitro* dynein-microtubule systems. Microtubules sparsely decorated with monomeric cytoplasmic dynein and processed by single-particle techniques have shown in 3D the geometry of attachment of the head domain (Mizuno *et al.*, 2007). In an alternative approach, microtubules polymerized in the presence of flagellar outer arm dyneins created pairs of helically arranged microtubules cross-linked by two rows of ordered dynein complexes (Oda *et al.*, 2007). Processed by helical methods this study showed new details of the 3D arrangement of the heads and tails of the three heavy chains in this species of outer arm. However, neither of these studies showed the stalks. This was most likely caused by their limited resolutions (~26 and ~27 Å, respectively). With larger data sets and perhaps with improvements in processing, these approaches offer greater hope of higher resolution and visualization of the entire dynein motor in contact with its microtubule track. A recent development in revealing the stalk of an intact dynein motor bound to the microtubule has been obtained by a technique termed cryo-positive staining (Ueno *et al.*, 2008). In this *in vitro* technique, cytoplasmic microtubules and purified dimeric outer arm flagellar dyneins were mixed in the presence of low concentrations of an electron dense stain (0.01–0.05% uranyl acetate) and then vitrified as for conventional cryo-EM. The stain helps to contrast fine structures such as the stalks, which were seen in great detail in 2D averages.

The first structural studies of isolated dynein molecules employed the freeze-etch, rotary replication technique (Goodenough and Heuser, 1984). These established the characteristic tail-head-stalk arrangement of domains (the tail known then as the "stem"). This technique lacks the resolution to see substructure within the head and is limited to visualizing the surface of the molecule rather than its internal densities. Other work has employed negative staining, and early images suggested that the head possessed a cavity or channel running through it (Amos, 1989; Marchese-Ragona *et al.*, 1988), a detail unseen in freeze-etch and early negative stain studies of dyneins in flagellar fragments (Avolio *et al.*, 1984).

More recent work using negatively stained molecules analyzed by single-particle image processing has led to 3D reconstructions of the ringlike head domain of a recombinant cytoplasmic dynein motor domain (Samsó and Koonce, 2004; Samsó *et al.*, 1998). In our own 2D studies of flagellar inner-arm dynein-c from

Chlamydomonas flagella (Burgess *et al.*, 2003; Burgess *et al.*, 2004a), we were able to describe the conformation of the molecule in two different nucleotide states thought to mimic the post- and pre-powerstroke conformations of the motor. In the absence of nucleotide (the "unprimed" or "postpowerstroke" conformation) the tail emerges near the base of the stalk. In the presence of ATP and vanadate, which forms a dynein-ADP. Vi complex that mimics the dynein-ADP.Pi conformation (the "primed" or "prepowerstroke" conformation), the tail emerges further away from the stalk base. These observations were interpreted to originate from the swinging of a rodlike structure (the linker) relative to the ring-like head. The linker has subsequently been confirmed in cytoplasmic dynein and identified as the N-terminal region of the motor domain using GFP- and BFP-tagged constructs by negative stain EM and FRET (Kon *et al.*, 2005; Roberts *et al.*, 2009).

II. Rationale

Cryo-EM, where samples are frozen in a thin layer of vitreous water, has revolutionized biological EM because the specimen is preserved in a near-native environment (Dubochet *et al.*, 1988). While high-resolution information can be obtained, a disadvantage of this technique is the low signal-to-noise ratio in the images. Analyses of macromolecules with low molecular weight (<500 kDa) lacking internal symmetry and possessing small, flexible domains (such as dynein's stalk) remain challenging.

By contrast, negative-stain EM is a high-contrast technique. With a resolution of only ~20 Å, it falls far short of the resolution achievable with X-ray crystallography and NMR (atomic resolution) and cryo-EM (which under favorable circumstances can reach sub-10 Å resolution, capable of revealing secondary structure). Negative staining relies on the molecule under investigation being adsorbed to a carbon film, stained, and then dried, with attendant concerns about distortion of its structure. Nevertheless, with appropriate caution and attention to potential artifacts, there is considerable evidence that this imaging technique can provide reliable structural information. For example, the stalk is revealed in detail, showing, for example, nucleotide-induced changes in its structure (Burgess *et al.*, 2003) and the preservation of its structure in cysteine mutants cross-linked to restrict helix–helix sliding within the stalk (Kon *et al.*, 2009). Moreover, a kink in the distal coiled coil revealed in the crystal structure of a stalk fragment chimera protein (Carter *et al.*, 2008) is clearly resolved by negative-stain EM (Fig. 1B and C), here in the context of the intact, fully functional motor (Roberts *et al.*, 2009).

The high contrast of negative-stain EM also makes this technique well suited for locating small subunits or subdomains within large macromolecules using, for example, specific labels to tag the protein. When no high-resolution structure of the macromolecule is available, such subdomain mapping can provide important insights into its structure and mechanism. Classically, antibody labels have been used. However, disadvantages of this technique include the possibility of incomplete labeling and disruption or perturbation of function (Lührmann and Stark, 2009). Below, we describe methods to locate green fluorescent protein (GFP)-based tags inserted at specific locations in dynein. Such genetic tagging has the advantage of 100% labeling efficiency, and the use of GFP and blue fluorescent protein (BFP) enables corroborating FRET studies to be carried out in solution.

III. Methods

Preparation of dynein samples suitable for imaging by negative-stain EM is covered elsewhere in this series (see 'Purification of Axonemal Dyneins and…' by King, volume 92 for flagellar dyneins and 'Protein Engineering Approaches to Study…' by Kon, Shima and Sutoh, volume 92 for recombinant cytoplasmic dyneins). It should be noted that dynein preparations should be made with a high degree of purity, since the presence of even a small amount of contaminant (e.g., a second isoform of flagellar dynein) could lead to erroneous interpretation.

A detailed description of all steps involved in collecting and processing EM images of flagellar dynein-c has been given (Burgess *et al.*, 2004b), including the preparation of grids, details of the negative staining method we use, the collection of images, and their processing. Briefly, we use the single-carbon film method in which molecules are applied to one surface of a continuous (i.e., nonholey) carbon film suspended over a copper EM grid. This differs from the so-called carbon sandwich technique, in which a second layer of carbon is applied to the first following staining (Frank, 2006). The resulting "carbon sandwich" more completely embeds the sandwiched molecules in the stain and therefore produces more accurate 2D projections of the 3D object, making it more suitable for 3D reconstruction of the imaged molecules. However, in having a second carbon film in the carbon sandwich, images are typically more noisy, especially when imaged under low-dose conditions, and there are reports that more flattening of the molecules occurs during drying, especially in molecules with large internal cavities (such as is likely for dynein) (Cheng *et al.*, 2006). To avoid these problems, and in particular to visualize the slender coiled-coil stalk domain, we use the single-carbon film method instead. This technique requires that we exercise caution when interpreting the resulting images since there is evidence that single-sided staining of the molecules occurs under these conditions. Such incomplete staining, in which parts of the molecules closest to the carbon film are fully embedded in stain while those furthest away are only partially embedded or not embedded at all, renders such images unsuitable for 3D reconstruction.

A. Preparation of Carbon Support Films

EM grids with a continuous carbon film suitable for negative staining can be purchased (e.g., from http://www.grid-tech.com, http://www.emsdiasum.com/, and others). We prefer to make our own, although specialist equipment is required. The carbon support film is prepared by evaporating carbon fiber (1-mm thick, Agar Scientific, Stansted, U.K.) onto freshly cleaved mica sheets (Agar Scientific, Stansted, U.K.) in a coating unit

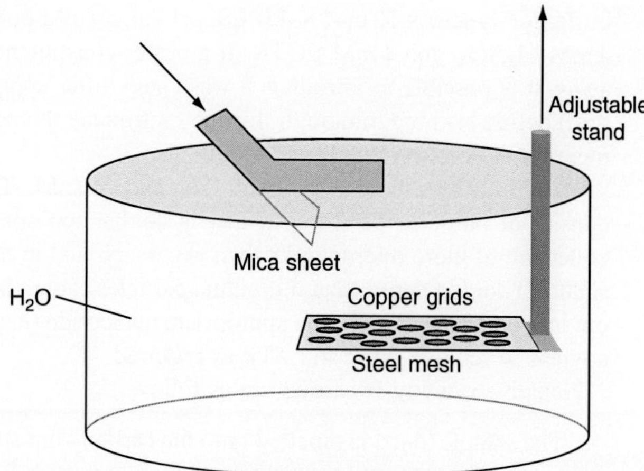

Fig. 2 Preparation of carbon-filmed grids for electron microscopy. Copper grids are washed once in acetone and twice in milliQ H_2O, and then laid shiny side up on a steel mesh resting on an adjustable stand under the surface of a bath of milliQ H_2O. Carbon films are floated onto the water surface above the grids by lowering the mica sheet into the water at an angle of ~45°. The carbon film is then collected onto the grids by gently raising the stand through the surface of the water while steering the carbon film using air jets (e.g., from an empty glass pipette or blowing) or an eyelash on a stick. The grids are dried at ambient temperature.

(e.g., Edwards E306A) under a vacuum of $\sim 10^{-6}$ mbar. We fix the carbon fiber between electrodes with a ~2-mm bow to ensure it is not under tension, and raise it ~170 mm above the mica to yield carbon films ~8 nm thick (Burgess *et al.*, 2004b). Current is applied rapidly through the carbon fiber, causing it to spark. Rapid evaporation produces carbon films of consistent thickness, which we mount onto 300 or 400 mesh copper grids (Agar Scientific; Fig. 2).

Before negative staining, carbon-filmed grids are treated to make them hydrophilic. A common way to do this is using glow discharge (Grassucci *et al.*, 2007), but we routinely use treatment with UV light for 30–50 min (Burgess *et al.*, 2004b). Though the surface chemistry is poorly understood, production of ozone by short-wavelength UV in the 182 nm range is critical for grids with good staining properties. We use an R-52 lamp (Ultra-Violet Products, Upland, CA, U.S.A.) modified with a custom bulb emitting UV radiation in this range. Grids are placed ~5 cm from the bulb with the carbon film facing up and irradiated while enclosed in a cardboard box to retain ozone and used immediately thereafter for negative staining.

B. Negative Staining

Purified dynein is diluted into an appropriate buffer before staining. Where possible we avoid reducing agents, detergents, sugars such as sucrose, glycerol, and high salt because they have a detrimental effect on the quality of staining. A typical dilution buffer for dynein is 10 mM K-PIPES, pH 7.0, 50 mM potassium acetate or 20 mM KCl, 4 mM $MgSO_4$, and 1 mM EGTA. If a buffer constituent cannot be avoided or diluted away, it is possible to introduce a wash step (after applying the diluted dynein to the grid) before staining, although this may introduce the possibility of protein rearrangement on the grid (Walker *et al.*, 2000).

We dilute dynein to a concentration of ~40 nM. This concentration produces a density of particles on the grid that is neither too sparse (which would require the collection of more micrographs than necessary, and in our experience leads to inferior staining) nor too dense (that individual particles cannot be resolved). Dilution is carried out into buffer containing the appropriate nucleotide (e.g., ADP, ATP plus vanadate) or apyrase to remove ADP and ATP as required.

Negative staining is carried out as follows:

i. The sample (5 μl) is pipetted onto the carbon-film side of a freshly UV-treated grid held in forceps, and molecules allowed to adsorb for ~30 s.
ii. Grids are then washed (if necessary) with two drops of wash buffer (including appropriate nucleotide, e.g., ATP and vanadate for the primed conformation). Drops are applied to the carbon side of the grid with a Pasteur pipette, then quickly flicked off to prevent wetting of the copper under side. Grids wetted on both sides typically yield poor staining.
iii. Molecules are then stained with two drops of 1% (w/v) of filtered aqueous uranyl acetate in the same way.
iv. Excess stain is wicked away from one edge of the grid on the carbon side using torn filter paper (Grade #1, Whatman).

C. Electron Microscopy

First we survey grids at 4000× magnification to locate negatively stained areas, and then examine critically the stain quality at 25,000–40,000× magnification. Choice of stain depth is critical for resolving fine parts of dynein's structure such as the stalk and inserted GFP/BFP tags. Deep staining obscures these features, whereas shallow staining results in poor overall contrast. Optimal imaging is typically found in intermediate areas, where molecules are delineated with a clear outline of stain (Burgess *et al.*, 2004b) (Fig. 3). Total electron dose is generally not controlled, but is typically high, ~50–1000 $e^-/Å^2$ (including time to focus). While low-dose imaging (10–30 $e^-/Å^2$) preserves high-resolution features (Unwin, 1974), stain migration upon radiation can improve the definition of fine protein structures, such as coiled coils (Walker *et al.*, 1991, 2000).

Raw micrographs of negatively stained molecules (Fig. 3) contain a wealth of structural information. However, they are noisy (low signal-to-noise ratio) owing to variations in the underlying carbon film and in the way stain crystallites accumulate around the molecules. To extract the maximum information from these images, it is necessary to apply image-processing techniques to reduce the noise. By aligning and averaging many images together into a single image, consistent information arising

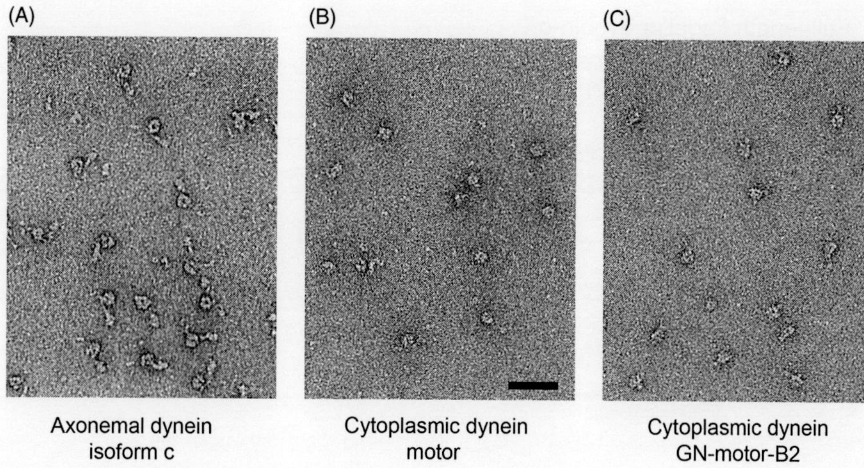

Fig. 3 Raw micrographs of negatively stained dynein molecules using the single-carbon film method. (A) Full-length inner-arm dynein-c purified from *Chlamydomonas* flagella. (B) Motor domain of recombinant cytoplasmic dynein from *Dictyostelium*. (C) Construct in (B) tagged with N-terminal GFP ("GN") and with BFP inserted just downstream of the second AAA+ module ("B2"). In these micrographs particles are optimally stained—the best stain depth shows particles as pale stain-excluding features on a dark background with a darker outline of stain surrounding each one. Noise in the images obscures details in the molecules in raw images; typically, stalks and the GFP and BFP moieties in (C) are difficult to detect. *Note*: the tails of full-length flagellar dynein-c (A) are clearly visible, but these have been deleted in the recombinant cytoplasmic proteins (B and C). Images were collected with a defocus of ~1 μm. Scale bar, 50 nm.

from the signal is reinforced while random noise is canceled. For individual, randomly oriented molecules such as dynein (Fig. 3) it is necessary first to computationally align the particles and then to classify them into homogeneous classes before calculating their averages and variances (Fig. 4). This is the technique called single-particle image processing.

D. Digitization and Image Preprocessing

Whether micrographs are collected on CCD camera or photographic emulsion, consideration should be made for the sampling on the object scale in the digitized images. A final pixel size at the specimen of ~5 Å is more than sufficient to retain information to ~20 Å, the resolution limit expected for molecules imaged in negative stain. For accurate measurements from digitized micrographs it is important to calibrate the magnification carefully. We calibrate our micrographs using a grid of negatively stained paramyosin filaments, recorded and digitized in the same way as the data. Paramyosin's axial spacing of 14.4 nm is unaltered by staining and dehydration (Elliott *et al.*, 1976).

Various software applications for single-particle image processing are available, including IMAGIC (van Heel *et al.*, 1996), EMAN (Ludtke *et al.*, 1999), and SPIDER

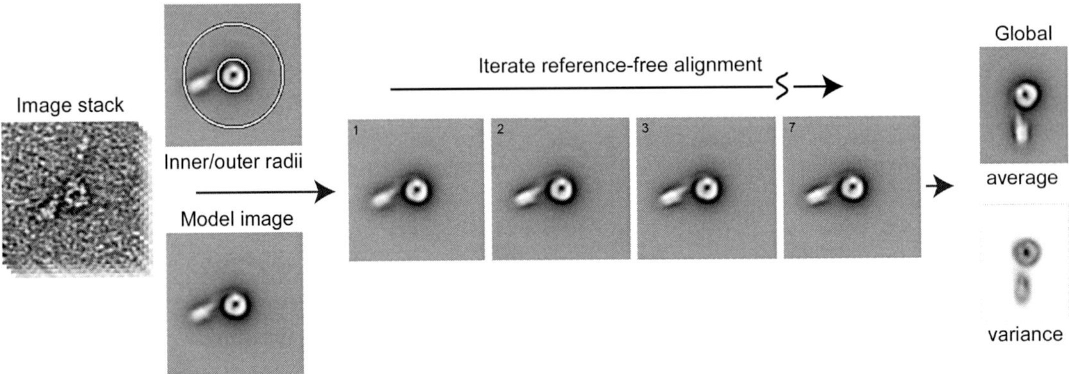

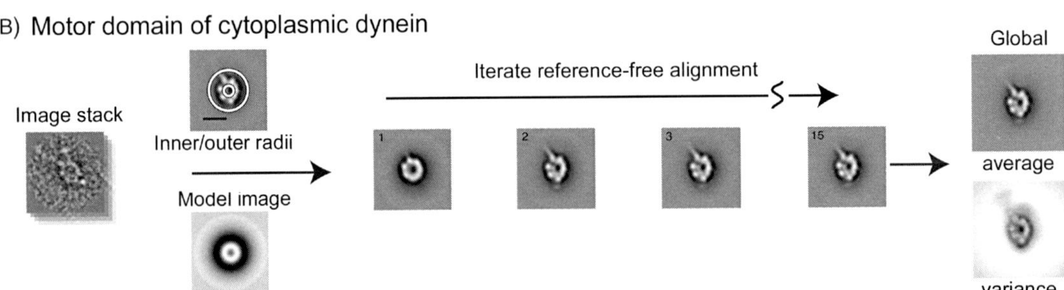

Fig. 4 Reference-free computational alignment of windowed dynein particles. Alignment strategy for (A) full-length flagellar dynein-c and (B) tailless recombinant cytoplasmic dynein motor domain. Circles show the inner and outer radii used for rotational alignment, overlaid on an image average of dynein [the inner radius in (B) (5 pixels) is the minimum permitted by the algorithm]. The black bar in (B) indicates the maximum allowed shift during translational alignment. Seven iterations were performed here for dynein-c (A), 15 for cytoplasmic dynein (B) using a model image (indicated) to maintain the particles centered within the window. After alignment, global average and variance images are calculated, often after an additional rotation and cropping of the window (as in A). The flexible stalk is not seen in its entirety in the global average in (B), but its location (emerging at the 11 o'clock position) is visible in the inverted variance image (black indicating high variance).

(Frank *et al.*, 1996). Image format conversion can be carried out using EM2EM (Image Science Software GmbH, Berlin, Germany), a free stand-alone application that can also convert between formats commonly used in single-particle image processing (e.g., TIFF, SPIDER, MRC, and IMAGIC). For 2D analysis of dynein, we have found SPIDER to be the most powerful because its scripting language allows the development of very sophisticated processing strategies for automation. In the text that follows, we refer to SPIDER commands using quotation marks (e.g., "WI" command).

Dynein particles in the digitized micrographs must first be identified. Particle coordinates, centered on dynein's head domain, can be selected manually using

SPIDER's graphical user-interface WEB (using the "Markers" function). A number of automated particle-picking methods are also available (reviewed in Zhu *et al.*, 2004). We routinely use the autoboxing feature in BOXER (Ludtke *et al.*, 1999). Autoboxing has the obvious advantage of dramatically increasing the speed of particle selection, though typically some manual screening must be applied to limit the false positives and false negatives that arise.

Particles are excised into individual windows (using SPIDER's "WI" command), sufficiently large to contain the (approximately) centered molecule with a sufficient region of surrounding background to allow image alignment (rotation and translation) errors to be restricted entirely to the background (rather than within the particle itself). Thus, for flagellar dynein-c, which contains head, stalk and tail domains, the window size is 140 × 140 pixels (70 × 70 nm). For the smaller motor domain (head and stalk only) of recombinant cytoplasmic dynein, we use a window size of 120 × 120 pixels (~60 × 60 nm). However, the appropriate window size depends upon the accuracy with which the "center" of the particle is recorded (larger for less accurate recording). In all cases we select the center of the head.

Once windowed we remove outlier pixels by thresholding pixel values in each window to a maximum of five standard deviations (5σ), either side of the mean ("TH" command). To remove gradual changes in stain depth across each windowed image, which can produce a very strong low-resolution feature that otherwise dominates image alignment, we then perform ramp subtraction ("RA" command) and then normalize each window to a mean pixel intensity of 0 and a standard deviation of 1 ("AR" command). Windowed particles are stored in a SPIDER image stack.

To obtain the best alignment and classification, it is necessary to start with sufficient numbers of particles. This is dependent on several factors, including the quality of staining, the range of orientations of the particles on the grid, and their conformational variability. A minimum number of particles would be typically ~2000. For dynein, in which there is considerable flexibility and more than one orientation, we start with about 10,000 particles, although this can increase to ~35,000 particles for more thorough investigations (Roberts *et al.*, 2009). When studying molecules, in different nucleotide-induced conformations, or tagged at different positions with GFP, we often combine all molecules together and perform a single grand alignment. This offers the advantages that all conformers and/or constructs are brought to a common alignment, and allows the possibility of removing misaligned or poorly stained particles (after the first round of classification) without bias (Figs. 5 and 6).

E. Image Alignment

Alignment of the molecules in a dataset must be as robust and reliable as possible to extract the most information in subsequent steps. Because the data contain some poorly stained molecules, and often molecules in several different orientations and even conformations, the reliability of a single alignment can be variable. We usually perform two alignments (Fig. 5). In the first we align the molecules as best as we can (Fig. 4) and then classify them (Fig. 5). From the resulting classes we remove bad particles

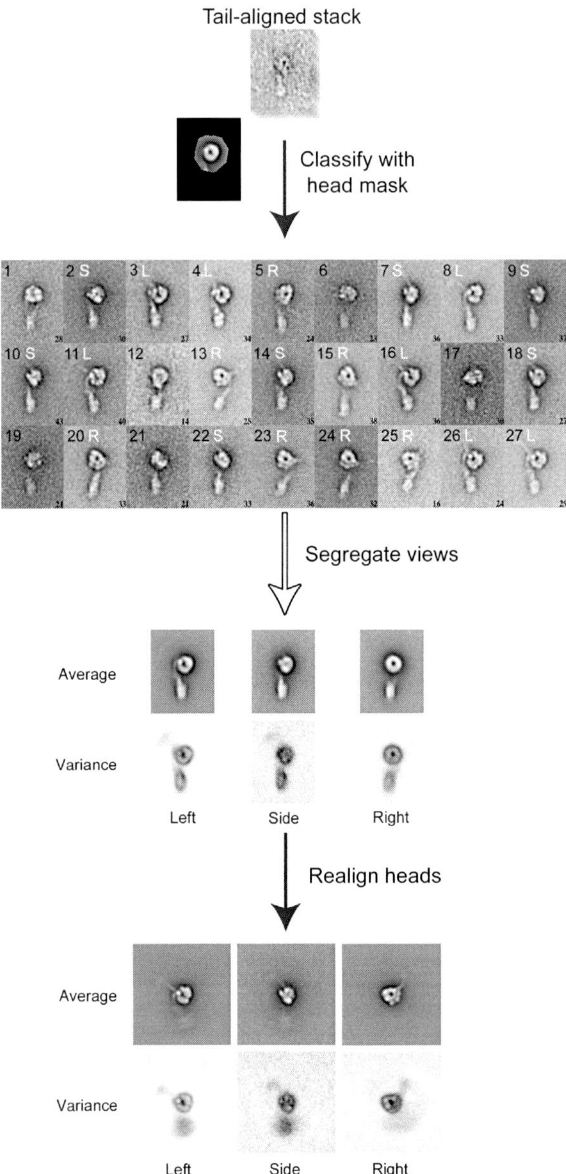

Fig. 5 Image classification for segregation of views of flagellar dynein-c. Classification of tail-aligned molecules using a head-shaped mask (indicated) produces numerous classes (27 of 300 classes shown). Among these, three characteristic views of dynein are seen: left views (indicated with "L"), side views ("S") and right views ("R"). Bad classes showing indeterminate structure (e.g., classes 1, 6, 12, 17, 19, and 21) are discarded at this stage. All molecules showing a particular view are combined for subsequent head-based realignment, independently, before further analysis. Segregation allows more detailed analysis, for example, of tail and stalk flexibility (indicated by high variance [dark] in the inverted variance images, lower row).

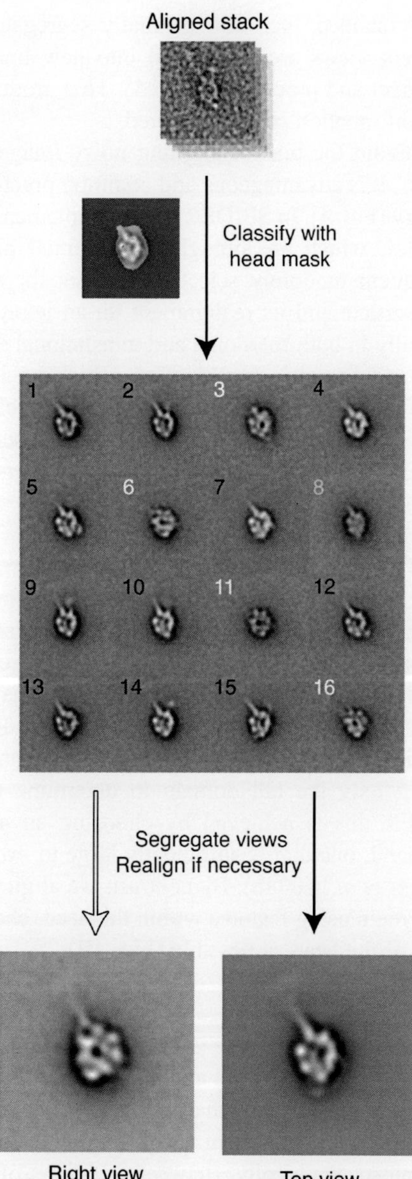

Fig. 6 Image classification for segregation of views of recombinant cytoplasmic dynein motor domains. Classification of all aligned molecules using a head-shaped mask (indicated) produces numerous classes (16 classes shown). Among these, two characteristic views are seen: top views (black numbers) and right views (white numbers). All images showing a particular view are combined for subsequent processing. Bad classes showing poor internal detail (class 8) are discarded at this stage. In some cases, images of a particular view are not aligned with respect to one other (e.g., class 3 vs 6, 11, and 16). These are realigned independently in a second round before further analysis.

(poorly stained, etc.) and manually segregate classes according to their orientation. Different views are segregated into new image stacks, for subsequent independent alignment and processing (Fig. 5). This produces superior alignments from which the most information can be gathered.

To avoid the bias of aligning noisy images of particles to one or more reference images, it is advantageous and common practice to employ a reference-free alignment strategy (Fig. 4). In SPIDER, this is implemented by selecting a random pair of starting particles, which are brought into mutual alignment, followed by the addition of subsequent randomly selected particles for alignment, followed by the removal of each particle and its realignment, in an iterative process. These steps are carried out internally in both rotational and translational steps (SPIDER's "AP RA" and "AP SA" commands) used in combination, which we incorporate into an alignment script. We have found for dynein, as well as for other elongated particles, that alignment is best achieved by starting with a rotational alignment followed by a translational alignment, whereas for more globular particles the reverse is true.

F. Whole Molecule Alignment

For full-length flagellar dynein-c (i.e., including the tail) we found it was helpful to align the particles first with the heads centralized and with the tails pointing downward (Fig. 4A). This enabled us to segregate different orientations of the particles (left, side, and right views, based on the relative emergence of the tail from the head) into different categories for subsequent independent processing (Fig. 5). This alignment is achieved by excluding the head domain from the rotational alignment (only), thereby leaving only the tail domain to determine the rotational orientation (Fig. 4A). In SPIDER this is achieved by choosing an appropriate inner radius in the AP RA command, one that is sufficiently large to *exclude* the central head (Fig. 4A; see also Burgess *et al.*, 2004b). By contrast, we align images of tailless recombinant cytoplasmic dynein using features within the head, choosing inner and outer radii to *include* as much of the head as possible (Fig. 4B).

G. Head Alignment

For obtaining the head alignment, particles are typically cropped (into 80×80 pixel windows) and treated with a soft-edged circular mask ("MA" command, 35 pixel radius, 5 pixel falloff). Alignment is typically iterated 15 times to ensure stability/convergence, determined empirically. Between iterations of reference-free alignment, it is useful to realign the image stack *en bloc* to a model image to prevent image drift during subsequent iterations of alignment. For initial model images, we either use an average image of the unaligned image stack ("AS R" command) rotationally averaged ("RO I" command) or the global average from a previous alignment attempt. Keeping the particles centered and rotationally stable between iterations (Fig. 4B) allows objective determination of alignment success, since the shifts and rotations obtained reduce as the alignment converges to a stable solution. We find that it also produces more stable alignments in general. The final alignment parameters (rotations and shifts) can then be applied to the unmasked, 120×120 pixel images and then cropped again to exclude artifacts around the image edges arising from the applied shifts and rotations. The pixel intensities in each image are then floated so that the minimum value is >0, as required for subsequent image classification. The aligned images are averaged ("AS R" command) to yield "global average" and "global variance" images (Fig. 4).

H. Head Classification to Segregate Different Views

Aligned dynein particles are classified into similar groups using multivariate statistical methods. We find that K-means clustering ("CL KM" command) outperforms hierarchical ascendant classification when molecules display continuous flexibility (Burgess *et al.*, 1997), such as dynein (Burgess *et al.*, 2003). Classification in SPIDER is performed using a mask to define the region within the image in which to apply multivariate statistical analysis. Classifications are first performed based on dynein's head domain (Figs. 5 and 6) by classifying all pixels within a custom-drawn mask or a mask generated by binary thresholding the global variance ("TH M" command). Images are classified several times into varying numbers of classes (average of 10–500 images per class) to determine empirically the optimum number of images per class. Images in each class are averaged ("AS R" command) to generate class averages and variances. Under our negative-staining conditions, dynein's head typically adopts preferred orientations on the EM grid and head classifications typically differentiate these characteristic views, as well as differences in stain depth and quality between images (Figs. 5 and 6). The "Markers" command within WEB is used to select well-stained class averages of a particular view. Individual particles from these classes are then compiled to generate documents listing all particles of a particular view. These lists are used to segregate the image stacks into new subsets for subsequent independent processing.

I. Visualizing Small Flexible Domains—The Stalk

Flexibility in dynein's stalk relative to the head means that it is not clearly resolved in image averages resulting from alignment and classification of the head (in full-length dynein-c, flexibility of the tail relative to the head introduces the same problem). Variance images, however, hint at the range of stalk positions in the data (e.g., Fig. 5, bottom row). Image classification can be used to group molecules with similar stalks by using a mask that encompasses all stalk positions while excluding the head entirely (Fig. 7). Such custom masks can be drawn in WEB ("Mask" function), using the variance image as a guide. When dealing with flexible structures it is advisable to test a range of masks and class sizes to ensure consistency in the resulting classes. Class averages can reveal the stalk's structure clearly, but this depends critically on the quality of staining in the original images. Stalk class averages sometimes also contain minor artifacts arising from classification of background noise.

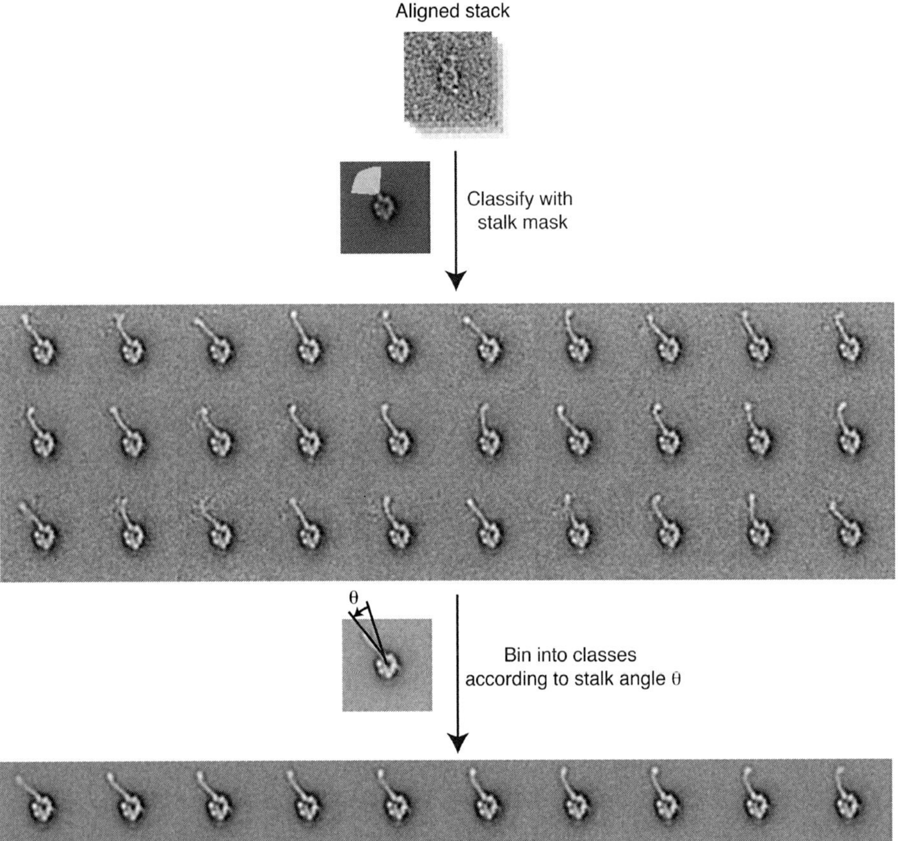

Fig. 7 Image classification of dynein's stalk domain. Classification using an appropriately shaped mask (white segment shape in upper left quadrant of image) shows the stalk in class averages (middle panel) at a range of angles. Typically, some contain artifacts from classification of noise and background features included within the mask and many do not have a visible stalk (not shown), owing to images in the dataset that were too deeply stained or of insufficient quality to resolve its fine structure. Ten artifact-free averages (lower panel) are created from all good stalk classes by binning their individual images according to their stalk angle (θ). This process reduces the artifacts arising from mask-based classification.

Where the quantity of stalk classes permits, artifact-free averages can be produced using the following procedure to group molecules solely by stalk angle:

i. The coordinates of the stalk base and head center are recorded using the "Markers" command in WEB, using the global average.
ii. In each clear stalk class, the coordinates of the MTBD are recorded.
iii. The angle of the stalk relative to a vector between the head center and stalk base is calculated for each class, using arithmetic commands in SPIDER.
iv. Each molecule in a given class is assigned this stalk angle.
v. Images are binned into classes on the basis of their stalk angle and averaged.

J. Mapping the Heavy Chain in the Motor by Locating Inserted GFP Tags

GFP-based tags (~27 kDa) inserted at various sites along the polypeptide chain of the motor can also be detected by negative-stain EM. Similar to the stalk, these tags are not fixed rigidly to dynein's head (owing to the flexible nature of GFP's N- and C-termini and to flexible inserts between GFP and the dynein sequence) and thus require specialized image processing to reveal their location (Figs. 8 and 9). We developed two steps to locate tags with positions peripheral to dynein's head, as outlined below.

First, to establish the limits to the range of positions that each tag adopts, we performed nine independent classifications of the same data to "scan" around the head periphery (Fig. 8). Subdividing the peripheral area into nine small segments (using overlapping wedge-shaped masks) produces superior class averages than using a single, large, annular-shaped mask, because the amount of background noise in each classification is greatly reduced. From each classification the class averages produced are compiled into a movie, and then all nine movies are concatenated to produce a "scanning classification" (Fig. 8). Careful study of such movies can reveal the position(s) of the tag even if they occur in only a

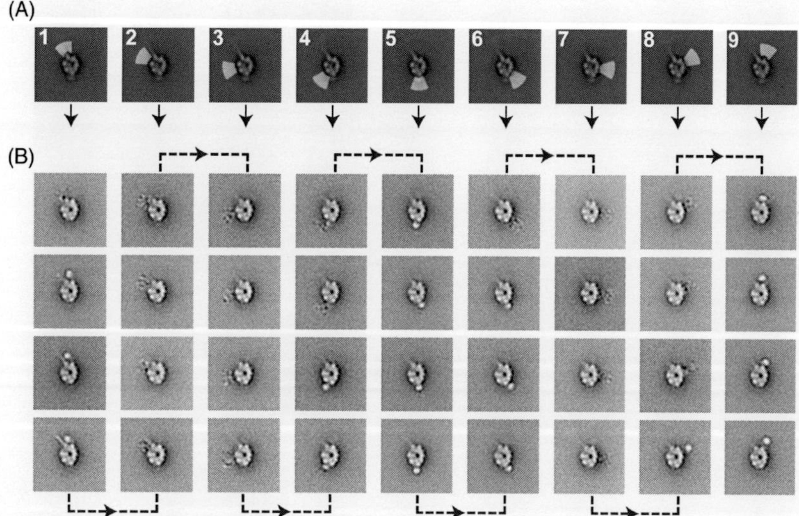

Fig. 8 Locating GFP/BFP tags by scanning classification. (A) Classification of dynein motor tagged with N-terminal GFP and BFP inserted just after the first AAA+ module. Nine distinct masks (white wedge-shaped features lying just outside the periphery of the head) are used to classify the images. Each mask covers a 45° wedge, with a 5° overlap with the preceding mask. (B) Typical examples of four classes (out of 72 each) for all nine mask-based classifications in (A). Small globular stain-excluding features can be seen in classification numbers 1 and 9 (corresponding to N-terminal GFP) and number 5 (corresponding to BFP). Individual classes from all nine classifications can be compiled into a single movie (indicated by the dashed lines) to produce a "scanning classification" of the head periphery.

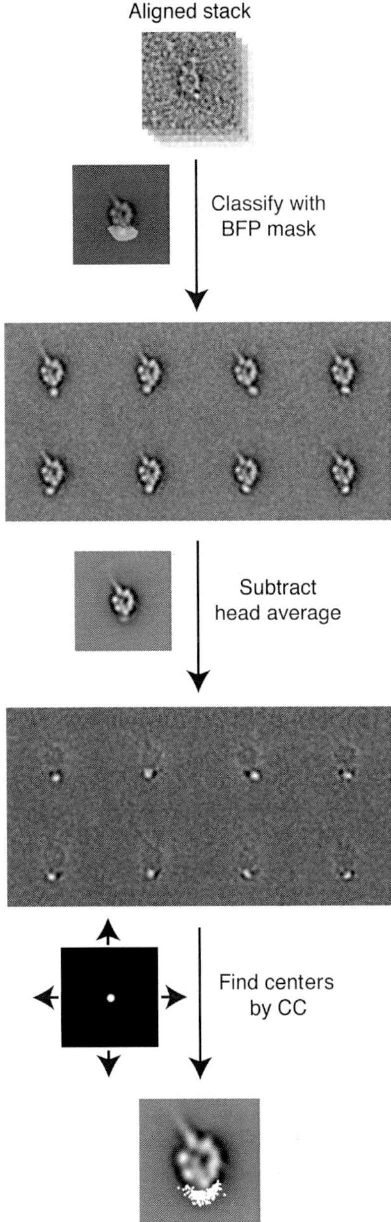

Fig. 9 Locating GFP/BFP tags by classification and autodetection. Informed by the "scanning classification" method (Fig. 8), tags are classified using a custom mask focussed on the region of interest. Tag coordinates are determined from class averages by subtracting a head average, then finding the remaining tag center by cross correlation with a model image (white disk on a black background). Coordinates are plotted as white spots on the head average (lower panel).

few classes. Scanning classifications can be viewed as movies at the following website: http://www.cell.com/supplemental/S0092-8674(08)01600-0

Second, based on the region of interest identified by scanning classification, a custom mask can be drawn for each tag to encompass all the positions found. The resulting class averages clearly resolve tags as globular densities peripheral to the head. Because the tags are typically peripheral to the head and sufficiently flexible that they disappear from the global average, it is possible to subtract ("SU" command) the global average (showing the head) from each class average (showing head plus tag) to isolate the tag itself as a discrete globular structure (Fig. 9). Before subtraction, images are normalized (mean pixel intensity $= 0$, $\sigma = 1$). Next, because each isolated tag is roughly circular in shape, its coordinates can be determined by cross-correlation with a model image ("CC N" command). The model image is a GFP-sized disc (created with the "MO" command, radius $= 3$ pixels) with a soft edge (produced by rotational averaging with the "RO I" command). Performing a peak search of the resulting cross-correlation function image (using "PK" command) obtains the coordinates of the center of each tag. Coordinates from images scoring above a given cross-correlation coefficient (to exclude images lacking a clear tag) can be weight-averaged (according to class size) to obtain the mean position of the tag (Fig. 9).

IV. Summary

EM in general has a great role to play in structural studies of dynein motor proteins because of their large size and extended structure. The inability to fully saturate microtubules with dynein motors precludes conventional helical methods that have been so successful in determining subnanometer density maps of kinesin-microtubules. Thus single-particle methods are especially relevant. Microtubule-attached states will be particularly challenging; for a complete picture we need to visualize the entire motor including its slender and flexible coiled-coil stalk and distal MTBD. Imaging conditions that visualize the entire motor (one-, two-, or three-headed) while simultaneously resolving the stalks themselves (which likely adopt a variety of positions along the microtubule lattice) will probably require a combination of EM approaches and large data sets. Labeling the MTBDs with electron dense tags may be a way forward. Some of the techniques described here may be applicable to cryo-EM data (e.g., to visualize the flexibility of stalk and tail in 2D). In the meantime, advances in our understanding of the dynein motor can be obtained by further studies of negatively stained dyneins, for example, by employing recombinant proteins modified by truncation or GFP-based tagging, which, in the absence of atomic models of dynein, can be used to broadly map the course of the polypeptide chain through this large motor. Indeed, such an approach has applications beyond the field of dynein motors, having the advantage that structural studies can be carried out in parallel with other biophysical studies taking advantage of FRET between GFP-based tags.

Acknowledgments

We thank Kazuhiro Oiwa and Hitoshi Sakakibara for purified inner-arm dynein-c, and the group of Kazuo Sutoh for recominant cytoplasmic dynein constructs, Matt Walker (MLW Consulting) for early EM contributions to this work, and Peter J. Knight for helpful discussions throughout this work. This work was supported by the BBSRC, HFSP (S.A.B.), and the Wellcome Trust (A.J.R.).

References

Amos L.A. (1989). Brain dynein crossbridges microtubules into bundles. *J. Cell. Sci.* **93**, 19–28.

Avolio, J., Lebduska, S., and Satir, P. (1984). Dynein arm substructure and the orientation of arm-microtubule attachments. *J. Mol. Biol.* **173**, 389–401.

Bui, K.H., Sakakibara, H., Movassagh, T., Oiwa, K., and Ishikawa, T. (2008). Molecular architecture of inner dynein arms in situ in Chlamydomonas reinhardtii flagella. *J. Cell Biol.* **183**, 923–932.

Burgess, S.A., Walker, M.L., Sakakibara, H., Knight, P.J., and Oiwa, K. (2003). Dynein structure and power stroke. *Nature* **421**, 715–718.

Burgess, S.A., Walker, M.L., Sakakibara, H., Oiwa, K., and Knight, P.J. (2004a). The structure of dynein-c by negative stain electron microscopy. *J. Struct. Biol.* **146**, 205–216.

Burgess, S.A., Walker, M.L., Thirumurugan, K., Trinick, J., and Knight, P.J. (2004b). Use of negative stain and single-particle image processing to explore dynamic properties of flexible macromolecules. *J. Struct. Biol.* **147**, 247–258.

Burgess, S.A., Walker, M.L., White, H.D., and Trinick, J. (1997). Flexibility within myosin heads revealed by negative stain and single-particle analysis. *J. Cell Biol.* **139**, 675–681.

Carter, A.P., Garbarino, J.E., Wilson-Kubalek, E.M., Shipley, W.E., Cho, C., Milligan, R.A., Vale, R.D., and Gibbons, I.R. (2008). Structure and functional role of dynein's microtubule-binding domain. *Science* **322**, 1691–1695.

Cheng, Y., Wolf, E., Larvie, M., Zak, O., Aisen, P., Grigorieff, N., Harrison, S.C., and Walz, T. (2006). Single particle reconstructions of the transferrin–transferrin receptor complex obtained with different specimen preparation techniques. *J. Mol. Biol.* **355**, 1048–1065.

DeLano, W.L. The PyMOL Molecular Graphics System (2002) DeLano Scientific, Palo Alto, CA, USA. http://www.pymol.org.

Dubochet, J., Adrian, M., Chang, J.J., Homo, J.C., Lepault, J., McDowall, A.W., and Schultz, P. (1988). Cryo-electron microscopy of vitrified specimens. *Q. Rev. Biophys.* **21**, 129–228.

Elliott, A., Offer, G., and Burridge, K. (1976). Electron microscopy of myosin molecules from muscle and non-muscle sources. *Proc R Soc Lond, B Biol. Sci.* **193**, 45–53.

Frank, J. (2006). "Three-Dimensional Electron Microscopy of Macromolecular Assemblies." Oxford University Press, New York.

Frank, J., Radermacher, M., Penczek, P., Zhu, J., Li, Y., Ladjadj, M., and Leith, A. (1996). SPIDER and WEB: Processing and visualization of images in 3D electron microscopy and related fields. *J. Struct. Biol.* **116**, 190–199.

Gee, M.A., Heuser, J.E., and Vallee, R.B. (1997). An extended microtubule-binding structure within the dynein motor domain. *Nature* **390**, 636–639.

Gibbons, I.R., Garbarino, J.E., Tan, C.E., Reck-Peterson, S.L., Vale, R.D., and Carter, A.P. (2005). The affinity of the dynein microtubule-binding domain is modulated by the conformation of its coiled-coil stalk. *J. Biol. Chem.* **280**, 23960–23965.

Goodenough, U.W., and Heuser, J.E. (1984). Structural comparison of purified dynein proteins with *in situ* dynein arms. *J. Mol. Biol.* **180**, 1083–1118.

Goodenough, U.W., and Heuser, J.E. (1982). Substructure of the outer dynein arm. *J. Cell Biol.* **95**, 798–815.

Grassucci, R.A., Taylor, D.J., and Frank, J. (2007). Preparation of macromolecular complexes for cryo-electron microscopy. *Nat. protoc.* **2**, 3239–3246.

Ishikawa, T., Sakakibara, H., and Oiwa, K. (2007). The architecture of outer dynein arms *in situ*. *J. Mol. Biol.* **368**, 1249–1258.

Kon, T., Imamula, K., Roberts, A.J., Ohkura, R., Knight, P.J., Gibbons, I.R., Burgess, S.A., and Sutoh, K. (2009). Helix sliding in the stalk coiled coil of dynein couples ATPase and microtubule binding. *Nat. Struct. Mol. Biol.* **16**, 325–333.

Kon, T., Mogami, T., Ohkura, R., Nishiura, M., and Sutoh, K. (2005). ATP hydrolysis cycle-dependent tail motions in cytoplasmic dynein. *Nat. Struct. Mol. Biol.* **12**, 513–519.

Koonce, M.P., and Samsó, M. (1996). Overexpression of cytoplasmic dynein's globular head causes a collapse of the interphase microtubule network in Dictyostelium. *Mol. Biol. Cell* **7**, 935–948.

Ludtke, S.J., Baldwin, P.R., and Chiu, W. (1999). EMAN: Semiautomated software for high-resolution single-particle reconstructions. *J. Struct. Biol.* **128**, 82–97.

Lührmann, R., and Stark, H. (2009). Structural mapping of spliceosomes by electron microscopy. *Curr. Opin. Struct. Biol.* **19**, 96–102.

Marchese-Ragona, S.P., Wall, J.S., and Johnson, K.A. (1988). Structure and mass analysis of 14S dynein obtained from Tetrahymena cilia. *J. Cell. Biol.* **106**, 127–132.

Mizuno, N., Narita, A., Kon, T., Sutoh, K., and Kikkawa, M. (2007). Three-dimensional structure of cytoplasmic dynein bound to microtubules. *Proc. Natl. Acad. Sci. USA* **104**, 20832–20837.

Mizuno, N., Toba, S., Edamatsu, M., Watai-Nishii, J., Hirokawa, N., Toyoshima, Y.Y., and Kikkawa, M. (2004). Dynein and kinesin share an overlapping microtubule-binding site. *EMBO J.* **23**, 2459–2467.

Nicastro, D., McIntosh, J.R., and Baumeister, W. (2005). 3D structure of eukaryotic flagella in a quiescent state revealed by cryo-electron tomography. *Proc. Natl. Acad. Sci. USA* **102**, 15889–15894.

Nicastro, D., Schwartz, C., Pierson, J., Gaudette, R., Porter, M.E., and McIntosh, J.R. (2006). The molecular architecture of axonemes revealed by cryoelectron tomography. *Science* **313**, 944–948.

Oda, T., Hirokawa, N., and Kikkawa, M. (2007). Three-dimensional structures of the flagellar dynein-microtubule complex by cryoelectron microscopy. *J. Cell Biol.* **177**, 243–252.

Roberts, A.J., Numata, N., Walker, M.L., Kato, Y.S., Malkova, B., Kon, T., Ohkura, R., Arisaka, F., Knight, P.J., Sutoh, K., and Burgess, S.A. (2009). AAA+ ring and linker swing mechanism in the dynein motor. *Cell* **136**, 485–495.

Samsó, M., and Koonce, M.P. (2004). 25 Angstrom resolution structure of a cytoplasmic dynein motor reveals a seven-member planar ring. *J. Mol. Biol.* **340**, 1059–1072.

Samsó, M., Radermacher, M., Frank, J., and Koonce, M.P. (1998). Structural characterization of a dynein motor domain. *J. Mol. Biol.* **276**, 927–937.

Sindelar, C.V., and Downing, K.H. (2007). The beginning of kinesin's force-generating cycle visualized at 9-A resolution. *J. Cell Biol.* **177**, 377–385.

Ueno, H., Yasunaga, T., Shingyoji, C., and Hirose, K. (2008). Dynein pulls microtubules without rotating its stalk. *Proc. Natl. Acad. Sci. USA* **105**, 19702–19707.

Unwin, P.N. (1974). Electron microscopy of the stacked disk aggregate of tobacco mosaic virus protein. II. The influence of electron irradiation of the stain distribution. *J. Mol. Biol.* **87**, 657–670.

Vale, R.D., and Milligan, R.A. (2000). The way things move: Looking under the hood of molecular motor proteins. *Science* **288**, 88–95.

Vale, R.D., Reese, T.S., and Sheetz, M.P. (1985). Identification of a novel force-generating protein, kinesin, involved in microtubule-based motility. *Cell* **42**, 39–50.

van Heel, M., Harauz, G., Orlova, E.V., Schmidt, R., and Schatz, M. (1996). A new generation of the IMAGIC image processing system. *J. Struct. Biol.* **116**, 17–24.

Walker, M.L., Burgess, S.A., Sellers, J.R., Wang, F., Hammer, J.A., Trinick, J., and Knight, P.J. (2000). Two-headed binding of a processive myosin to F-actin. *Nature* **405**, 804–807.

Walker, M.L., Knight, P.J., and Trinick, J. (1991). Properties of the myosin molecule revealed by negative staining. *Micron. Microsc. Acta* **22**, 413–422.

Zhu Y., Carragher B., Glaeser R.M., Fellmann D., Bajaj C., Bern M., Mouche F., de Haas F., Hall R.J., Kriegman D.J., Ludtke S.J., Mallick S.P., Penczek P.A., Roseman A.M., Sigworth F.J., Volkmann N., Potter C.S. (2004) Automatic particle selection: results of a comparative study. *J. Struct. Biol.* **145**, 3–14.

CHAPTER 3

Immunogold Labeling of Flagellar Components *In Situ*

Stefan Geimer

Zellbiologie/Elektronenmikroskopie NWI/B1, Universität Bayreuth, 95440 Bayreuth, Germany

Abstract

Immunogold electron microscopy is a classic high-resolution method for the selective localization of macromolecules in the context of cells and subcellular structures. Specific antibodies are used to affix small particles of colloidal gold, which are easily visible in the electron microscope, to the macromolecule of interest. There are different immunogold-labeling techniques; in the postembedding immunogold-labeling

978-0-12-374973-4
DOI: 10.1016/S0091-679X(08)91003-7

technique, the biological material is first fixed, dehydrated, and embedded in resin and the antibody reactions are done on the sectioned material. In the preembedding immunogold-labeling technique, the antibody reactions are carried out prior to fixation, dehydration, and resin embedding of the biological specimen. The whole-mount immunogold-labeling technique does not involve resin embedding at all; the material is applied to an electron microscopy grid and the antibody reactions are carried out on the grid. The aim of this chapter is to describe in detail postembedding and preembedding techniques applicable for the immunogold labeling of components of *Chlamydomonas* flagella and basal bodies. Special emphasis is put on the flat embedding of *Chlamydomonas* cells, which allows the analysis of individual flagella along their whole length, a method especially suitable to the study of intraflagellar transport (IFT). Depending on the fixation protocol and resin used, such flat embeddings can be utilized for the localization of components of the IFT machinery by postembedding immunogold labeling or the ultrastructural analysis of the IFT complex by standard electron microscopy or electron tomography.

I. Introduction

Chlamydomonas reinhardtii is one of the most powerful model organisms to study the cell biology of cilia and flagella. Cilia from this organism have been used since the early days of electron microscopy for analysis of the complex ultrastructure of this organelle. To understand the inner workings of such a complex organelle, a detailed knowledge of its ultrastructure is not enough; also needed is an understanding of the specific macromolecules that contribute to the structures visualized. The method of choice for high-resolution localization of macromolecules in the structural context of cells and organelles is immunogold electron microscopy. In this method an antibody raised against an antigenic determinant is used to specifically detect the antigen and the antigen–antibody complex is visualized by small gold particles. This technique involves two conflicting requirements; on the one hand, the specimen must be processed in such a way that the fine structure is preserved as close as possible in the native state, on the other hand, the antigenicity of the macromolecule to be localized must be retained as much as possible.

When localizing flagellar membrane-associated components, such as membrane proteins or components of intraflagellar transport (IFT), good ultrastructural preservation of the flagellar membrane is crucial. Analysis of IFT trains by electron microscopy indicates that they are connected to the outer doublet microtubules and the flagellar membrane (Kozminski *et al.*, 1993, 1995; Pazour *et al.*, 1998; Pedersen *et al.*, 2006). Common protocols for the isolation of IFT particles rely on the disintegration of the flagellar membrane by either nonionic detergent or freeze thawing (Cole *et al.*, 1998; Qin *et al.*, 2004). For the localization of components of IFT, the preembedding technique, which would involve the permeabilization of the flagellar membrane, is not a good approach as any disturbance of the flagellar membrane would likely alter the ultrastructure of the IFT trains. Postembedding immunogold labeling of flat embedded *Chlamydomonas* cells is more suitable for the localization of IFT components (Fig. 2B–F; Pedersen *et al.*, 2006). The postembedding technique does not require the flagellar membrane to be permeabilized as the exposure of the antigens is achieved by ultrathin sectioning of the resin-embedded material. When *Chlamydomonas* cells are pelleted and then processed for electron microscopy, the flagella orient randomly and it is extremely laborious to find longitudinal sections over a long stretch of a flagellum. To overcome this problem, the technique of flat-embedding *Chlamydomonas* cells can be used. Live *Chlamydomonas* cells encountering a solid surface adhere to it starting with their flagellar tips. Flagellar beating ceases and finally both flagella stick to the surface along nearly their full lengths and the cell begins to glide (Mitchell *et al.*, 2004). When *Chlamydomonas* cells are fixed and embedded in that position, longitudinal sections through almost the whole length of flagella can be obtained (Mitchell and Nakatsugawa, 2004), an ideal situation for their analysis (Figs. 1D and 2A).

Flagellar components that are not extracted by detergents (the axonemal microtubules, radial spokes, dynein arms, etc.) and components of the basal bodies can be localized by either postembedding or preembedding immunogold labeling (Fig. 3; Geimer and Melkonian, 2005; Keller *et al.*, 2009; Pedersen *et al.*, 2003). Both techniques are described in this chapter and are well suited for such labeling. However, the postembedding technique has the disadvantage that only protein epitopes directly exposed on the section surface are accessible to the antibodies (Stierhof *et al.*, 1991). Structures may be clearly visible within the depth of the section but are not labeled with colloidal gold because they are not exposed at the surface (Fig. 2G). In the preembedding technique, the antibody reactions are completed before resin embedding, allowing the antibody to reach all accessible antigens, thus providing a three-dimensional localization of protein epitopes.

Another method for the localization of components that are retained in isolated cytoskeletons/axonemes is the whole-mount immunogold-labeling technique (Johnson *et al.*, 1994; Lechtreck and Geimer, 2000). For whole-mount immunogold labeling, isolated cytoskeletons axonemes are settled onto an electron microscope grid, immunogold labeled, and negatively stained. This method is fast and only a minimal amount of material is needed. Such immunogold-labeled whole mounts can show excellent labeling densities as fixation with aldehydes, dehydration, and embedding in plastic resin are omitted. This technique will not be described in this chapter but a detailed practical introduction of how to apply this method to isolated *Chlamydomonas* flagellar axonemes is given in Johnson (1995). Also immunogold scanning electron microscopy of isolated axonemes has been used to localize flagellar components (Sloboda and Howard, 2007).

For additional background on immunogold-labeling techniques, there are several excellent articles (Morphew, 2007; Schwarz and Humbel, 2007; Stirling, 1990) and books (Griffiths *et al.*, 1993; Newman and Hobot, 1993).

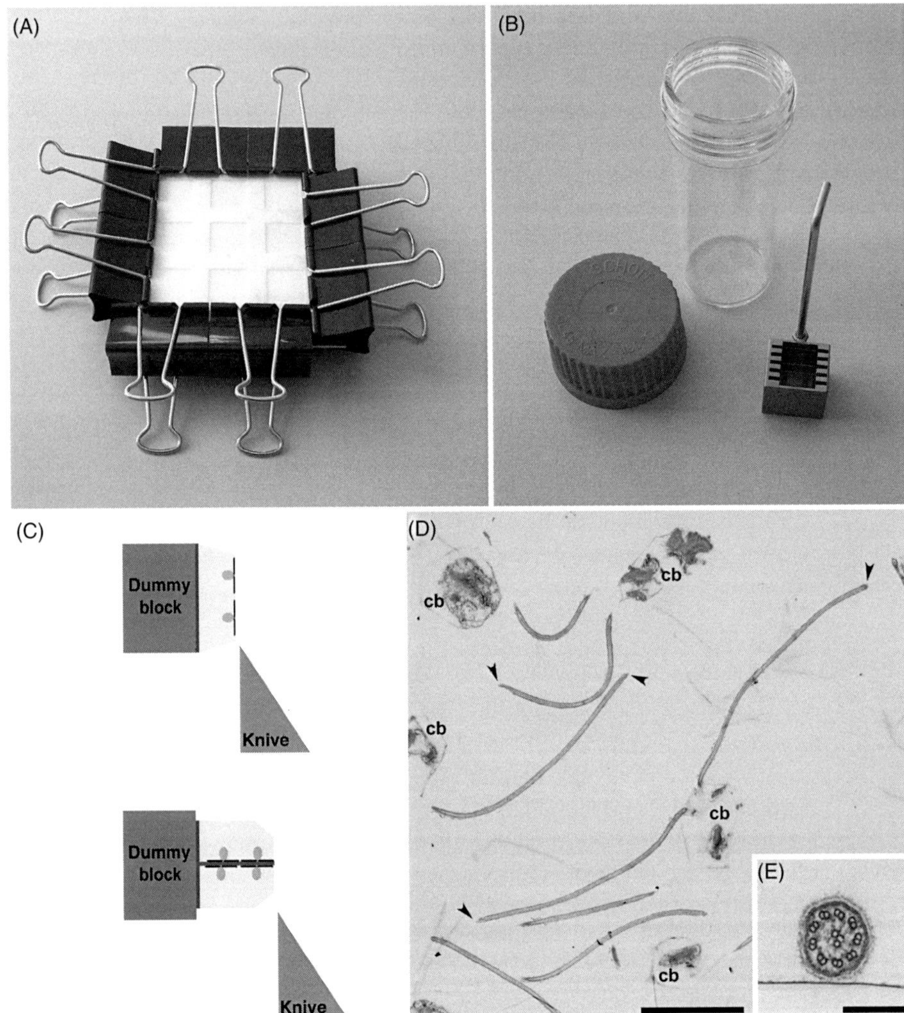

Fig. 1 Flat embedding of *Chlamydomonas* cells. (A) Assembly for polymerizing LR Gold flat embeddings in a Chang monolayer mold (for a detailed description, see Section II.A, step 10). (B) Custom-made coverslip staining rack holding five 18 mm × 18 mm coverslips and suitable glass container with screw cap. (C) Mounting of flat embeddings for ultrathin sectioning. To get longitudinal sections of the flagella, ultrathin sections are cut parallel to the surface of the flat embedding (upper cartoon). To get cross sections of the flagella, two pieces of the flat embedding are first glued together and then this sandwich is glued onto a dummy block (lower cartoon). This way ultrathin sections perpendicular to the surface of the flat embedding are cut and in such sections it is easy to find cross sections of flagella. (D, E) Electron micrographs of flat embedded (Epon embedding) *Chlamydomonas* cells. (D) Longitudinal sections of flagella. Several flagella can be seen in longitudinal section along almost their entire length, from the flagellar tip (arrowheads) to their proximal end where entering the cell body (cb). (E) Cross section of a flagellum. Note the thin black line in the lower part of the picture, which represents the boundary between the material of the flat embedding and the coverslip (before its removal by hydrofluoric acid). Scale bar: D, 5 μm; E, 200 nm.

II. Methods for Immunogold Labeling of Flagellar Components

A. Flat Embedding of *Chlamydomonas* Cells in LR Gold

1. Introductory Remarks

The coverslips used as a substrate for flat embedding need to be extensively soaked in water before use. They are often factory precleaned and may be contaminated with traces of detergent that can damage the flagellar membrane. Incubate the coverslips for a few days in distilled water (with several changes) and then overnight in growth medium MI. Squares of Aclar film, which can be easily removed from all sorts of polymerized resin, can be used instead of glass coverslips. The Aclar film should also be soaked extensively before use. To date, however, the results I have obtained using glass coverslips are somewhat superior to those obtained with Aclar, but this needs to be further explored.

1. Collect cells from 100–200 ml of an actively growing *Chlamydomonas* culture by centrifugation (50–80 × *g*, 20 min, 18°C). Gently resuspend the cells in 10–20 ml growth medium (MI) using a large-bore pipette.
2. Overlay a coverslip (18 mm × 18 mm) with a 0.1% (w/v) poly-L-lysine solution and incubate for 2–5 min. For this and the following incubation steps, the coverslip can rest on a few pieces of dental wax that are cut smaller than the coverslip. The dental wax is secured with a piece of double-sided adhesive tape inside a Petri dish and is used to elevate the coverslip, making it easier to grip with the forceps.
3. Wash the coverslip briefly by dipping it into MI. Immediately overlay with *Chlamydomonas* cells and allow the cells to settle and adhere for 10–30 min.
4. Decant excess cells and gently wash the coverslip by dipping it into MI.
5. Immediately overlay the coverslip with MI containing 0.25–0.5% glutaraldehyde, 0.1–0.2% tannic acid, pH 7.2, and fix for 30–40 min at room temperature.
6. Decant the fixative and wash the coverslip by putting it for 2 min in a Petri dish with MI.
7. Overlay the coverslip with 0.05–0.1% osmium tetroxide and incubate for 20 min at 4°C. Decant the osmium tetroxide and wash the coverslip by putting it for 5 min in a Petri dish with MI, repeat once.
8. Dehydrate the specimen using a graduated series of ethanol according to the following scheme: 30 and 50% at 4°C; then 70 and 95% ethanol at −25°C, 15 min each. Dehydration and the following infiltration are done using a coverslip-staining jar (or coverslip-staining rack and suitable container).
9. For infiltration use LR Gold containing 0.4% benzil (simply called "LR Gold" in the following) for all steps. Bubble the LR Gold for 10 min with gaseous nitrogen before use, as oxygen will inhibit the polymerization (this being especially important for the LR Gold used for the final embedding of the sample). Transfer the specimen from 95% ethanol into a 1:1 mixture of 100% ethanol and LR Gold and incubate for 2 h at −25°C. Transfer the specimen

into LR Gold and infiltrate with two changes, 2 h each, at −25°C, followed by overnight incubation at −25°C. On the next day, transfer the specimen into fresh LR Gold and incubate for 3 h at −25°C. Once the specimen is in pure LR Gold, care has to be taken that the temperature does not fall below −25°C because the LR Gold will freeze.

10. For flat embedding, place a Chang monolayer mold onto a piece of sheet metal (about 12 cm × 12 cm) and fill the cavities with LR Gold. Place the coverslips with

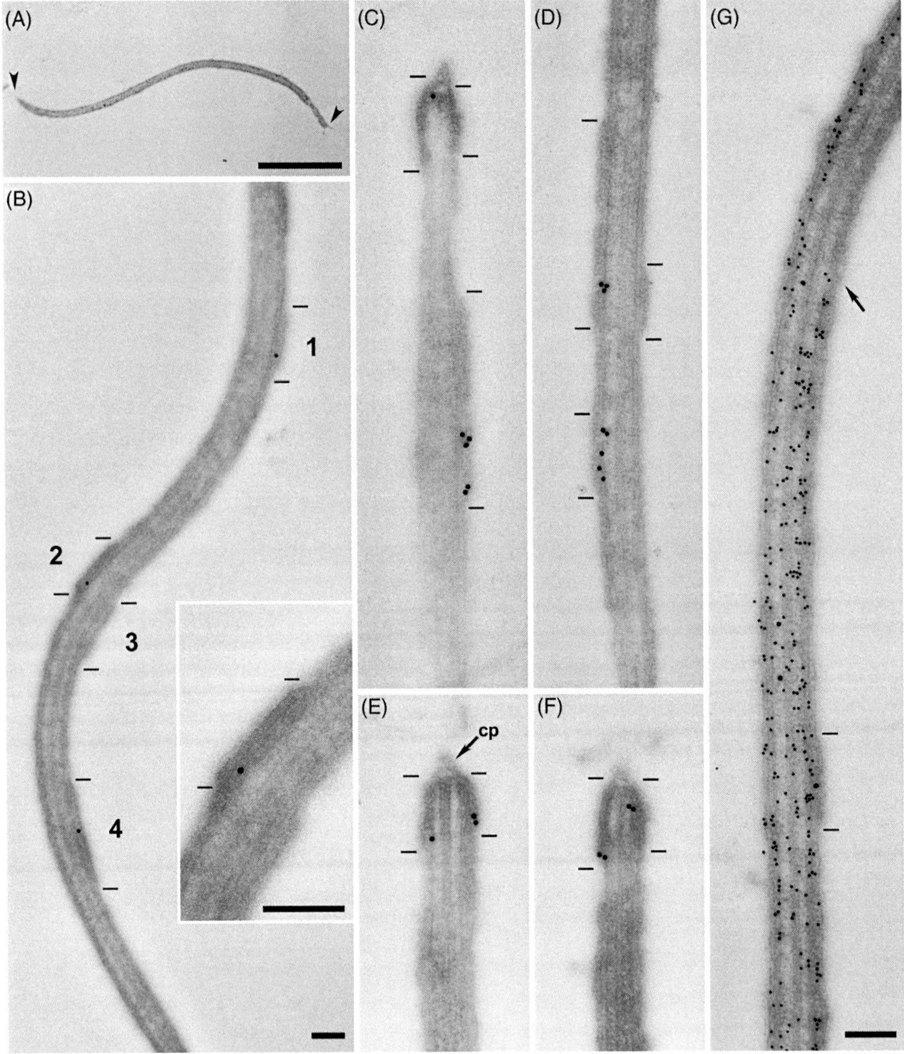

Fig. 2 (*Continued*)

the cells facing up in the cavities and overfill with LR Gold. Cover the Chang monolayer mold with a piece of Aclar film (no air bubbles should be trapped) and put a glass plate with the size of the mold (or a little bigger, thickness about 2 mm) on top of it. Clamp the whole sandwich together with eight 51-mm foldback clips (Fig. 1A). Polymerization is carried out at −25°C under UV light for at least 48 h.

11. To remove the coverslip from the polymerized LR Gold layer of the flat embedding, a little mechanical manipulation might be sufficient. If the coverslip does not come off easily, it can be detached by immersion into liquid nitrogen. As polymerized LR Gold is quite brittle, the resin layer will crack into smaller pieces, but this poses no problem as only small pieces of the flat embedding are needed.

B. Embedding of Isolated Cytoskeletons from *Chlamydomonas* in LR Gold

1. Isolate cytoskeletons from *Chlamydomonas* according to the protocol of Wright *et al.* (1985).
2. Pellet isolated cytoskeletons suspended in $MT^{Mg^{2+}}$ buffer in a 1.5-ml microfuge tube at about $16{,}000 \times g$ (10 min, 4°C). The pellet should be approximately 1 mm thick. The pellet size can be adjusted by resuspending the pellet and removing or adding cytoskeletons.
3. Fix the pellet in $MT^{Mg^{2+}}$ buffer containing 0.1–0.25% glutaraldehyde and 2–3% formaldehyde, pH 7.2, for 40–60 min at 15°C. After 15 min of incubation try to release the pellet from the wall of the microfuge tube. A glass Pasteur pipette with its tip drawn out and melted shut is a good tool for this purpose.
4. Wash the pellet with two changes of $MT^{Mg^{2+}}$ buffer, 5 min each.
5. Dehydrate with a graduated series of ethanol as described in A, step 8.
6. Infiltrate the pellet with LR Gold as described in A, step 9. Depending on the size of the pellet, the last infiltration step should be extended to up to 24 h with at least two changes in between.

Fig. 2 Postembedding immunogold labeling of flagellar components on flat embedded *Chlamydomonas* cells. (A, G) Longitudinal sections of flagella labeled with a monoclonal antibody directed against acetylated-α-tubulin (purified immunoglobulin 1:100; clone 6-11B-1; Sigma-Aldrich; Piperno and Fuller, 1985) and goat antimouse IgG conjugated to 10 nm colloidal gold (1:30; British BioCell). (A) Longitudinal section of a flagellum from its tip (right arrowhead) to almost its proximal end (left arrowhead) where the flagella enters the cell body. (G) Longitudinal section of a flagellum showing dense labeling of outer doublet and central pair microtubules with gold particles. Note the disappearance of the labeling from the right outer doublet and the central pair microtubules in the right upper part (arrow) of the flagellum. This is most probably caused by the microtubules leaving the surface of the section, thus not being accessible to the antibodies. An IFT train (marked by short lines) is visible. (B–F) Longitudinal sections of flagella labeled with a polyclonal antibody against IFT46 (1:50; Hou *et al.*, 2007), an IFT complex B protein, and goat antirabbit IgG conjugated to 15 nm colloidal gold (1:30; British BioCell). IFT trains are marked by short lines. (B) Four IFT trains are visible (numbered 1–4), three of which are labeled with gold particles. Inset: higher magnification view of IFT train no. 2, the characteristic ultrastructure of the IFT train is clearly visible. (C–F) Longitudinal sections showing IFT trains labeled with colloidal gold. In C, E, and F IFT trains at the flagellar tip can be seen. In E, the central plug (cp), a structure associated with the distal end of the central pair microtubules, is visible. Scale bar: A, 2 μm; B and inset, 200 nm; C–G, 200 nm.

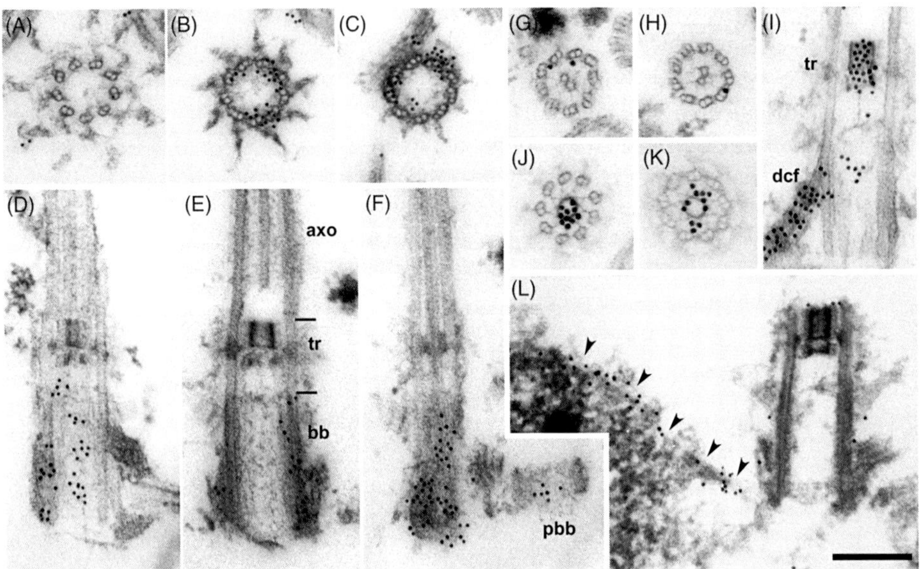

Fig. 3 Postembedding and preembedding immunogold labeling of isolated cytoskeletons. (A–F) Postembedding immunogold labeling of isolated cytoskeletons with a polyclonal antibody against POC1 (1:250; Keller *et al.*, 2009) and goat antirabbit IgG conjugated to 12 nm colloidal gold (1:30, Jackson ImmunoResearch). (A–C) Three consecutive serial cross sections (from distal to proximal) through a basal body. At the most distal end of the basal body no labeling is visible; more proximal triplet microtubules of the basal body are heavily labeled with colloidal gold. (D–F) Three consecutive longitudinal serial sections through a basal body (bb), the transitional region (tr), and proximal end of the axoneme (axo). The triplet microtubules of the basal body and also probasal body (pbb) show dense labeling. (G–K) Postembedding immunogold labeling of isolated cytoskeletons with a polyclonal antibody against centrin (1:100; Höhfeld *et al.*, 1994) and goat antirabbit IgG conjugated to 15 nm (G, H, J, K; 1:30; British BioCell) or 10 nm colloidal gold (I; 1:30, Sigma Aldrich). (G, H) Random cross sections through axonemes showing gold particles associated primarily with the A tubules of the outer doublets. (J, K) Random cross sections through the flagellar transitional region with dense centrin labeling, mostly over the central ring-like hub. (I) Longitudinal section through a basal body, transitional region (tr), and proximal portion of the axoneme. Dense centrin labeling is found in the stellate structure of the transitional region (tr) and the distal connecting fiber (dcf). Also the lumen of the basal body shows centrin labeling. (L) Preembedding immunogold labeling of isolated cytoskeletons with a polyclonal antibody against centrin (1:100; Höhfeld *et al.*, 1994) and goat antirabbit IgG conjugated to 10 nm colloidal gold (1:30; Sigma Aldrich). The nucleus basal body connector, which interconnects the basal body and the cell nucleus, is labeled. Note that in this preembedding labeling the transitional region and lumen of the basal body do not show labeling with gold particles. This is most probably caused by the antibodies being not able to penetrate into the lumen of the transitional region and basal body. Scale bar: A–L, 200 nm.

7. For embedding, transfer the pellet into a gelatin capsule. As is characteristic of acrylic resins, oxygen inhibits polymerization of LR gold. It is unavoidable to trap some air when closing the capsule, but the trapped air bubble should be kept as small as possible by overfilling the capsule with LR Gold to convexity before closing it. Polymerization is carried out at −25°C under UV light for at least 48 h.

C. Postembedding Immunogold-Labeling Technique

1. Introductory Remarks

Nickel grids are often used because of their high resistance to chemical damage during the procedures. Gold-gilded copper grids are a good alternative, as the magnetic properties of nickel grids can be inconvenient. If nickel grids are used, nonmagnetic forceps are needed and it may be necessary to demagnetize the grids with a demagnetizing coil to prevent extreme astigmatism in the transmission electron microscope.

All incubations are done by floating the grids (section side down) on drops of the appropriate buffers. Incubations of the grids on drops of antibody solutions are performed on a piece of clean parafilm in a Petri dish. A drop of 50 μl is large enough for the incubation of up to six grids. When dealing with a precious primary antibody, the volume of the drop can be even reduced to as little as 2 μl, which is enough for the incubation of one grid. During the entire labeling procedure, care must be taken to prevent the sections from drying out. One way to do this is by placing the Petri dish with drops and floating grids in a humid chamber constructed from a box with a snug-fitting lid and some wet paper towels.

Washing steps should be done on drops of a larger volume. For this purpose, larger drops of the appropriate buffer are placed on parafilm or more conveniently, the wells of a 96-well multititer plate filled to convexity. The grids can be transferred from drop to drop by using a wire loop with a slightly larger diameter than a grid (most standard inoculation loops are well suited for this purpose). When dunked into the buffer, moved below the grid, and then lifted out, the loop traps a drop of buffer with the grid floating on top. This is a very convenient and fast way to transfer the grids from drop to drop and it minimizes the chance of damaging the grids by handling them with forceps.

For labeling sections of flat embedded flagella, it is not recommended to incubate overnight at 4°C as prolonged incubations can reduce the ultrastructural preservation of the embedded material (Berryman and Rodewald, 1990), an effect I also observed with preparations of flat embedded flagella.

1. After sectioning, transfer the grids to drops of a freshly prepared saturated solution of sodium metaperiodate and incubate for 2–4 min. It has been found that treatment with sodium metaperiodate can partially restore protein antigenicity on thin sections of osmicated material (Bendayan and Zollinger, 1983).
2. Wash the grids extensively by transferring them across four changes of distilled water and two changes of Na-PBS (sodium phosphate-buffered saline), 5 min each.
3. Block the grids for 30–60 min with blocking buffer.
4. Transfer the grids to drops of the primary antibody diluted in blocking buffer and incubate for about 90 min at room temperature.
5. Wash the grids by transferring them across five changes of Na-PBS, 5–10 min each.
6. Incubate the grids on drops of the secondary antibody conjugated to colloidal gold diluted in blocking buffer (1:25–1:100) for about 90 min at room temperature.

7. Wash the grids by transferring them across five changes of Na-PBS, 5–10 min each.
8. Incubate the grids for 8 min on drops of Na-PBS containing 1% glutaraldehyde. This step is to chemically fix bound antibodies in place as they might be released by the low pH of the uranyl acetate solution used for poststaining.
9. Wash the grids by transferring them across four changes of distilled water, 5 min each. Blot the grids dry by touching the edge to a piece of filter paper.
10. Poststain the grids with 2% uranyl acetate in distilled water for about 20 min and lead citrate for 2–4 min.

D. Preembedding Immunogold Labeling of Isolated Cytoskeletons from *Chlamydomonas* and Flat Embedding in Epon

1. Collect cells from 50–100 ml of an actively growing *Chlamydomonas* culture of a cell wall-deficient mutant (preferably strain cw-2) by centrifugation (50–80 × *g*, 20 min, 18°C).
2. Resuspend the cells in 10–20 ml $MT^{Mg^{2+}}$ buffer and pellet as above. Resuspend in about 5 ml $MT^{Mg^{2+}}$ buffer and put on ice for 5 min.
3. Lyse the cells by adding an equal volume $MT^{Mg^{2+}}$ buffer containing 2–3% Triton X-100. Lysis of the cells should be completed after 5 min, check in the light microscope.
4. Add an equal volume of $MT^{Mg^{2+}}$ containing 0.2–0.5% glutaraldehyde and 6% formaldehyde, pH 7.2, to the suspension of cytoskeletons and incubate for 10 min on ice.
5. Overlay a coverslip with a 0.1% (w/v) poly-L-lysine solution and incubate for 2–5 min.
6. Wash the coverslip briefly by dipping it into $MT^{Mg^{2+}}$ buffer and immediately overlay with the suspension of isolated cytoskeletons. Allow the cytoskeletons to settle and adhere for about 30 min.
7. Decant the cytoskeletons and wash the coverslip by putting it in a Petri dish with Na-PBS for 5–10 min, repeat three to five times. Overlay the coverslip with blocking buffer and incubate for 30–60 min at room temperature.
8. Decant the blocking buffer and gently wash the coverslip by dipping it into Na-PBS. Overlay the coverslip with the primary antibody diluted in blocking buffer and incubate for 90 min at room temperature.
9. Wash the coverslip by putting it in a Petri dish with Na-PBS for 5–10 min, repeat three to five times. Overlay the coverslip with the secondary antibody conjugated to colloidal gold diluted in blocking buffer (1:25–1:100) and incubate for 90 min at room temperature.
10. Wash the coverslip by putting it in a Petri dish with Na-PBS for 5 min, repeat three to five times. Overlay the coverslip with $MT^{Mg^{2+}}$ containing 2.5% glutaraldehyde, pH 7.2, and incubate for 40 min at 4°C.
11. Decant the fixative and wash the coverslip by putting it for 5 min in a Petri dish with distilled water.

12. Overlay the coverslip with 1% osmium tetroxide and incubate for 30–40 min at 4°C. Decant the osmium tetroxide and wash the coverslip by putting it for 5 min in a Petri dish with distilled water, repeat twice.
13. Dehydrate the specimen using a graduated series of ethanol according to the following scheme: 30 and 50% at 4°C; then 70%, 90%, and two times 100% ethanol; and a 1:1 mixture of 100% ethanol and propylene oxide and two times propylene oxide at −25°C, 15 min each. Dehydration and the following first step of infiltration are done using a coverslip-staining jar (or coverslip-staining rack and suitable container).
14. Transfer the coverslip to a 1:1 mixture of propylene oxide and Epon and incubate overnight at −25°C.
15. With the lid closed, let the coverslip-staining jar warm up to room temperature. Fill a small glass Petri dish with a 1:1 mixture of propylene oxide and Epon to a height of a few millimeters (mark liquid level at the outside of the Petri dish) and put the coverslip in the Petri dish (side with adhered cytoskeletons up).
16. Place the Petri dish with its lid half opened in a hood and let the propylene oxide evaporate. Complete evaporation of the propylene oxide should take 6–8 h, regulate rate of evaporation by closing the lid of the Petri dish every hour or so for about 30 min. Evaporation is complete when the resin volume is reduced by about 50%.
17. Decant the Epon as completely as possible from the coverslip, put it on a piece of parafilm placed in a Petri dish, overlay with Epon, and incubate overnight. Repeat once over the following day.
18. For flat embedding, decant the Epon as complete as possible from the coverslip. There should be as little Epon as possible on the back side (the side without cytoskeletons) of the coverslip. Excess Epon can be removed by wiping the back side of the coverslip over the edge of a microscope slide. Place a piece of Aclar film on a glass plate. As spacers, place two strips (about 4 mm × 25 mm) of Aclar film 14 mm apart on top of the Aclar on the glass plate and put a small drop of Epon in between. Place the coverslip (cytoskeletons facing down) on the spacers. Be careful not to trap air bubbles under the coverslip. Polymerize for 24–48 h at 65°C.
19. Removal of the coverslip from the Epon layer of the flat embedded material is done with hydrofluoric acid. Sand away all Epon residue on the back of the coverslip. Put the coverslip into a 50-ml plastic tube containing enough hydrofluoric acid to completely cover it and incubate until the coverslip dissolves off the Epon layer (usually takes between 10 and 20 min, swirling helps). Transfer the Epon layer of the flat embedding into a 50-ml plastic tube with distilled water and incubate for 10 min, repeat three times. The incubation in hydrofluoric acid softens the Epon layer. To reharden the flat embedding, cover a microscope slide with Aclar film. Put the flat embedding on top of the Aclar film and as spacers place two strips of Aclar film on opposite sides of the flat embedding. Place a second piece of Aclar film (the size of the microscope slide) and another microscope slide on top. Press the whole sandwich together using two 51-mm foldback clips and reharden for several days at 65°C.

The procedure described above can be also used to flat embed *Chlamydomonas* cells for structural analysis of the flagella by standard electron microscopy (Pedersen *et al.*, 2006) or double tilt-axis electron tomography (Pigino *et al.*, 2009). For this method, allow *Chlamydomonas* cells to settle and adhere as described in A, steps 1–4. Fix the cells with MI containing 2.5% glutaraldehyde, pH 7.2, for 30 min at room temperature. Decant fixative and fix for another 30 min with MI containing 2.5% glutaraldehyde and 0.2–0.4% tannic acid, pH 7.2, at room temperature. Then proceed as described above, starting with step 11.

E. Tips for the Ultramicrotomy of Flat Embeddings

Once the glass coverslip is removed, the side of the flat embedding where the *Chlamydomonas* cells or isolated cytoskeletons are located should be handled with great care to not scratch or contaminate the surface. This is very important as a glassy surface is absolutely essential for an exact alignment of the block face with the knife edge in preparation for sectioning. A good way to handle the flat embeddings is to put them (cell/cytokeleton side up) onto a piece of removable adhesive tape which is fixed onto a microscope slide. Having the flat embedding mounted this way is very convenient as it is held in place when cutting smaller pieces and the microscope slides can be stored dust free in Petri dishes. Areas with a suitable density of cells/cytoskeletons and a glassy surface are cut out of the flat embedding under a dissecting microscope using a scalpel. Small pieces of the flat embedding (about 1–2 mm × 1–2 mm) are mounted onto dummy blocks using a two-component adhesive with an epoxy resin base (the material also can be mounted so that cross sections of the flagella are obtained; for mounting see Fig. 1C). After hardening the adhesive overnight at room temperature, a block face of a suitable size is trimmed. For trimming, the use of an ultramicrotome and glass knives is recommended, as such mechanical trimming will give a precisely shaped block face. It is advisable to keep the block face as small as possible; the only limit to the minimum area of the block face is the ease of handling the sections once they are cut. A good size for the block face is about 0.2–0.3 mm×0.2–0.3 mm. As the flagella were attached to the coverslip and have only a diameter of about 220 nm, they are contained only within the first few hundred nanometes of the flat embedding (Fig. 1E). This makes it necessary to be extremely accurate in aligning the block face with the knife edge; the first or second section should be a full or nearly full section. It is not possible to just go on and take sections until the full area of the block face is being cut, because the material of interest will be gone after a few sections.

III. Materials

Suppliers for electron microscopy:

Electron Microscopy Sciences (EMS), Hatfield, PA, USA
Polysciences, Warrington, PA, USA
Ted Pella, Redding, CA, USA

Suppliers for immunogold reagents:

Aurion, Wageningen, The Netherlands
British BioCell International, Cardiff, UK
Jackson ImmunoResearch, West Grove, PA, USA
Nanoprobes, Yaphank, NY, USA
Sigma-Aldrich, St. Louis, MO, USA

Materials are listed according to method sections; recurrent materials are only listed once.

A. Flat Embedding of *Chlamydomonas* Cells in LR Gold

Chlamydomonas cells grown in MI
Growth medium MI (Sager and Granick, 1953)
Glass coverslips (#1 thickness)
Poly-L-lysine solution, 0.1% (w/v) in distilled water (e.g., Sigma-Aldrich, Cat# P 8920)
Dental wax sheets (available from most suppliers for electron microscopy)
Double-sided adhesive tape
Glutaraldehyde, 25%, EM grade or equivalent
Tannic acid (Mallinckrodt Baker, Phillipsburg, NJ, USA; Cat# 1674-02)
Osmium tetroxide, 1% (w/v) in distilled water
Absolute ethanol
LR Gold™ resin and benzil as catalyst for light polymerization (available from most suppliers for electron microscopy)
Coverslip-staining jars with plastic screw caps (e.g., EMS, Cat# 72242-21). Several incubations (dehydration and infiltration) are done in coverslip-staining jars that hold four 18 mm × 18 mm or 22 mm × 22 mm coverslips. When processing larger numbers of samples, it is convenient to use coverslip-staining racks and suitable glass containers with screw caps. However, to my knowledge, only coverslip staining racks holding 22 mm × 22 mm coverslips are commercially available (e.g., EMS, Cat# 72240). For those interested in custom-made coverslip staining racks (Fig. 1B), I can provide a mechanical drawing with exact measurements.
Chang monolayer mold with nine cavities, 20 mm × 20 mm × 1 mm (e.g., EMS, Cat# 70920)
Piece of sheet metal (about 12 cm × 12 cm, 1–1.5 mm thick)
Aclar® film (7.8 mil (199 μm) thickness)
Glass plate (about 10 cm × 10 cm, 1–2 mm thick)
Foldback clips (51 mm)
UV lamp mounted in a deep freezer for polymerizing the LR Gold. See instructions of the manufacturer for details.
A two-component adhesive on epoxy resin base, for example, UHU plus endfest 300 (UHU GmbH and Co KG., Buehl/Baden, Germany)
Removable adhesive tape, for example, Scotch Removable Magic Tape #811 (3M, St. Paul, MN, USA)

B. Embedding of Isolated Cytoskeletons from *Chlamydomonas* in LR Gold

$MT^{Mg^{2+}}$ buffer (30 mM HEPES, 5 mM Na-EGTA, 15 mM KCl, 5 mM $MgSO_4$, pH 7.2)
Formaldehyde, 16%, EM grade or equivalent
Gelatin capsules

C. Postembedding Immunogold-Labeling Technique

Pioloform-covered gold-gilded copper grids (gold-gilded copper grids are available from EMS)
Plastic box with snug-fitting lid
Platinum wire loop (e.g., inoculation loop)
Saturated solution of sodium metaperiodate
Na-PBS (8.1 mM Na_2HPO_4, 1.5 mM NaH_2PO_4, 150 mM NaCl, pH 7.4)
Bovine serum albumin (BSA), fraction V
Gelatin from cold water fish skin (fish gelatin) (e.g., Sigma-Aldrich, Cat# G7765)
Tween 20
Blocking buffer: Na-PBS containing 1–2% (w/v) BSA, 0.1% (v/v) fish gelatin, 0.05% (v/v) Tween 20, pH 7.4

Primary antibody. Prior to their use in immunogold EM, antibodies should be tested in immunofluorescence microscopy, but it should be kept in mind that getting a signal with an antibody in immunofluorescence is no guarantee for also getting a signal in immunogold EM. The primary antibody should be used at a final concentration of 1–5 μg/ml of specific IgG. The optimal concentration of the antibody can be determined by series of dilutions, a good starting point is the dilution that gave good results in immunofluorescence. The criterion for optimal antibody concentration is a high labeling density without background (Schwarz and Humbel, 2007).

Secondary antibody conjugated with colloidal gold. The 10 nm gold particles are a good compromise as they are easily visible in the electron microscope and give a good labeling density. As the gold particle size decreases, the density of gold label increases (Humbel *et al.*, 1998; Yokota, 1988). The use of very small gold markers (1 nm) such as ultrasmall gold (Aurion) or the gold compound Nanogold™ (Nanoprobes) and their enlargement by either deposition of metallic silver or gold (silver or gold enhancement) is described elsewhere (Stierhof, 2009, Stierhof *et al.*, 1995). Gold probes can be conjugated to protein A, protein G, protein A/G, or IgG. Staphylococcal protein A and streptococcal protein G bind to the Fc portion of IgG molecules from human and many animal species. Human, rabbit, pig, and guinea pig IgGs are excellent protein A binders. Protein G binds strongly to all human IgG subclasses and IgG molecules from rabbit, goat, sheep, horse, and cow. Protein A/G is a recombinant protein which displays the immunoglobulin Fc-binding sites for both, staphylococcal protein A and streptococcal protein G, presenting a broader spectrum toward IgGs of different species and subclasses (Ghitescu *et al.*, 1991). Gold probes conjugated to IgG are species specific and have a higher sensitivity but may show higher nonspecific binding than protein A, protein G, or protein A/G (Slot *et al.*, 1989).

2% (w/v) uranyl acetate in distilled water
Lead citrate according to Reynolds (1963)

D. Preembedding Immunogold Labeling of Isolated Cytoskeletons from *Chlamydomonas* and Flat Embedding in Epon

Cell wall-deficient *Chlamydomonas* mutant, preferably cw2 (strain CC-851 cw2 mt+, Chlamydomonas Center culture collection, Duke University, Durham, NC, USA; strains can be ordered online at http://www.chlamy.org/strains.html)

Triton X-100
Propylene oxide
An Epon-812 (Luft, 1961) replacement like EMbed 812 (EMS) or Poly/Bed® 812 (Polysciences). In this chapter the resin is called Epon.
Small glass Petri dishes (diameter about 40 mm)
Polymerization oven (65°C)
Sand paper (220 grit or similar)
Hydrofluoric acid, 48 wt% in water

IV. Summary and Discussion

This chapter describes the application of different immunogold-labeling techniques for the *in situ* localization of components of *Chlamydomonas* flagella and basal bodies. The presented methods rely on prior knowledge of advanced electron microscopy techniques but can be conducted in every standard electron microscopy laboratory. The fixatives used and their concentrations should be understood as guiding principles only. As with any immunocytochemical method that involves chemical fixation, it cannot be predicted how a particular fixation will affect specific antigens; this must be tested empirically. In the methods described here, glutaraldehyde and formaldehyde as well as tannic acid and osmium tetroxide are used as fixatives. Low concentrations of tannic acid enhance overall cell morphology and membrane contrast, while preserving sufficient antigenicity (Berryman *et al.*, 1992). The use of osmium tetroxide as fixative in immunocytochemical studies is critical. At temperatures greater than 0°C, osmium tetroxide acts proteolytically (Maupin and Pollard, 1983; Tanaka *et al.*, 1989) and antigenicity is often destroyed (Roth *et al.*, 1981). However, low concentrations of osmium tetroxide have been successfully used as fixative for postembedding immunogold electron microscopy (Lechtreck *et al.*, 1999; Pedersen *et al.*, 2006). Also the use of other fixatives might be explored. For example, uranyl acetate has been successfully used as postfixative in immunocytochemical studies enhancing membrane ultrastructure without apparent effects on the antigens studied (Berryman and Rodewald, 1990; Valentino *et al.*, 1985). A very promising preparation method would be cryofixation (preferably by high-pressure freezing) of *Chlamydomonas* cells adhered to a substrate, followed by freeze-substitution and flat embedding. Compared to the more conventional EM methods presented here, such a preparation most likely will improve ultrastructural preservation and also preservation of antigenicity, but special equipment is needed.

The most significant disadvantage of the postembedding technique is that only protein epitopes directly exposed on the section surface are accessible for the antibodies (Stierhof *et al.*, 1991), thus producing a two-dimensional localization of the protein epitopes. Pre-embedding immunogold techniques can provide a three-dimensional localization of protein epitopes but requires that the membrane be permeabilized to allow access of the antibodies. Because of the close association of the IFT trains with the flagellar membrane, an ideal labeling technique would be one that does not require permeabilization of the flagellar membrane. For this purpose, recently developed techniques for correlative light and electron microscopy might be applicable. These techniques are based on oxygen radicals generated by a fluorescent compound under illumination, which are used to drive the oxidation of diaminobenzidine (DAB) into an electron-dense precipitate that can be visualized by electron microscopy. In the FlAsH/ReAsH technique, biarsenical labeling reagents become fluorescent when they bind to recombinant proteins containing a specific tetracysteine motif and can be used to generate oxygen radicals for the oxidation of DAB (Gaietta *et al.*, 2002). In a method termed GRAB (GFP [green fluorescent protein] recognition after bleaching), oxygen radicals are generated by bleaching green fluorescent protein (Grabenbauer *et al.*, 2005). The spatial resolution achievable is expected to be similar or even somewhat higher compared to immunogold-labeling techniques (Grabenbauer *et al.*, 2005). The successful application of these techniques would permit the labeling of appropriately tagged IFT proteins within intact flagella. Subsequent analysis of thick sections by electron tomography could localize components of the IFT system in three-dimensional volumes of flagella. However, it has to be noted that these methods have not yet been successfully combined with protocols involving cryofixation followed by freeze substitution.

Acknowledgments

I thank Michael Melkonian, Joel Rosenbaum, Dennis Diener, Heinz Schwarz, Lara Perasso, Michaela Wilsch-Bräuninger, Pietro Lupetti, Gaia Pigino, and others for numerous and helpful conversations over the years regarding EM technique and Dennis Diener and Michaela Wilsch-Bräuninger for critically reading the manuscript. Special thanks to Benedikt Westermann for his ongoing support.

References

Bendayan, M., and Zollinger, M. (1983). Ultrastructural localization of antigenic sites on osmium-fixed tissues applying the protein A-gold technique. *J. Histochem. Cytochem.* **31**, 101–109.

Berryman, M.A., Porter, W.R., Rodewald, R.D., and Hubbard, A.L. (1992). Effects of tannic acid on antigenicity and membrane contrast in ultrastructural immunocytochemistry. *J. Histochem. Cytochem.* **40**, 845–857.

Berryman, M.A., and Rodewald, R.D. (1990). An enhanced method for post-embedding immunocytochemical staining which preserves cell membranes. *J. Histochem. Cytochem.* **38**, 159–170.

Cole, D.G., Diener, D.R., Himelblau, A.L., Beech, P.L., Fuster, J.C., and Rosenbaum, J.L. (1998). *Chlamydomonas* kinesin-II-dependent intraflagellar transport (IFT): IFT particles contain proteins required for ciliary assembly in *Caenorhabditis elegans* sensory neurons. *J. Cell Biol.* **141**, 993–1008.

Gaietta, G., Deerinck, T.J., Adams, S.R., Bouwer, J., Tour, O., Laird, D.W., Sosinsky, G.E., Tsien, R.Y., and Ellisman, M.H. (2002). Multicolor and electron microscopic imaging of connexin trafficking. *Science* **296**, 503–507.

Geimer, S., and Melkonian, M. (2005). Centrin scaffold in *Chlamydomonas reinhardtii* revealed by immunoelectron microscopy. *Eukaryot. Cell* **4**, 1253–1263.

Ghitescu, L., Galis, Z., and Bendayan, M. (1991). Protein AG-gold complex: An alternative probe in immunocytochemistry. *J. Histochem. Cytochem.* **39**, 1057–1065.

Grabenbauer, M., Geerts, W.J., Fernadez-Rodriguez, J., Hoenger, A., Koster, A.J., and Nilsson, T. (2005). Correlative microscopy and electron tomography of GFP through photooxidation. *Nat. Methods* **2**, 857–862.

Griffiths, G., Burke, B., and Lucocq, J. (1993). "Fine Structure Immunocytochemistry." Springer, Berlin, New York.

Höhfeld, I., Beech, P.L., and Melkonian, M. (1994). Immunolocalization of centrin in *Oxyrrhis marina*. *J. Phycol.* **30**, 474–489.

Hou, Y., Qin, H., Follit, J.A., Pazour, G.J., Rosenbaum, J.L., and Witman, G.B. (2007). Functional analysis of an individual IFT protein: IFT46 is required for transport of outer dynein arms into flagella. *J. Cell Biol.* **176**, 653–665.

Humbel, B.M., de Jong, M.D., Muller, W.H., and Verkleij, A.J. (1998). Pre-embedding immunolabeling for electron microscopy: An evaluation of permeabilization methods and markers. *Microsc. Res. Tech.* **42**, 43–58.

Johnson, K.A. (1995). Immunoelectron microscopy. *In* "Methods in Cell Biology" (W. Dentler, and G. Witman, eds.), Vol. 47, pp. 153–162. Academic Press, San Diego.

Johnson, K.A., Haas, M.A., and Rosenbaum, J.L. (1994). Localization of a kinesin-related protein to the central pair apparatus of the *Chlamydomonas reinhardtii* flagellum. *J. Cell Sci.* **107**, 1551–1556.

Keller, L.C., Geimer, S., Romijn, E., Yates, J., III, Zamora, I., and Marshall, W.F. (2009). Molecular architecture of the centriole proteome: The conserved WD40 domain protein POC1 is required for centriole duplication and length control. *Mol. Biol. Cell* **20**, 1150–1166.

Kozminski, K.G., Beech, P.L., and Rosenbaum, J.L. (1995). The *Chlamydomonas* kinesin-like protein FLA10 is involved in motility associated with the flagellar membrane. *J. Cell Biol.* **131**, 1517–1527.

Kozminski, K.G., Johnson, K.A., Forscher, P., and Rosenbaum, J.L. (1993). A motility in the eukaryotic flagellum unrelated to flagellar beating. *Proc. Natl. Acad. Sci. USA* **90**, 5519–5523.

Lechtreck, K.F., and Geimer, S. (2000). Distribution of polyglutamylated tubulin in the flagellar apparatus of green flagellates. *Cell Motil. Cytoskeleton* **47**, 219–235.

Lechtreck, K.F., Teltenkötter, A., and Grunow, A. (1999). A 210 kDa protein is located in a membrane-microtubule linker at the distal end of mature and nascent basal bodies. *J. Cell Sci.* **112**, 1633–1644.

Luft, J.H. (1961). Improvements in epoxy resin embedding methods. *J. Biophys. Biochem. Cytol.* **9**, 409–414.

Maupin, P., and Pollard, T.D. (1983). Improved preservation and staining of HeLa cell actin filaments, clathrin-coated membranes, and other cytoplasmic structures by tannic acid-glutaraldehyde-saponin fixation. *J. Cell Biol.* **96**, 51–62.

Mitchell, B.F., Grulich, L.E., and Mader, M.M. (2004). Flagellar quiescence in *Chlamydomonas*: Characterization and defective quiescence in cells carrying sup-pf-1 and sup-pf-2 outer dynein arm mutations. *Cell Motil. Cytoskeleton* **57**, 186–196.

Mitchell, D.R., and Nakatsugawa, M. (2004). Bend propagation drives central pair rotation in *Chlamydomonas reinhardtii* flagella. *J. Cell Biol.* **166**, 709–715.

Morphew, M.K. (2007). 3D immunolocalization with plastic sections. *In* "Methods in Cell Biology" (J. R. McIntosh, ed.), Vol. 79, pp. 493–513. Elsevier Inc., San Diego.

Newman, G.R., and Hobot, J.A. (1993). "Resin Microscopy and on-Section Immunocytochemistry." Springer-Verlag, Berlin, New York.

Pazour, G.J., Wilkerson, C.G., and Witman, G.B. (1998). A dynein light chain is essential for the retrograde particle movement of intraflagellar transport (IFT). *J. Cell Biol.* **141**, 979–992.

Pedersen, L.B., Geimer, S., and Rosenbaum, J.L. (2006). Dissecting the molecular mechanisms of intraflagellar transport in *Chlamydomonas*. *Curr. Biol.* **16**, 450–459.

Pedersen, L.B., Geimer, S., Sloboda, R.D., and Rosenbaum, J.L. (2003). The Microtubule plus end-tracking protein EB1 is localized to the flagellar tip and basal bodies in *Chlamydomonas reinhardtii*. *Curr. Biol.* **13**, 1969–1974.

Pigino, G., Geimer, S., Lanzavecchia, S., Paccagnini, E., Cantele, F., Diener, D.R., Rosenbaum, J.L., and Lupetti P. (2009). Electron-tomographic Analysis of Intraflagellar Transport Particle Trains in Situ. *J. Cell Biol.* Oct 5; **187**(1), 135–148.

Piperno, G., and Fuller, M.T. (1985). Monoclonal antibodies specific for an acetylated form of alpha-tubulin recognize the antigen in cilia and flagella from a variety of organisms. *J. Cell Biol.* **101**, 2085–2094.

Qin, H., Diener, D.R., Geimer, S., Cole, D.G., and Rosenbaum, J.L. (2004). Intraflagellar transport (IFT) cargo: IFT transports flagellar precursors to the tip and turnover products to the cell body. *J. Cell Biol.* **164**, 255–266.

Reynolds, E.S. (1963). The use of lead citrate at high pH as an electron-opaque stain in electron microscopy. *J. Cell Biol.* **17**, 208–212.

Roth, J., Bendayan, M., Carlemalm, E., Villiger, W., and Garavito, M. (1981). Enhancement of structural preservation and immunocytochemical staining in low temperature embedded pancreatic tissue. *J. Histochem. Cytochem.* **29**, 663–671.

Sager, R., and Granick, S. (1953). Nutritional studies with *Chlamydomonas* reinhardi. *Ann. NY Acad. Sci.* **56**, 831–838.

Schwarz, H., and Humbel, B.M. (2007). Correlative light and electron microscopy using immunolabeled resin sections. *In* "Methods in Molecular Biology" (J. Kuo, ed.), Vol. 369, pp. 229–256. Humana Press, Totowa.

Sloboda, R.D., and Howard, L. (2007). Localization of EB1, IFT polypeptides, and kinesin-2 in *Chlamydomonas* flagellar axonemes via immunogold scanning electron microscopy. *Cell Motil. Cytoskeleton* **64**, 446–460.

Slot, J.W., Posthuma, G., Chang, L.Y., Crapo, J.D., and Geuze, H.J. (1989). Quantitative aspects of immunogold labeling in embedded and in nonembedded sections. *Am. J. Anat.* **185**, 271–281.

Stierhof, Y.-D. (2009). Immunolabeling of ultra-thin sections with enlarged 1 nm gold or Qdots. *In* "Handbook of Cryo-Preparation Methods for Electron Microscopy" (A. Cavalier, D. Spehner, and B.M. Humbel, eds.), pp. 587–616. CRC Press, Boca Raton.

Stierhof, Y-D., Hermann, R., Humbel, B.M., and Schwarz, H. (1995). Use of TEM, SEM, and STEM in imaging 1 nm colloidal gold particles. *In* "Immunogold-Silver Staining: Principles, Methods, and Applications" (M.A. Hayat, ed.), pp. 97–118. CRC Press, Boca Raton.

Stierhof, Y-D., Schwarz, H., Dürrenberger, M., Villiger, W., and Kellenberger E. (1991). Yield of immunolabel compared to resin Sections and thawed cryosections. *In* "Colloidal Gold: Principles, Methods, and Applications" (M. A. Hayat, ed.), Vol. 3, pp. 87–115. Academic Press, San Diego.

Stirling, J.W. (1990). Immuno- and affinity probes for electron microscopy: A review of labeling and preparation techniques. *J. Histochem. Cytochem.* **38**, 145–157.

Tanaka, K., Mitsushima, A., Kashima, Y., Nakadera, T., and Osatake, H. (1989). Application of an ultrahigh-resolution scanning electron microscope (UHS-T1) to biological specimens. *J. Electron. Microsc. Tech.* **12**, 146–154.

Valentino, K.L., Crumrine, D.A., and Reichardt, L.F. (1985). Lowicryl K4M embedding of brain tissue for immunogold electron microscopy. *J. Histochem. Cytochem.* **33**, 969–973.

Wright, R.L., Salisbury, J., and Jarvik, J.W. (1985). A nucleus-basal body connector in *Chlamydomonas reinhardtii* that may function in basal body localization or segregation. *J. Cell Biol.* **101**, 1903–1912.

Yokota, S. (1988). Effect of particle size on labeling density for catalase in protein A-gold immunocytochemistry. *J. Histochem. Cytochem.* **36**, 107–109.

CHAPTER 4

Scanning Electron Microscopy to Examine Cells and Organs

Jovenal T. SanAgustin[*], **John A. Follit**[*], **Gregory Hendricks**[†], *and* **Gregory J. Pazour**[*]

[*]Program in Molecular Medicine, University of Massachusetts Medical School, Worcester, Massachusetts 01605

[†]Department of Cell Biology, University of Massachusetts Medical School, Worcester, Massachusetts 01655

Abstract

Scanning electron microscopy is an excellent method for viewing the surface of cells and organs, and provides exquisite detail of surface projections. This method has a long history of use in the analysis of eukaryotic cilia and flagella. In this chapter, we provide methods used by our group to examine mouse kidneys and the embryonic node. The methods provided here can be used with little modification to examine other mammalian organs or used as a starting point to develop methods for use in other organisms.

METHODS IN CELL BIOLOGY, VOL. 91

978-0-12-374973-4
DOI: 10.1016/S0091-679X(08)91004-9

I. Introduction

The surface detail provided by scanning electron microscopy (EM) reveals features that are not visible by any other technique. In particular, this approach provides exquisite detail of cellular projections such as cilia and flagella. Scanning EM provided key evidence of the prevailing view that nearly every vertebrate cell type has either primary or motile cilia projecting from their surface. For example, the detailed scanning EM study of the kidney by Andrews and Porter in 1974 noted cilia on all cell types of the tubule except for the intercalated cells (Andrews and Porter, 1974). This documentation predated the understanding of the importance of kidney primary cilia by almost 30 years. In addition to this paper, numerous publications have used scanning EM to document primary cilia on a multitude of cell types in the organs of vertebrate animals. The primary cilium homepage (http://www.bowserlab.org/primarycilia/ciliumref.html) has an extensive list of older publications that document primary cilia, many of which used scanning EM. In addition, the book *Tissues and Organs: A Text Atlas of Scanning Electron Microscopy* (Kessell and Kardon, 1979) is required viewing for anyone interested in scanning EM and cilia.

The analysis of genetically modified mice with defects in cilia-related genes has led to a resurgence of interest in scanning EM. In this work, we describe methods used in our laboratory for the examination of cilia in the kidney (Fig. 1) and on the embryonic node (Fig. 2) of mice. These protocols can be used with either no or little modification to look at other mammalian organs and the methods can be used as a starting point to look at other nonmammalian organisms like planaria (see Chapter 4 by Rompolas *et al.*, volume 93). In addition, scanning EM can be combined with more advanced techniques such as immunogold labeling for further characterization of the surface (Agematsu *et al.*, 1997; Goldberg, 2008).

II. Materials

1. Cacodylate buffer: 0.1 M sodium cacodylate, pH 7.4. Make from sodium cacodylate trihydrate (Electron Microscopy Sciences, Hatfield, PA, USA).
   Note: Cacodylate is an arsenic-containing compound and should be handled with care. Use disposable labware for preparing and storing cacodylate buffers.
2. Fixative: 2.5% glutaraldehyde in 0.1 M sodium cacodylate, pH 7.4. Make from 25% stock (Electron Microscopy Sciences).
   Note: Glutaraldehyde can cause burns and eye damage, use caution when handling.
3. Ethanol solutions (10, 30, 50, 70, 85, 95, and 100%).
4. 1% osmium tetroxide. Make from 4% stock (Electron Microscopy Sciences).
   Note: Osmium tetroxide is a highly dangerous *volatile* chemical. Exposure to low concentrations of fumes can cause blindness and even death.

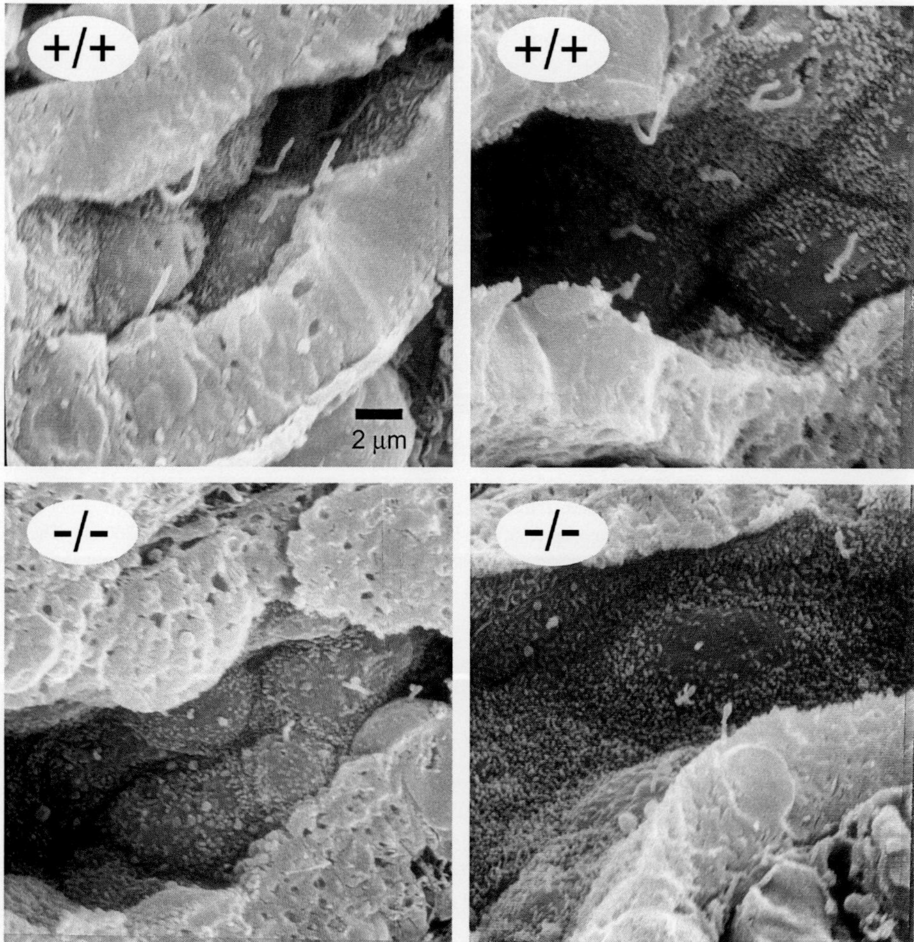

Fig. 1 Scanning EM of mouse kidney. Primary cilia in the kidney of Tg737 mutant mice are shorter than normal. Kidneys from 4-day-old pups were fixed with glutaraldehyde, freeze fractured, metal impregnated, and examined by scanning EM. Numerous cilia were found on the epithelial cells in the tubules and collecting ducts of the wild-type mice (+/+). Cilia were also found in the homozygous mutant (–/–) pups, but they were usually <2-μm long and most were only short stubs. Originally published as Fig. 6A in Pazour *et al.* (2000). Used with permission from *The Journal of Cell Biology.*

5. Liquid nitrogen.
   Note: Wear protective gear to prevent frostbite.
6. Microporous specimen capsules (Electron Microscopy Sciences #701880).
7. Aluminum specimen mount (slotted head, 12.7-mm diameter, with 7-mm-long pin, Electron Microscopy Sciences #75200).
8. Double-sided carbon conductive tape (Electron Microscopy Sciences #77816).

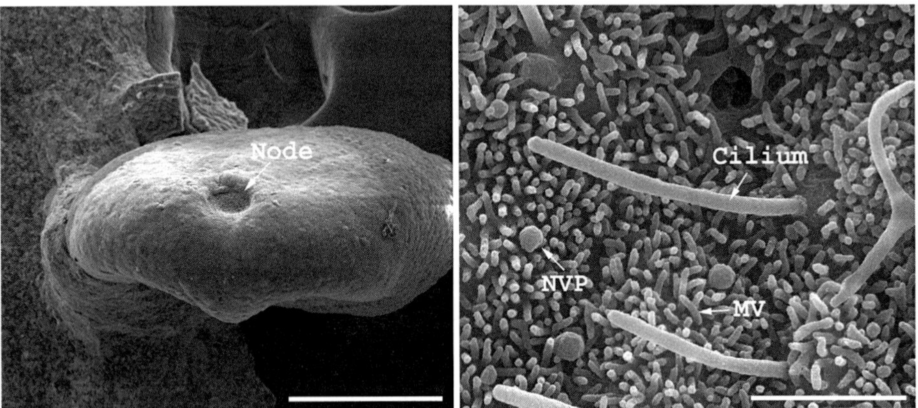

Fig. 2 Scanning EM of nodal cilia. The embryonic node is a pit on the ventral surface of the early mouse embryo. The cells in the bottom of the pit contain motile primary cilia and are involved in breaking left–right symmetry of the early embryo (Nonaka *et al.*, 1998; Tanaka *et al.*, 2005). The left panel shows a low magnification image of a mouse embryo with the node in the center. Scale bar is 200 μm. The right panel shows a high magnification image of cells within the node. Note the cilia, microvilli (MV), and nodal vesicular parcels (NVP). Scale bar is 2 μm.

9. Colloidal silver paste (Electron Microscopy Scicnccs #12640).
10. O-F solution: 1% osmium tetroxide, 1.5% potassium ferrocyanide trihydrate in 0.1 M sodium cacodylate, pH 7.4. $K_4Fe(CN)_6 \cdot 3H_2O$ was purchased from Sigma Chemical Company (#P-9387).
11. 2% tannic acid in 0.1 M sodium cacodylate, pH 7.4. Tannic acid was purchased from Electron Microscopy Sciences (#02170).
12. 1% aqueous hydroquinone. Hydroquinone was purchased from Eastman Kodak Company (Rochester, NY, USA, #105 0350).

III. Metal-Impregnation Method for Scanning EM Analysis of Mouse Kidneys (modified from McManus *et al.*, 1993)

1. Euthanize animals by an approved method and remove the kidneys. Cut the kidneys of mature animals (6 weeks and older) in half. Immerse kidneys in the fixative for 2 h at room temperature or overnight at 4°C.

   Note: Preservation of the tissue is better if the animal is perfusion fixed prior to the removal of the organs. Methods for perfusion fixation (adapted for mice) are given in Sprando (1990). The mice are perfused for 20 min with the fixative.
2. Place a small (i.e., 10 cm × 10 cm × 5 cm) metal block in an ice bucket and cover it with liquid nitrogen.

3. Remove the kidneys from the fixative and flash-freeze them in the liquid nitrogen. Save the fixative for step 5.
4. Position the kidneys on the submerged metal block and fracture them into ~4-mm fragments by striking with a hammer.
5. Collect the kidney fragments and thaw them in the saved fixative.
6. Wash the kidney fragments three times in cacodylate buffer, 15 min per wash at room temperature.
7. Postfix the fragments in O-F solution for 2 h at room temperature. Steps involving the use of osmium tetroxide must be carried out in a fume hood.
8. Transfer the kidney fragments from the O-F solution into the 2% tannic acid solution and incubate for 3 h at room temperature. Tannic acid enhances the metal deposition in the tissue.
9. Repeat steps 7 and 8. Then transfer the kidney fragments into 1% hydroquinone and leave overnight. This will complete the reduction of the O-F components to an insoluble metallic complex. No washes are needed between the transfers.
10. Wash the metal-impregnated kidney fragments twice with distilled water.
11. Dehydrate the metal-impregnated kidneys through a graded ethanol series. This is done by moving the tissue through the following ethanol concentrations: 10, 30, 50, 70, 85, 95, 100, and 100% with the tissue incubating in each solution for 10 min.
12. Collect the kidney fragments into microporous specimen capsules immersed in 100% ethanol. The kidney fragments are ready for critical point drying.
13. Fill the process chamber of a CO_2 critical point dryer (i.e., Autosamdri-815, Series A, Tousimis Research Corp, Rockville, MD, USA) with 100% ethanol, and quickly transfer the capsules containing the kidney fragments into the chamber. The samples must not be allowed to dry out. The kidney samples then undergo a process where the ethanol is replaced by liquid CO_2. Once all the ethanol has been purged and replaced with liquid CO_2 at −15°C, the samples are carried through the "critical point" by increasing the pressure in the chamber to 1100 psi and the temperature to above 31°C.
14. Store the capsules in a desiccator after the critical point drying step. The experiment may be stopped at this point.
15. Attach a piece of double-sided carbon tape on the head of a specimen mount.
16. Select kidney fragments showing internal fractures and position them on the carbon-taped specimen mount with the fractures facing up.
17. Paint silver paste round the edges of the kidney fragments, and then trace a line from the fragments to a portion of the head of the specimen mount not covered by the carbon tape. This ensures the specimen is properly grounded to the metal stud. The O-F and tannic acid treatments caused significant amounts of metal to be deposited in the material, making an additional surface coat of metal unnecessary.
18. Examine the fractured surfaces of the kidney in a scanning electron microscope (FEI Quanta 200 FEG SEM, Hillsboro, Oregon, USA.).

IV. Method for Scanning EM Visualization of Mouse Nodal Cilia

1. Euthanize a timed-pregnant female by an approved method and dissect out the uterus. Mouse development can vary by genetic background but dissections between 9 and 11 p.m. on day postcoital day 7 works best for our mouse colony. Separate the decidua and dissect each embryo (Hogan *et al.*, 1994). It is best to do the dissections in phosphate-buffered saline (PBS) to keep the tissues from drying out.
2. Cut the embryo in half along the anterior/posterior axis. Place the half of the embryo containing the ventral surface and the node immediately into fix. The other half of the embryo can be used for genotyping.
3. Fix embryos 12–16 h at 4°C without agitation.
4. Wash the fixed embryos three times in cacodylate buffer, 5 min per wash at room temperature.
5. Postfix in 1% aqueous osmium tetroxide for 1 h at room temperature. Steps involving the use of osmium tetroxide must be carried out in a fume hood. The tissue turns black during this step.
6. Wash three times in cacodylate buffer as in step 4.
7. Dehydrate the osmicated embryos in a graded ethanol series. This is done by moving the tissue through the following ethanol concentrations: 10, 30, 50, 70, 85, 95, 100, and 100% with the tissue incubating in each solution for 5 min.
8. Place the embryos into microporous specimen capsules immersed in 100% ethanol. The embryos are ready for critical point drying.
9. Fill the process chamber of a CO_2 critical point dryer (i.e., Autosamdri-815, Series A, Tousimis Research Corp.) with 100% ethanol, and quickly transfer the capsules containing the embryos into the chamber. The samples must not be allowed to dry out. The samples then undergo a process where the ethanol is replaced by liquid CO_2. Once all the ethanol has been purged and replaced with liquid CO_2 at −15°C, the samples are carried through the "critical point" by increasing the pressure in the chamber to 1100 psi and the temperature to above 31°C.
10. Store the capsules in a desiccator after the critical point drying step. The experiment may be stopped at this point.
11. Attach a piece of double-sided carbon tape on the head of a specimen mount.
12. Use an eyelash mounted on a wooden handle to move the embryo onto the specimen mount. Position the embryo on the carbon-taped specimen mount with the ventral node facing up (cut side down against the carbon tape).
13. Paint silver paste around the base of the embryo, and then trace a line from the fragments to a portion of the head of the specimen mount not covered by the carbon tape. This ensures the specimen is properly grounded to the metal stud.
14. Sputter coat the mounted embryos with iridium to a thickness of 3 nm (Cressington 208 HR Sputter Coater, Ted Pella, Redding, CA, USA).
15. Examine the surface of the embryo in a scanning electron microscope (FEI Quanta 200 FEG SEM).

V. Summary

Two different methods are provided for the visualization of cilia in mice by scanning EM. Either method should work for most tissues of interest. However, because the fractured surfaces of kidneys tend to have cracks and discontinuities in their surfaces, metal coatings can leave large portions of the surfaces unusable because they are not properly grounded. By metal impregnating these surfaces, a continuous ground path can be achieved over the entire specimen. When studying other tissues of interest, one may try both methods to determine which is optimal.

Visualization of the surface of cells and organs by scanning EM provides information that complements other microscopy approaches such as transmission EM and light microscopy. The ability of scanning EM to scrutinize the fine structure of the surface of the cilium suggests that this methodology will become increasingly useful as we begin to study mutations that do not simply block ciliary assembly but have more subtle effects on ciliary structure.

Acknowledgments

The Pazour laboratory is supported by grant GM060992 from the National Institutes of Health. The EM laboratory is supported by the Diabetes Endocrinology Research Center grant DK32520.

References

Agematsu, H., Sawada, T., Watanabe, H., Yanagisawa, T., and Ide, Y. (1997). Immuno-scanning electron microscope characterization of large tubules in human deciduous dentin. *Anat. Rec.* **248**, 339–345.

Andrews, P.M., and Porter, K.R. (1974). A scanning electron microscopic study of the nephron. *Am. J. Anat.* **140**, 81–116.

Goldberg, M.W. (2008). Immunolabeling for scanning electron microscopy (SEM) and field emission SEM. *Methods Cell Biol.* **88**, 109–130.

Hogan, B., Beddington, R., Costantini, F., and Lacy, E. (1994). "Manipulating the Mouse Embryo: A Laboratory Manual." Cold Spring Harbor Laboratory Press, Plainview, NY.

Kessell, R.G., and Kardon, R.H. (1979). "Tissues and Organ: A Text-Atlas of Scanning Electron Microscopy." W. H. Freeman, San Francisco.

McManus, W.R., McMahon, D.J., and Oberg, C.J. (1993). High-resolution scanning electron microscopy of milk products: A new sample preparation procedure. *Food Struct.* **12**, 475–482.

Nonaka, S., Tanaka, Y., Okada, Y., Takada, S., Harada, A., Kanai, Y., Kido, M., and Hirokawa, N. (1998). Randomization of left-right asymmetry due to loss of nodal cilia generating leftward flow of extraembryonic fluid in mice lacking KIF3B motor protein. *Cell* **95**, 829–837.

Pazour, G.J., Dickert, B.L., Vucica, Y., Seeley, E.S., Rosenbaum, J.L., Witman, G.B., and Cole, D.G. (2000). *Chlamydomonas IFT*88 and its mouse homologue, polycystic kidney disease gene *Tg737*, are required for assembly of cilia and flagella. *J. Cell Biol.* **151**, 709–718.

Sprando, R.L. (1990). Perfusion of the rat testis through the heart using heparin. *In* "Histological and Histopathological Evaluation of the Testis" (Russel, L.D., Ettlin, R.A., Sinha Hikim, A.P., and Clegg, E.D., eds.), pp. 277–280. Cache River Press, Clearwater, FL.

Tanaka, Y., Okada, Y., and Hirokawa, N. (2005). FGF-induced vesicular release of Sonic hedgehog and retinoic acid in leftward nodal flow is critical for left-right determination. *Nature* **435**, 172–177.

CHAPTER 5

X-ray Fiber Diffraction Studies on Flagellar Axonemes

Kazuhiro Oiwa[*,†], **Shinji Kamimura**[‡], *and* **Hiroyuki Iwamoto**[Ï]

[*]Kobe Advanced ICT Research Center, National Institute of Information and Communications Technology, 588-2 Iwaoka, Nishi-ku, Kobe 651-2492, Japan

[†]Graduate School of Life Science, University of Hyogo, Harima Science Park City, Hyogo 678-1297, Japan

[‡]Department of Biological Sciences, Faculty of Science and Engineering, Chuo University, 1-13-27 Kasuga, Bunkyo, Tokyo 112-8551, Japan

[Ï]Research and Utilization Division, SPring-8, Japan Synchrotron Radiation Research Institute, Hyogo 679-5198, Japan

METHODS IN CELL BIOLOGY, VOL. 91

978-0-12-374973-4
DOI: 10.1016/S0091-679X(08)91005-0

Abstract

Eukaryotic cilia and flagella are highly ordered and precisely assembled cellular organelles. Here, to understand the mechanism of the orderly undulations of cilia and flagella, we shall draw a blueprint of their core structures and supporting scaffolds, that is, axonemes, and we shall describe the dynamic structural changes of components of the organelles. Small-angle X-ray scattering and diffraction are among the principal tools used to study protein polymers. These methods are now well established as indispensable tools that complement electron microscopy, providing information on the structure and dynamics of biological materials at atomic resolution in near-physiological environments. For instance, X-ray diffraction studies of skeletal muscles have contributed greatly to our understanding of the structure and molecular mechanisms of muscles. However, owing to the minute size and low diffracting power of axonemes, few attempts at X-ray diffraction of axonemes have been reported. The advent of third-generation synchrotron radiation facilities now makes these attempts feasible, because we now have stable and intense X-rays that enable us to obtain diffractions from the axonemes. In this chapter, we provide a concise practical guide to this new avenue for structural analysis of axonemes.

I. Introduction

The axoneme, with a diameter of ~150 nm, is a highly ordered and precisely assembled superstructure with more than 250 constituent proteins (Dutcher, 1995; Pazour *et al.*, 2005). The most widespread form of the axoneme has a 9 + 2 arrangement of microtubules: nine doublets surrounding a pair of singlets (the central pair microtubules), with radial spokes extending from each of the peripheral doublets toward the central pair. Coordinated beating and bend propagation of cilia and flagella are generated by active sliding of the peripheral doublet microtubules, driven by ensembles of various types of dyneins under the regulation of axonemal components. In an axoneme, the activity of the dynein molecules is propagated through linear or two-dimensional arrays of dynein arms closely packed on the peripheral doublet microtubules. Axonemal dyneins and other constituent proteins thus show large-scale integrated behavior that is responsible for the beating of flagella and for wave propagation. Detailed information on the molecular architecture of axonemes and the structural changes in the constituent proteins during motion is thus indispensable for understanding the mechanism of waveform generation by flagella and cilia.

As described in Chapters 1 and 2 in this volume, electron microscopy has been practically the only means used to obtain structural information on axonemal components since the 9 + 2 architecture was first described in the 1950s (e.g., Afzelius, 1959), with the exception of a few pioneering X-ray diffraction studies (Silvester, 1964; Yamaguchi *et al.*, 1972). Advanced electron microscopic techniques have provided three-dimensional structural details of axonemal components through, for example, electron tomography of metal replicas of rapidly frozen and cryofractured sperm axonemes from the dipteran *Monarthropalpus flavus* (Lupetti *et al.*, 2005), and cryoelectron tomography of sea urchin sperm (Nicastro *et al.*, 2005) and *Chlamydomonas* flagella (Bui *et al.*, 2008; Ishikawa *et al.*, 2007; Nicastro *et al.*, 2005, 2006).

Recently, however, intense X-ray beams generated by the third-generation synchrotron radiation facilities have made it feasible to analyze axonemal structure by means of small-angle X-ray scattering/diffraction technique (SAXS) (e.g., Kamimura *et al.*, 2007; Sugiyama *et al.*, in press; Toba *et al.*, 2007). The major advantage of the SAXS technique over electron microscopy is that it can be applied to native, functional axonemal samples. Because of this, data can be collected free of fixation-induced artifacts. The effects of biologically active chemicals such as ATP and calcium can also be tested on the same sample. Even the time course of structural change can be followed in time-resolved measurements (e.g., Wakayama *et al.*, 2004). Moreover, because all the parameters needed for analysis—such as X-ray wavelength, specimen-to-detector distance, and the pixel size of the detector—can be accurately defined, the SAXS technique is especially effective in precisely analyzing periodical structures: many axonemal components are known to have axial periodicities of multiples of 8 nm, that is, the basic axial periodicity of α–β tubulin dimers.

As stated earlier, recordings of high-quality diffraction patterns from eukaryotic ciliary/flagellar axonemes were made possible by the intense, well-oriented X-ray beams generated by third-generation synchrotron radiation facilities. The use of weaker sources such as laboratory-scale rotating-anode generators is impractical. In the following part, therefore, we focus on studies that use SAXS beamlines in these facilities.

II. Rationale

The following is a very brief introduction to the SAXS technique, including the principle of X-ray scattering in general and specific applications to protein crystals, muscle, and axonemes. For more details, see textbooks such as those by Holmes and Blow (1965) or Cantor and Schimmel (1980). Innumerable textbooks have been published on protein crystallography. For diffraction from muscle, the textbook by Squire (1981) may be consulted. Details of the mathematical treatment of diffraction from axonemes have also been published (Iwamoto, 2008).

A. X-ray Scattering from Sample

X-rays are electromagnetic radiation occupying the spectrum from 10^{-3} to 10 nm in wavelength. For studies of the structure of proteins and their complexes, the wavelengths used are confined to the approximate range of 0.05–0.3 nm. X-rays, like visible light, exhibit a wave–particle duality. X-rays propagate with the speed of light $c = 2.998 \times 10^8\,\mathrm{m \cdot s^{-1}}$. The wavelength λ and frequency ν are related by

$$\lambda = \frac{c}{\nu} \tag{1}$$

A beam of X-rays is regarded as a stream of photons. A photon is characterized by its energy E and momentum p, which are related to the wavelength λ and frequency ν by

$$E = h\nu \tag{2}$$

$$p = \frac{h}{\lambda} \tag{3}$$

where h is Planck's constant $6.626 \times 10^{-34}\,\mathrm{J \cdot s}$. Equation (3) is called the *de Broglie's* relation. The wavelength λ of X-ray is precisely given from Eqs. (1) and (2).

In visible light or X-rays, photons are scattered by the electrons of the sample. Because of their properties as waves, the scattered photons interfere with each other and are deflected in various ways, creating a scattering or diffraction pattern. Mathematically, the scattering/diffraction pattern is a Fourier transform of the electron density distribution of the sample expressed in reciprocal space, in which lengths are inversely related to the corresponding lengths in real space. If a second Fourier transform is performed on the pattern, it restores the original structure of the sample in real space, often at a different magnification. In the case of visible light, the second Fourier transform is achieved by a lens, but in the case of X-ray, usually the first Fourier transform is directly subjected to analysis because of the difficulty in manufacturing a usable lens.

B. X-ray Scattering from Periodically Arranged Objects

Samples such as protein crystals, muscles, or axonemes contain large numbers of identical repeating units that are arranged with specific periodicities. The Fourier transform of a single repeating unit (often called a unit cell) is a weak, continuous function called a structure factor. The periodic arrangements of many repeating units strengthen and modify the function in complex ways, and the resulting scattering/diffraction pattern looks very different from the original structure factor. In the following, by taking protein crystals, muscle, and axonemes as examples, we describe how the original structure factor is modified by periodic arrangement.

Before proceeding to biological specimens, we briefly describe the basic principles of Fourier transform of periodically arranged objects. Figure 1 compares the Fourier transform of a single dynein molecule and that of periodically arranged dynein molecules. In mathematical terms, the periodically arranged molecules are considered as the convolution of two functions, that is, molecular shape and periodicity (first row). On the other hand, the Fourier transform of the dynein arrays is the product of the Fourier transforms of the two functions stated above. This demonstrates an important

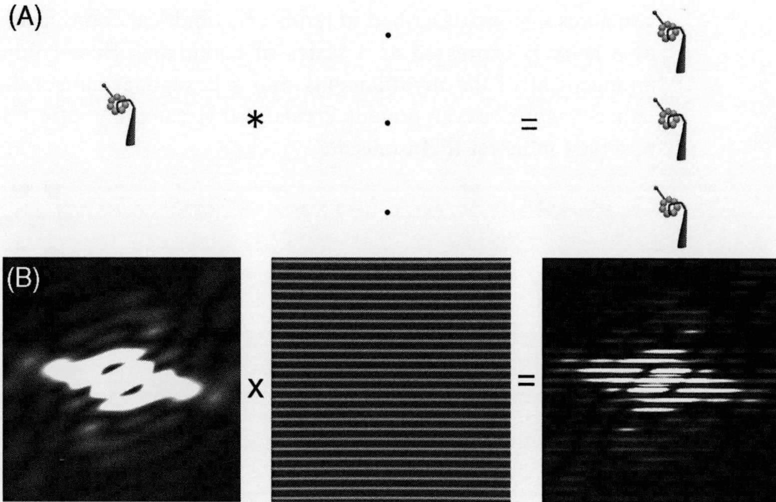

Fig. 1 X-ray scattering from periodically arranged objects. (A) objects in real space, (B) scattering pattern from the objects in (A). An array of three dynein molecules (A, right) is considered to be the convolution (asterisk) of a single dynein molecule (A, left) and three dots representing the periodicity (A, middle). The scattering pattern from the three dynein molecules (B, right), on the other hand, is the product of the scattering pattern (Fourier transform) from a single dynein molecule (B, left) and that from the three dots (B, middle).

principle called the convolution theorem, stating that the Fourier transform of the convolution of two functions is the product of the Fourier transforms of the two functions.

Next, we show in Fig. 2 how the convolution theorem works in three biological specimens, that is, protein crystal, muscle, and axoneme. The former two are very popular materials for X-ray diffraction studies, and the comparison with these materials will give us a better understanding of the principle of diffraction from axonemes.

The structure of a protein crystal (Fig. 2A) is regarded as the convolution of the molecular shape and a regular three-dimensional lattice, and the Fourier transform of the latter gives a collection of discrete, infinitesimally small peaks. As a result, the continuous function of the molecular transform is reduced to a collection of discrete spots (reflections). Despite this disrupting effect of the lattice (lattice sampling), it is possible to record a large number of reflections owing to the high regularity of the lattice, and the molecular shape can be restored at an atomic resolution. To record as many reflections as possible, the protein crystals must be rotated to maximize the number of lattice planes that meet the "Bragg condition" for producing reflections.

In muscle (Fig. 2B), contractile proteins are arranged helically in myofilaments. The myofilaments are in turn arranged into a two-dimensional hexagonal lattice. Because of this, the structure of muscle can be regarded as the convolution of three functions, that is, molecular shape, helix, and hexagonal lattice. As a result, the Fourier transform of a muscle is the product of the Fourier transforms of the three functions. As in other structures that are described in terms of cylindrical coordinates, the Fourier transform of a helix is expressed as a series of continuous Bessel functions. Because of the arrangement of the myofilaments into a hexagonal lattice, the Bessel functions are lattice-sampled as in protein crystals; this sampling effect is most conspicuously observed in insect flight muscle.

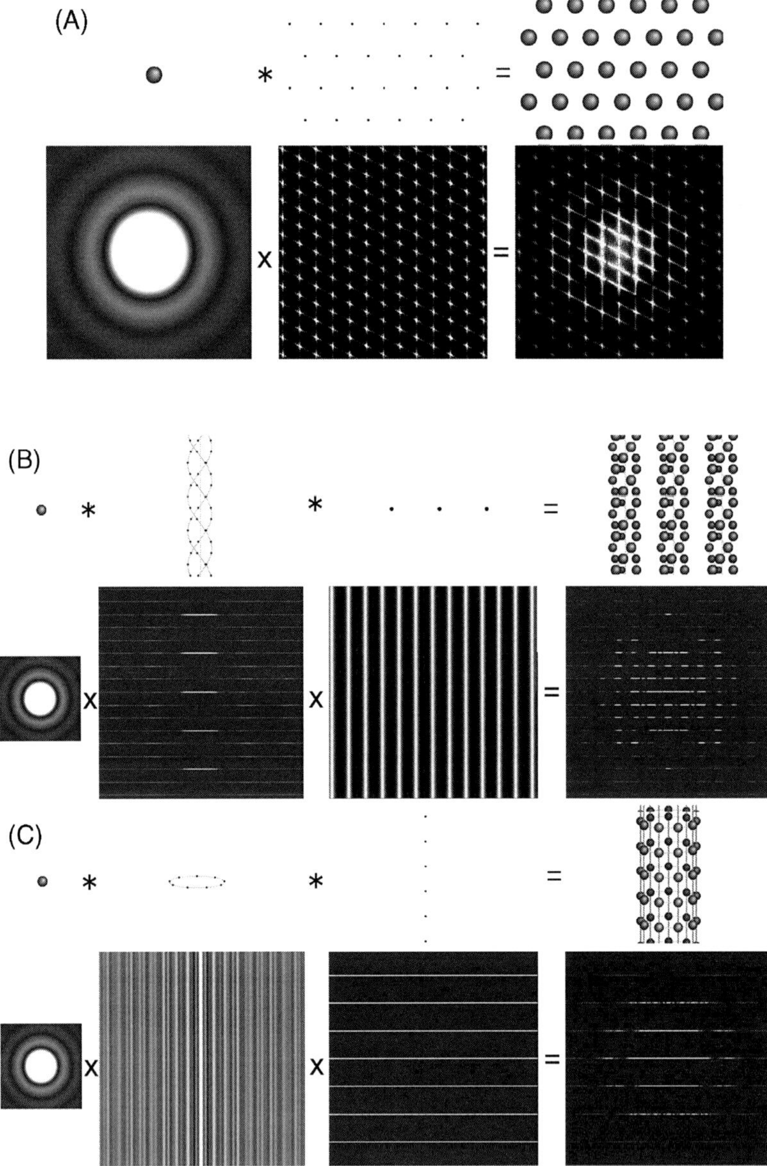

Fig. 2 *(continued)*

Diffraction from muscle is different from that from protein crystals in two ways: (1) The filament lattice in muscle is far less regular, and the number of reflections available for structure analysis is much more limited. Reflecting this, the extent of lattice sampling is variable, making some reflections sharply disrupted but leaving others continuous. This variable effect of lattice sampling often makes data analysis complicated. (2) Because of the large number of myofibrils with randomly rotated lattice planes, the diffraction pattern from a muscle is a rotary-averaged one, and all the reflections are simultaneously observable without the need to rotate the muscle around its long axis.

An axoneme (Fig. 2C) is a cylindrical structure best described in terms of cylindrical coordinates, and its reflections are naturally expressed by Bessel functions. A single axoneme diffracts X-rays only very weakly. Therefore, a large number of axonemes are prepared, and after they have been oriented by using the proper techniques, as described below, diffraction patterns are recorded *en masse*. The X-ray photons scattered by an axoneme are not expected to interfere with those scattered by other axonemes in the bulk (they diffract incoherently), so the scattering intensities are simply summed. Therefore, the diffraction pattern from the oriented mass of axonemes can be treated as if it were from a single axoneme (again after rotary averaging, because of the random rotations around the axonemal axes). In the analysis of diffraction patterns from axonemes, this offers a great advantage: both equatorial and

Fig. 2 Schematic diagrams of diffraction patterns from protein crystal (A), muscle (B), and axoneme (C). For simplicity, the shape of the repeating unit (protein molecule) is represented as a sphere, and the arrangement of the repeating units is reduced to two-dimensional planes. In each of the (A)–(C) panels, the top panels represent the objects in the real space, and the bottom panels represent the diffraction/scattering patterns from the objects above. A protein crystal (A, right) is considered to be the convolution of the molecule (left) and a large, very regular lattice (middle). The Fourier transform of the lattice is also a lattice, consisting of numerous spots (middle). The resulting diffraction from the crystal is also a collection of spots (reflections), and the intensity of each spot reflects the Fourier transform of the single molecule. In muscle (B), we consider myosin filaments arranged in a hexagonal lattice. In each myosin filament, the myosin heads are arranged in a three-start helix (here we ignore the filament backbone). Then the whole array of myosin filaments is considered to be the convolution of three functions, that is, molecular shape, the helical arrangement of myosin heads, and the hexagonal arrangement of filaments (simplified as three dots). Although the Fourier transform of helices is the sum of continuous Bessel functions (layer-line reflections, second pattern from the left), it is disrupted (lattice-sampled) in the final diffraction pattern from muscle in an incomplete manner (rightmost pattern) due to the hexagonal filament lattice that is far less regular than in protein crystals. Because of this, it is not immediately clear if a peak of a reflection comes from Bessel function or from lattice sampling. In axoneme (C), we ignore the microtubules and consider only the periodically arranged structures (here an outer dynein arm represented as a sphere). Then, the whole axoneme is considered to be the convolution of the molecular shape, circular arrangement of nine molecules at a single level, and the periodicity of molecules along the axonemal axis (24-nm interval). This is a nonhelical arrangement of dyneins but helical arrangements are also possible (see Iwamoto, 2008). In any case, the final diffraction pattern (rightmost) consists of a number of layer-line reflections, each of which is the product of Bessel functions and the molecular transform. Neighboring axonemes in the bulk sample do not interfere with each other, so that the pattern is free of lattice sampling. In the actual diffraction patterns, the situation is more complicated than this, because of the asymmetrical molecular shapes of constituent proteins (this also applies to muscle).

layer-line reflections are expressed as pure sums of Bessel functions. This process is free of complications due to lattice sampling. [The Miller indices (*h*,*k*,*l*), which are so familiar in protein crystals and muscle, are irrelevant here.] This allows straightforward interpretations of diffraction data.

C. The Generation of Axonemal Diffraction

Figure 3 explains how the components of an axoneme contribute to the diffraction pattern. An axoneme consists of nine peripheral doublet microtubules, a central pair of singlet microtubules, and other periodically arranged structures such as dynein arms and radial spokes. For simplicity we ignore the central pair. As in the usual microtubules, tubulin monomers in doublet microtubules are arranged in a helical manner, which gives rise to layer-line reflections at a spacing of $1/4.0\,\text{nm}^{-1}$ and beyond. Macroscopically, however, the doublet microtubules are elongated tubes with uniform densities. In this case, all the diffraction intensities are concentrated on the equator and no other reflections are observed (Fig. 3A). Because the mass of the doublet microtubules is substantial, the equatorial reflections are very strong.

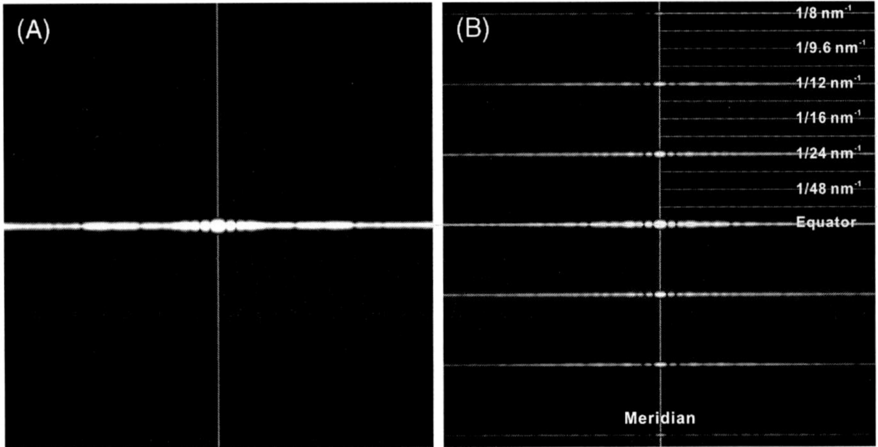

Fig. 3 Contributions of doublet microtubules and outer dynein arms to the diffraction pattern from an axoneme. (A) diffraction pattern from doublet microtubules alone. If the microtubules are ideally uniform tubes, they generate only equatorial reflections. (B) Diffraction pattern from outer dynein arms alone (nonhelical case as in Fig. 2C). They generate a number of layer-line reflections, as well as equatorial reflections which are much weaker than in (A). The actual diffraction pattern is not a simple combination of (A) and (B), but these calculations provide ideas as to the relative contributions of microtubules and dyneins (and other periodically arranged structures) to equatorial and layer-line reflections. The densities of microtubules and dyneins are based on the map data from cryoelectron microscopy tomography (Ishikawa *et al.*, 2007), and the patterns were calculated by using Axolotl2 (Iwamoto, 2008).

Next, we consider the reflections arising from structures periodically arranged on the doublet microtubules, such as dyneins and radial spokes. As has been shown in Fig. 2C, these proteins give rise to a series of evenly spaced off-equatorial layer-line reflections as well as equatorial reflections (Fig. 3B). However, the contribution of these proteins to the equatorial reflections is much smaller than that of microtubules. The repeat of the structures is either 24 or 96 nm, so that many layer-line reflections occur inside the tubulin reflections at $1/4.0\,\mathrm{nm}^{-1}$. An important conclusion drawn here is that the off-equatorial layer-line reflections inside the tubulin reflections are exclusively ascribable to these periodically arranged structures. The actual diffraction pattern from the axonemes is not the simple sum of the pattern from doublet microtubules (Fig. 3A) and that from periodically arranged structures (Fig. 3B), but for a first approximation, the generation of axonemal diffraction may be described in this way.

III. Methods

A. Specimen Preparation

For the SAXS studies, we have used flagella of sea urchin sperm and *Chlamydomonas*. Large quantities of axonemes are easily available from sea urchin sperm. For SAXS, there are many advantages to the use of *Chlamydomonas* flagella, since the organism is easy to grow synchronously in large quantities in simple defined media and has a large repertory of flagellar mutants with defined lesions. These mutant flagella are useful for defining the origins of each reflection in X-ray diffraction patterns. However, there are no strong limitations on the choice of samples if they can be prepared in high purity, high concentrations, and large quantities.

1. Sea Urchin Sperm

Sea urchin sperm is collected by injecting 0.55 M KCl into the body cavities of male sea urchins (*Hemicentrotus pulcherrimus* or *Anthocidaris crassispina*). Collected sperm is kept refrigerated and used for the experiment within 3 days. The following procedures to prepare axonemes are based on the previous report (Kamimura *et al.*, 1985). Ten milliliters of the collected sperm is first diluted with 0.55 M NaCl. The cell membrane is then extracted for 1 min with a 100- to 200-ml solution containing 0.08% Triton X-100, 0.4 M K-acetate, 1 mM ethylene glycol-bis-(beta-aminoethyl ether)-*N,N'*-tetraacetic acid (EGTA), 0.1 mM ethylenediamine tetraacetic acid (EDTA), 1 mM dithiothreitol (DTT), and 10 mM Tris-HCl (pH. 8.3) on ice.

After demembranation, spermatozoa are collected by centrifugation (4000 × *g*, 20 min, R14AF, Hitachi, Tokyo, Japan), and 30–40 slow pestle strokes in a Dounce homogenizer are given to remove the sperm heads mechanically. After removal of the sperm heads by 60–45–30% sucrose density gradient centrifugation (9000 × *g*, 30 min, R10S, Hitachi, Tokyo, Japan), the axonemal fragments are collected with

8000 × *g* centrifugation for X-ray diffraction analysis. For all the experiments using sea urchin sperm axonemes, a solution containing 0.4 M K-acetate, 1 mM EGTA, 0.1 mM EDTA, 1 mM DTT, and 20 mM Tris-HCl (pH 8.3) is used. For X-ray diffraction analysis, the precipitated samples resulting from centrifugation (9000 × *g*, 15 min) are resuspended in the same volume of buffer with 2% methylcellulose (M0512, Sigma, St. Louis, MO, USA).

2. *Chlamydomonas* Flagella

Ten to twenty liters of *Chlamydomonas* culture in TAP medium are grown as described by (Gorman and Levine, 1965). The cultures reach $\sim 2 \times 10^6$ cells ml^{-1}, and the volume required depends on the mutants used. Care should be taken that healthy cells grow and are not contaminated because debris from dead cells is a major cause of background scattering of the diffraction. To harvest these large cultures, we use centrifugations with a high-capacity rotor that can process 12 l of sample in a single run (Sorvall RC12BP with H-12000 rotor, 750 × *g* for 9 min, Kendro Laboratory Products, Newton, CT, USA) instead of a tangential-flow filtration system. Flagellar axonemes of *Chlamydomonas* are prepared basically according to the method described by King (1995). Once the cells are harvested and washed, we deflagellated them with dibucaine. Twenty liters of the culture provides flagella of ~ 2.5 mg ml^{-1} × 2 ml. To reduce background scattering from the sample, we demembranate the flagella. The axonemes are suspended and washed in HMDEKP solution, containing 30 mM HEPES-K, 5 mM $MgSO_4$, 1 mM EGTA, 1 mM dithiothreitol, 50 mM K-acetate, and 0.5% poly(ethylene glycol) (pH 7.4) at a final concentration of 5–10 mg ml^{-1}. Removal of the membranes is carried out by addition of 0.2% (final) Nonidet P40 (octylphenoxy polyethoxy ethanol) to the suspension. The resultant suspension is washed twice with HMDEKP buffer. For X-ray diffraction analysis, samples precipitated by centrifugation (3000 × *g*, 10 min) are resuspended in the same volume of buffer with 2% methylcellulose (M0512, Sigma, St. Louis, MO, USA).

B. Apparatus for Shear-Flow Alignment

In a suspension of axonemes, individual axonemes are randomly oriented and as a consequence, the suspension exhibits circular diffraction patterns when being irradiated by X-rays (see Fig. 5A). Although there are no specific diffraction spots, a number of relatively sharp diffraction peaks are superimposed on a broad, diffuse scattering in the rotary-averaged intensity profile. These peaks are a mixture of meridional and equatorial reflections, which report longitudinal and transverse regularity of protein arrangement, respectively. However, if axonemes are aligned in a uniform orientation, the two sets of reflections start to separate from each other in horizontal and vertical directions (Fig. 5B). Such a diffraction pattern can be treated as a fiber diffraction, as those from naturally oriented biological (e.g., collagen, muscle) and synthetic (e.g., nylon) polymers.

As mentioned above, axonemes must be aligned to record X-ray fiber diffraction patterns from a flagellar suspension. So far, two techniques have mainly been used to align filamentous samples: slow sedimentation with low-gravity centrifugations or/and biased Brownian motion under very strong magnetic fields (Bras *et al.*, 1998; Oda *et al.*, 1998; Popp *et al.*, 1987; Stubbs, 1999; Torbet *et al.*, 1981; Yamashita *et al.*, 1998). Successful alignments have been shown for bacterial flagella (Yamashita *et al.*, 1998), microtubules (Bras *et al.*, 1998), tobacco mosaic virus (Namba *et al.*, 1989; Stubbs, 1999), actin filaments (Oda *et al.*, 1998; Popp *et al.*, 1987), and other biological fibers (Torbet and Maret, 1981; Torbet *et al.*, 1981). However, these methods require a long time (from several hours to a few weeks) to accomplish alignment, and the results are not always reproducible under various experimental conditions. More inconveniently, these techniques are not suitable for many other types of biological filaments, particularly for those which easily become inactive or denatured, or those in which physiologically active components are easily degraded during the aligning procedures. We have therefore developed a quick and reproducible technique for the shear-flow alignment of biological filaments. The technique employs a parallel rotating disc instrument usually used for rheological studies (Sugiyama *et al.*, in press).

The apparatus used for the shear-flow alignment of axonemes is made of two stainless tubes (inner diameter, 16–17 mm, Nogata Denki Kogyo, Tokyo, Japan), one of which is connected to a DC motor with a rubber belt (Fig. 4). Before the experiments, the rate of disc rotation should be calibrated with a speed meter (testo 465, Testo AG, Lenzkirch, Germany). In the experimental shear-flow alignment protocol, the two tubes are placed with their center axes aligned, leaving a narrow space (0.1–0.35 mm, the asterisk in Fig. 4), where a coverslip (16 or 18 mm in diameter, No. 1, Matsunami, Tokyo, Japan) or a thin Kapton film (50H, 12.5 μm thick, DuPont, Wilmington, DE, USA) was glued to the opening of the tubes previously. The coverslips or Kapton films are thus used as two parallel discs. Coverslips give better results, as judged from the quality of fiber alignment, probably because of superior mechanical stability. However, the Kapton films should be used instead of coverslips in cases where X-rays of 0.15 nm wavelength are used.

After the specimen suspension (50–100 μl) was placed in the narrow space between the two discs (the asterisk in Fig. 4), one of the discs is rotated at a constant rate (5–30 rotations · s^{-1}), giving a stable gradient of flow velocity (shear flow) to the suspension of specimens. When the gap between the two parallel discs is 0.15–0.25 mm, the calculated shear rate is 750–7500 s^{-1} around the area 6 mm away from the center of rotation of the axis.

From a series of X-ray diffraction experiments so far, we found that the use of a parallel disc rotary rheometer was effective in aligning sea urchin sperm axonemes and *Chlamydomonas* flagellar axonemes suspended in physiological solutions. By giving high shear flow (1000–5000 s^{-1}) to these axonemes suspended in a methylcellulose-containing (~2%) solution, we found that fiber alignment was accomplished within 5 s (Fig. 5).

Specimen thickness (i.e., the length of the X-ray beam path through the specimen suspension) was restricted to <0.25 mm for practical reasons. To get a higher

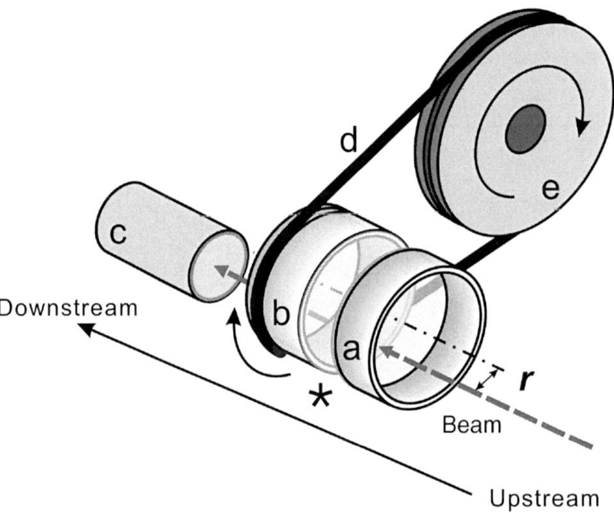

Fig. 4 A photograph and a schematic drawing of the apparatus used for the shear-flow alignment of biological filaments. For clarity, the apparatus in the drawing has the opposite handedness to that in the photograph. The X-ray beam passes through an area with 6 mm off the center (*r*) of a pair of tubes (a and b). Suspension of axonemes (2–5 mg/ml) is placed in the space (*, 0.1–0.35 mm) between the two parallel discs (coverslips or Kapton films, glued on the opening of the tubes). One of the discs (b) is rotated by a DC motor (e) and a rubber belt (d). The X-ray diffraction was measured downstream through a vacuum chamber (c). The deflection angle of scattered X-ray is so small with respect to the incident beam that an X-ray detector is placed far downstream of the sample position to enlarge the diffraction pattern to an observable size. A vacuum chamber is placed in the space between the sample and the detector and air is evacuated from the chamber to reduce the scattering/absorption of the X-rays. The maximum specimen-to-detector distance ranges from 3 to 10 m, depending on the facility.

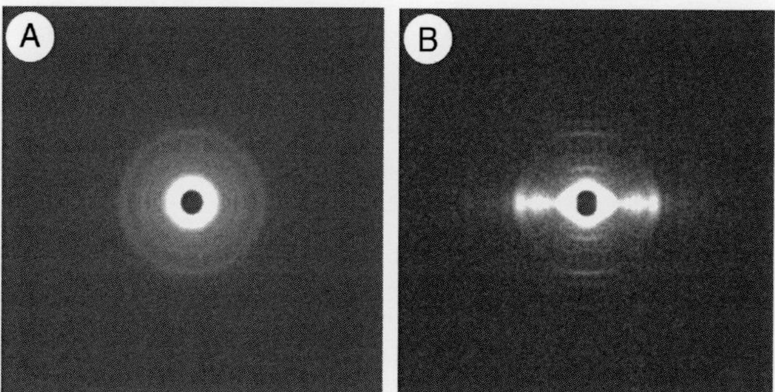

Fig. 5 Diffraction patterns observed with the suspension of axonemes isolated from wild-type *Chlamydomonas* flagella (protein concentration, 2.5 mg ml^{-1}). The medium contains 1% methylcellulose. Raw diffraction images obtained with a CCD detector are shown with no background subtraction and enhancement. (A) The diffraction obtained before application of the shear flow. (B) The diffraction at 4 s after application of the shear flow to the specimen (~30 rotations s^{-1}). Direction of shear applied is parallel to the vertical axis in each diffraction pattern. Difference between reflections in the meridional and equatorial directions is obvious. The wavelength is 0.09 nm and the camera length is 3.34 m. The exposure time at each pattern was 800 ms.

signal-to-noise ratio of the diffraction signals, it is necessary to use a thicker sample (>0.5 mm), but the spin rate of the rheometer disc should increase to more than 40 rotations s^{-1}. This causes the major problem of bubble formation inside the medium suspending the flagellar axonemes during the high-rate disc rotation.

Use of a higher shear rate (>1000 s^{-1}) and methylcellulose with higher molecular weight gives better results. Polyvinylpyrrolidone (10%) works similarly (Sugiyama *et al.*, in press). Such aqueous polymers could have several complicating effects on the medium (e.g., enhancing viscosity, reducing flow turbulence, and non-Newtonian flow-thinning effects). All of these medium properties might be working to stabilize the creeping flows produced by rotation of the disc. In the case of axonemes, application of high shear rates of 1000–6000 s^{-1} for up to 15 min gave no apparent change in the X-ray diffractions, indicating that the shearing forces caused no apparent mechanical damage to the structure. The estimated shear force on the axonemes was less than the order of 10 nN; this would have had little effect on the dynein motors of the doublet microtubules in the axonemes.

C. SAXS Measurements

As stated earlier, recordings of high-quality diffraction patterns from eukaryotic ciliary/flagellar axonemes were made possible by the intense, well-oriented X-ray beams generated by third-generation synchrotron radiation facilities. The use of weaker sources such as laboratory-scale rotating-anode generators is impractical. In the following part, therefore, we focus on how to use the SAXS beamlines in these facilities.

1. Beamlines

Figure 6 shows the standard arrangement of the components of a typical SAXS beamline in a third-generation synchrotron radiation facility. Synchrotron radiation is generated when the path of electrons or positrons traveling at near-light speed is bent by a magnetic field. To force the electrons or positrons to travel along a circular orbit, a synchrotron storage ring has a number of bending magnets, and each of them generates synchrotron radiation. The second-generation facilities make use of the radiation from these bending magnets, but the storage rings of the third-generation facilities have a number of "insertion devices" in addition to the bending magnets. An insertion device, called an undulator or wiggler, consists of usually two rows of magnets with alternating polarities. As electrons or positrons pass through the gap between the two rows of magnets, they literally "undulate," and emit radiation that is more intense and better oriented than that generated by the bending magnets. Synchrotron radiation is basically polychromatic, but the radiation from an undulator is to some extent monochromatic, and the peak wavelength can be tuned by changing the gap between the two rows of magnets. The radiation is further monochromatized by a pair of silicon or diamond monocrystals, and the final bandwidth of the X-rays available for experiments is typically ~0.01%. X-ray wavelengths can usually be tuned in the range of ~0.15 and ~0.08 nm, but the exact range depends on facilities. The monochromatized X-ray beams are focused by a pair of bent mirrors to obtain a small beam size at the sample position (typically ~100 μm).

Unlike in protein crystallography, where the interatomic distances of interest are comparable to the X-ray wavelength, the distances of interest in axonemes are much longer. The basic axial repeat of axonemal components is 24 nm (outer dynein arms) or 96 nm (inner dynein arms, radial spokes, etc.). The first Bessel peak on the equatorial reflection appears at a distance corresponding to a space of >100 nm. These values are much greater than the X-ray wavelength, meaning that the deflection angle of scattered X-rays is very small with respect to the incident beam (hence the small-angle scattering). To enlarge the diffraction pattern to an observable size, an X-ray detector is

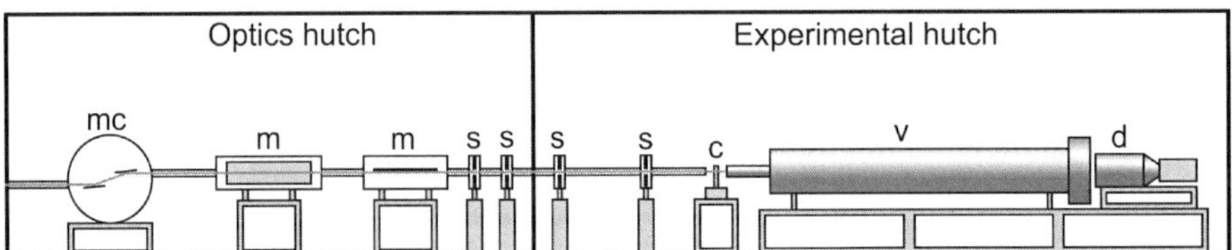

Fig. 6 Schematic diagram of a typical SAXS beamline in a third-generation synchrotron radiation facility. The X-ray beam is generated by an undulator located far left of the diagram (not shown). A SAXS beamline typically consists of an optics hutch, which contains a monochromator (mc) and vertical and horizontal bent mirrors (m). The experimental hutch, where the specimens are irradiated, contains the stage for the specimen chamber (c), a vacuum path (v), and a detector (d). Before the beam reaches the specimen, it passes through a number of slits (s), which define the beam size and also remove parasitic scattering from upstream components.

placed far downstream of the sample position, and the space between the sample and the detector is evacuated to reduce the scattering/absorption of the X-rays by air. The maximum specimen-to-detector distance ranges from 3 to 10 m, depending on the facility.

2. Detectors

Various types of X-ray area detectors have been developed, including imaging plates (IPs), semiconductor-based detectors, and wire detectors. IPs are high-sensitivity substitutes for X-ray films. An IP is a plate coated with phosphor that retains its excited state until it is hit by red laser beams and emits luminescence. It has a large area, a high dynamic range, and has no distortion of the image, but it has a long readout time. Therefore, it is suitable for highly accurate static measurements. The Rigaku R-Axis is an integrated IP-based detector with built-in readout and eraser. It has a higher throughput than stand-alone IPs, but its use is still limited to static measurements.

Semiconductor-type detectors, represented by cooled charge-coupled device (CCD) area detectors, are the detectors most widely used in synchrotron radiation facilities. They have faster readout, and the fastest end of the products can be used for millisecond time-resolved measurements. To increase their sensitivity, they are often used with fiber- or optics-coupled scintillators, giving them photon-counting capabilities and ~100% quantum efficiencies. The shortcoming is that they often suffer from some distortion in image, lower dynamic range, and accumulation of dark current. Another problem is that the detector area is limited. A large detector area may be achieved by the use of multiunit detectors, but some dead areas exist because of the gaps between the detector units.

Wire detectors apply high voltage between wire anodes and cathodes and count the gas ionization events caused by X-ray photons. Detectors of this type are very fast and free of saturation. However, they suffer from larger pixel sizes, lower counting efficiency compared with CCD area detectors, nonuniform sensitivity, and complex electronics. For these reasons, they are not widely used in beamlines.

3. Data Processing

Whether they are IP-based or of the semiconductor-type, commercially available detectors usually come with data acquisition software that can store data in formats readable by common image-processing software packages. Naturally, no axoneme-specific data-processing software has been released, so those who wish to do more quantitative analyses are currently forced to write the program codes themselves. Those who are not familiar with X-ray data processing should seek advice from beamline scientists or experts in fiber diffraction.

The usual data processing of diffraction patterns from axonemes includes the determination of the spacings, integrated intensities, and intensity profiles of reflections. These operations are usually done after subtraction of background scattering. At present, it is impossible to restore the original axonemal structure directly from these processed data, so that an approach is taken to build model structures, calculate diffraction patterns from them, and test which model best explains the observed data. A model can be built from cryoelectron microscopy map data: a segment of densities is excised from the map data to create the "repeating unit" (as explained above). After the unit has been properly oriented, it is possible to calculate a diffraction pattern, for example, by using the "Axolotl2" software provided as supplementary material by Iwamoto (2008).

D. Getting Started at Synchrotron Facilities

1. Choice of Facilities

As stated earlier, the only practical means to conduct SAXS experiments on flagellar axonemes is to visit one of the third-generation synchrotron radiation facilities. If you already have a technique for or perspective on the preparation of suitable axonemal samples, the first thing to do is to collect information about these facilities. The largest facilities include SPring-8 in Hyogo, Japan, APS in Chicago, IL, USA, and ESRF in Grenoble, France. However, smaller but newer, high-performance facilities have been, or are being, constructed all over the world, and they should also be taken into consideration. There are several factors to think about before selecting a facility.

1. Consider the travel distance to the facility. Geometrical proximity is very important, not only because it affects travel expenses but also because you may have to carry or ship fragile materials or experimental equipment to the facility.
2. Make sure that the facility has at least one beamline that is dedicated to SAXS and open to visiting users.
3. Make sure that the beamline is of the "undulator" type. "Bending-magnet" beamlines deliver much weaker beams and should be avoided.
4. Check the tunabilities of camera lengths (specimen-to-detector distances) and X-ray wavelengths. These parameters affect the SAXS resolution, which defines the maximum spacing that can be resolved in real space. If you wish to record a peak at a spacing of 96 nm, a combination of a ~3-m camera and a wavelength of 0.15 nm would be sufficient.
5. Check what kind of detectors are available. Silicon-based detectors have great variations in specifications, such as speeds of exposure and readout, area size, sensitivity, and pixel number and size. Make sure that the beamline has the right detector for your experimental purposes.
6. It is very important to have a beamline scientist who is familiar with both SAXS and biological materials. Today, a substantial proportion of users of SAXS beamlines study nonbiological samples such as synthetic polymers.
7. Finally, the presence of off-line infrastructure, such as wet labs, is also an important factor. If you want to do the final steps of sample preparation immediately before the X-ray measurements, you may need to use a centrifuge on the spot.

2. Submitting a Proposal

After the right beamline is found in the right synchrotron facility, the next step is to write a proposal. The proposal is reviewed by the committee of the facility in light of the scientific merit and technical feasibility of experiments. In SPring-8, calls for proposals are made twice a year, and once a proposal has been accepted it is effective for 6 months. Before you write a proposal, it is advisable to contact the beamline scientist and discuss the specifics of your experiments.

IV. Results and Discussion

A. Flow-Induced Alignment of Axonemes

After the specimen was placed between the discs, one of the discs was rotated at 5–30 rotation/s. The estimated shear rate was 750–7500 s^{-1}. Under conditions of shear flow, rigid slender bodies suspended in a medium should be aligned to the flow, in accordance with the theoretical investigations by Jeffery (1922) and the experimental demonstrations by Stover *et al.* (1992). However, our experiments without methylcellulose showed poor alignment in the case of axonemes, even if a very high shear rate was given ($>1000\ s^{-1}$). Therefore, applying shear flow alone is not enough to align the axonemes.

Although the exact mechanism is unclear, adding methylcellulose to the axonemal suspension medium was highly effective in facilitating flow-induced alignment. Under our present conditions, alignment was accomplished within 5 s, as shown in Fig. 5. Uniform orientation with a small angular deviation ($<5°$) enabled us to execute a detailed structural analysis of the axonemes by SAXS. On the other hand, as the diffraction pattern became gradually obscured after the cessation of disc rotation, continuous application of shear flow was crucial.

B. X-ray Diffraction Patterns

For the observation of X-ray diffraction patterns, we used a synchrotron radiation X-ray beam of the beamline 45XU [beam size = 0.1 (vertical) × 0.2 (horizontal) mm, $\lambda = 0.09$ or 0.15 nm)]at SPring-8 (JASRI, Hyogo, Japan). Diffraction signals were recorded with a Hamamatsu cooled CCD camera (C4880, Hamamatsu Photonics, Hamamatsu, Japan) with 0.1- to 5.0-s exposures. A camera length of 1–3 m was chosen to suit the range of signals we needed. Control observations (background) were carried out with buffer medium containing methylcellulose. We performed the background subtraction and image averaging ($n = 10$–120 diffraction patterns) by using custom-built software or ImageJ (ver.1.38x, Wayne Rasband, NIH, Bethesda, MD, USA).

A typical X-ray diffraction pattern of axonemes of *Chlamydomonas* is shown in Fig. 7. The axonemes in the medium are aligned by a shear flow applied to the direction parallel to the vertical axis of the pattern. This pattern is obtained after accumulation of a series of 50 diffraction patterns (exposure time, 800 ms each),

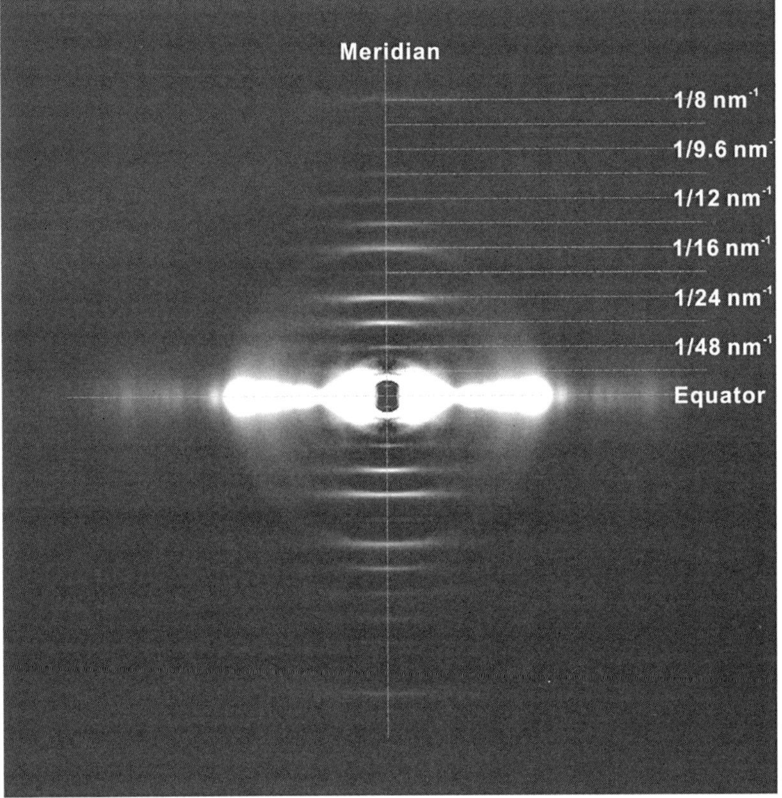

Fig. 7 An example of the observed X-ray diffraction pattern obtained from wild-type axonemes of *Chlamydomonas* flagella in the presence of 1 mM ADP and 50 μM vanadate ($\lambda = 0.09$ nm, camera length = 3.34 m, total exposure time = 40 s). The pattern was obtained with the synchrotron radiation X-ray at SPring-8, BL45XU. The diffraction pattern showed several distinct meridional and equatorial reflections.

subtraction of background scattering and then the folding of the four quadrants of the resultant pattern.

Number of reflections are found both on the meridian and the equator. As mentioned in Section II, since the doublet microtubules are macroscopically elongated tubes with uniform densities, the diffraction intensities derived from the doublet microtubules are concentrated on the equator (Fig. 3A). Strong but broad peaks are found on the equator, which are at 1/27, 1/23, 1/21, 1/19, 1/17, and 1/13 nm^{-1}. These peaks reflect structural regularity of axonemal components in the radial direction, such as the diameter of doublet microtubules and spacing within/among the doublets or other axonemal components.

Besides the strong equatorial reflections, a series of evenly spaced off-equatorial layer-line reflections is observed. These clear layer-line signals represent the longitudinal regularity of axonemal structures. Although the first-order layer line at $1/96\,nm^{-1}$ is too close to the equator to identify, the space between the layer lines provides the basic repeat of 96 nm. Reflections from the 2nd to the 12th order are clearly found inside the 4-nm tubulin reflections except 5th, 10th, and 11th reflections. Note that the 4th ($1/24\,nm^{-1}$) and 8th ($1/12\,nm^{-1}$) layer lines are intensified, reflecting the 24-nm repeat of the outer arms and that the 5th, 10th, and 11th reflections are absent or very weak. Theoretical consideration described in Section II suggests that the emergence of meridional intensity depends on whether the axoneme is helical or not, and if it is, on the number of helical strands. Absence of the 5th, 10th, and 11th reflections allows us to identify the type of helix of the axoneme.

Comparisons of diffraction patterns obtained from several types of flagella, each prepared from wild-type strains, and mutants lacking the whole outer arm (*oda1*), and spoke (*pf14*) provide useful information about the origin of the meridional reflections. For instance, the $1/24\text{-}nm^{-1}$ meridional reflection distinct in the wild type becomes faint in *oda1*, suggesting that this reflection originated mainly from the axial repeat of the outer dynein arms. The weaker $1/32\text{-}nm^{-1}$ meridional reflection of *pf14* than in the wild type suggests that this reflection comes mainly from the repeat of the spoke.

V. Summary

With third-generation synchrotron sources, improved specimen preparation methods, and new data-processing algorithms, X-ray fiber diffraction studies make the structure determination of large macromolecular complexes at high resolutions feasible. The rheological method described here is one of these advances. It has advantages against the alignment methods previously used. First, it provides simple and easy way to align the specimen in seconds, which allows the application of this method to physiologically unstable filaments. Second, the quality of alignment is better than or almost equivalent to that achieved with other conventional methods. Third, only a small volume (50–100 μl) of the specimen is required. The combination of this alignment method and the intense, well-oriented X-ray of synchrotron radiation promise a great possibility of structural analysis by means of X-ray diffraction in the field of life science as well as structural analysis of the axonemes.

Acknowledgments

We thank Drs Shiori Toba and Hitoshi Sakakibara for providing data on X-ray diffractions of *Chlamydomonas* flagellar axonemes and for their technical support. This work was supported by the Grant-in-Aid for Scientific Research on the Priority Area "Regulation of Nano-systems in Cells" (#16083207 to KO; #1704911, #19037010 to SK) by the Ministry of Education, Science, and Culture of Japan.

Most of the data used here were from the experiments approved by the SPring-8 Proposal Review Committee (2005B0075, 2005B0331, 2005B0384, 2006A1175, 2006A1322, 2006A1329, 2006B1418, 2006B1498, 2007A1187, 2007A1191, 2007A1590, 2007B1448, 2007B1470, 2008A1020, 2008A1091, 2008A1123, 2008A1544, 2008A1444).

References

Afzelius, B. (1959). Electron microscopy of the sperm tail; results obtained with a new fixative. *J. Biophys. Biochem. Cytol.* **5**, 269–278.

Bras, W., Diakun, G.P., Diaz, J.F., Maret, G., Kramer, H., Bordas, J., and Medrano, F.J. (1998). The susceptibility of pure tubulin to high magnetic fields: A magnetic birefringence and x-ray fiber diffraction study. *Biophys. J.* **74**, 1509–1521.

Bui, K.H., Sakakibara, H., Movassagh, T., Oiwa, K., and Ishikawa, T. (2008). Molecular architecture of inner dynein arms in situ in Chlamydomonas reinhardtii flagella. *J. Cell Biol.* **183**, 923–932.

Cantor, C.R., and Schimmel, R.R. (1980). "Biophysical Chemistry. Part II: Techniques for the Study of Biological Structure and Function." W.H. Freeman & Co., New York.

Dutcher, S.K. (1995). Flagellar assembly in two hundred and fifty easy-to-follow steps. *Trends Genet.* **11**, 398–404.

Gorman, D.S., and Levine, R.P. (1965). Cytochrome f and plastocyanin: Their sequence in the photosynthetic electron transport chain of Chlamydomonas reinhardi. *Proc. Natl. Acad. Sci. USA* **54**, 1665–1669.

Holmes, K.C., and Blow, D.M. (1965). "The Use of X-ray Diffraction in the Study of Protein and Nucleic Acid Structure." Interscience, John Wiley & Sons, New York.

Ishikawa, T., Sakakibara, H., and Oiwa, K. (2007). The architecture of outer dynein arms *in situ*. *J. Mol. Biol.* **368**, 1249–1258.

Iwamoto, H. (2008). Theory of diffraction from eukaryotic flagellar axonemes. *Cell Motil. Cytoskeleton* **65**, 563–571.

Jeffery, G.B. (1922). The motion of ellopsoidal particles immersed in a viscous fluid. *Proc. R. Soc. Lond. Ser. A.* **102**, 161–179.

Kamimura, S., Iwamoto, H., and Fujisawa, T. (2007). Analysis of small-angle X-ray diffractions from the flow-aligned axonemes of sea-urchin spermatozoa. *Biophys. J.* **92** (Abst. issue), 500a.

Kamimura, S., Yano, M., and Shimizu, H. (1985). ATP hydrolysis coupled to microtubule sliding in sea-urchin sperm flagella. *J. Biochem.* **97**, 1509–1515.

King, S.M. (1995). Large-scale isolation of Chlamydomonas flagella. *Methods Cell Biol.* **47**, 9–12.

Lupetti, P., Lanzavecchia, S., Mercati, D., Cantele, F., Dallai, R., and Mencarelli, C. (2005). Three-dimensional reconstruction of axonemal outer dynein arms in situ by electron tomography. *Cell Motil. Cytoskeleton* **62**, 69–83.

Namba, K., Pattanayek, R., and Stubbs, G. (1989). Visualization of protein-nucleic acid interactions in a virus. Refined structure of intact tobacco mosaic virus at 2.9 A resolution by X-ray fiber diffraction. *J. Mol. Biol.* **208**, 307–325.

Nicastro, D., McIntosh, J.R., and Baumeister, W. (2005). 3D structure of eukaryotic flagella in a quiescent state revealed by cryo-electron tomography. *Proc. Natl. Acad. Sci. USA* **102**, 15889–15894.

Nicastro, D., Schwartz, C., Pierson, J., Gaudette, R., Porter, M.E., and McIntosh, J.R. (2006). The molecular architecture of axonemes revealed by cryoelectron tomography. *Science* **313**, 944–948.

Oda, T., Makino, K., Yamashita, I., Namba, K., and Maeda, Y. (1998). Effect of the length and effective diameter of F-actin on the filament orientation in liquid crystalline sols measured by x-ray fiber diffraction. *Biophys. J.* **75**, 2672–2681.

Pazour, G.J., Agrin, N., Leszyk, J., and Witman, G.B. (2005). Proteomic analysis of a eukaryotic cilium. *J. Cell Biol.* **170**, 103–113.

Popp, D., Lednev, V.V., and Jahn, W. (1987). Methods of preparing well-orientated sols of F-actin containing filaments suitable for X-ray diffraction. *J. Mol. Biol.* **197**, 679–684.

Silvester, N.R. (1964). The cilia of Tetrahymena pyriformis: X-ray diffraction by the ciliary membrane. *J. Mol. Biol.* **8**, 11–19.

Squire, J.M. (1981). "The Structural Basis of Muscular Contraction." Plenum Press, New York.

Stover, C.A., Koch, D.L., and Cohen, C. (1992). Observations of fiber orientation in simple shear flow of semi-dilute suspensions. *J. Fluid Mech.* **238**, 277–296.

Stubbs, G. (1999). Developments in fiber diffraction. *Curr. Opin. Struct. Biol.* **9**, 615–619.

Sugiyama, T., Miyashiro, D., Takao, D., Iwamoto, H., Sugimoto, Y., Wakabayashi, K., and Kamimura, S. (2009). Quick shear-flow alignment of biological filaments for X-ray fiber diffraction under physiological conditions. *Biophys. J.* in press.

Toba, S., Iwamoto, H., Fujisawa, T.S.H., and Oiwa, K. (2007). Conformational changes of flagellar axonemes revealed by fiber diffraction study of flagella from *Chlamydomonas* strains. *Biophys. J.* **92** (Abst. issue), 500a.

Torbet, J., Freyssinet, J.M., and Hudry-Clergeon, G. (1981). Oriented fibrin gels formed by polymerization in strong magnetic fields. *Nature* **289**, 91–93.

Torbet, J., and Maret, G. (1981). High-field magnetic birefringence study of the structure of rodlike phages Pf1 and fd in solution. *Biopolymers* **20**, 2657–2669.

Wakayama, J., Tamura, T., Yagi, N., and Iwamoto, H. (2004). Structural transients of contractile proteins upon sudden ATP liberation in skeletal muscle fibers. *Biophys. J.* **87**, 430–441.

Yamaguchi, T., Hayashi, M., Wakabayashi, K., and Higashi-Fujime, S. (1972). X-ray and optical diffraction studies on the outer fibres of sea-urchin sperm tail flagella. *Biochim. Biophys. Acta* **257**, 30–36.

Yamashita, I., Suzuki, H., and Namba, K. (1998). Multiple-step method for making exceptionally well-oriented liquid-crystalline sols of macromolecular assemblies. *J. Mol. Biol.* **278**, 609–615.

CHAPTER 6

Markers for Neuronal Cilia

Jacqueline S. Domire *and* **Kirk Mykytyn**

Department of Pharmacology, Department of Internal Medicine Division of Human Genetics,
College of Medicine, The Ohio State University, Columbus, Ohio 43210

Abstract

Primary cilia were first detected on neurons in the mammalian brain over 40 years ago using electron microscopy. However, this approach is very labor intensive and has inherent limitations that restrict its utility for studying neuronal cilia. While the study

978-0-12-374973-4
DOI: 10.1016/S0091-679X(08)91006-2

of cilia in other tissues was greatly facilitated by the identification of specific ciliary markers, historically there have been no markers for neuronal cilia. Fortunately, recent developments make the study of neuronal cilia more practical. First, specific proteins have been shown to selectively localize to neuronal cilia and can serve as markers by immunolabeling. Second, neurons have been shown to possess cilia in culture, which allows for the use of additional approaches, such as live-cell imaging of neuronal cilia. This chapter provides an overview of the current techniques for visualizing neuronal cilia in tissue as well as fixed and living cells. These approaches allow for the identification of additional neuronal ciliary proteins and provide a basis for future functional studies.

I. Introduction

Investigation of neuronal cilia has been hindered by the fact that the marker commonly used for visualizing primary cilia, acetylated α-tubulin, is not specific for neuronal cilia and instead localizes throughout the cell body and processes of neurons. The discovery that immunolabeling for somatostatin receptor 3 (Sstr3) selectively labels neuronal cilia provided the first marker of neuronal cilia and revealed that these organelles are abundant and widely distributed throughout the rodent brain (Handel *et al.*, 1999). As Sstr3 is a G protein-coupled receptor (GPCR) activated by neuropeptides, this finding also suggested that neuronal cilia are sensory organelles that detect neuromodulators in the brain. This is further supported by the subsequent findings that other GPCRs, including serotonin receptor 6 (5-Ht6) and melanin-concentrating hormone receptor 1 (Mchr1), selectively localize to neuronal cilia (Berbari *et al.*, 2008a,b; Brailov *et al.*, 2000). Regardless of the precise functions of these receptors at the ciliary membrane, they serve as useful markers of neuronal cilia. There are currently commercial antibodies to Sstr3 and Mchr1 (Table I) that effectively label subsets of cilia in the brain (Berbari *et al.*, 2008b). A commercial antibody to 5-Ht6 that labels cilia has not yet been reported. Currently, the most extensive marker of neuronal cilia is type III adenylyl cyclase (ACIII) and immunolabeling with a commercial antibody to ACIII (Table I) labels the greatest number and widest distribution of neuronal cilia in the mouse and rat brain (Bishop *et al.*, 2007). There is also a commercial antibody to an unknown epitope that labels cilia throughout the rat brain

Table I
Summary of Commercial Antibodies That Label Mouse Neuronal Cilia

Antibody	Vendor	Catalog Number	Species	Dilution
Melanin-concentrating hormone receptor 1	Santa Cruz Biotechnology	sc-5534	Goat	1:250–1:500
Somatostatin receptor 3	Gramsch Laboratories	ss-830	Rabbit	1:500–1:1000
Somatostatin receptor 3	Santa Cruz Biotechnology	sc-11617	Goat	1:250–1:500
Adenylyl cyclase III	Santa Cruz Biotechnology	sc-588	Rabbit	1:500–1:1000

only (Fuchs and Schwark, 2004). Yet, it is likely that some neuronal cilia are not positive for any of these markers and have not been detected. Here, we provide detailed descriptions of several immunocytochemical approaches for labeling neuronal cilia, including the advantages of each approach, the potential pitfalls, and the alternative methods. We also present approaches for enabling live-cell imaging of neuronal cilia. Although the methods below are described for the examination of mouse neuronal cilia, they should also be applicable to other mammalian species.

II. Immunoenzymatic Labeling of Neuronal Cilia in Brain Sections

A. Rationale

In this approach, the primary antibody is coupled to an enzyme that is then visualized by the conversion of a chromogen into a pigmented dye. We typically use the glucose oxidase-diaminobenzidine (DAB)-nickel method (Shu *et al.*, 1988). This is the standard indirect peroxidase antiperoxidase method in which the primary antibody, through a series of immunological steps, is bound to a tertiary antibody that is conjugated to horseradish peroxidase (Sternberger *et al.*, 1970). The major difference is that its substrate, hydrogen peroxide, is generated by glucose oxidase activity on glucose, which then catalyzes the oxidation of DAB to form an insoluble brown precipitate. This precipitate is then turned black by the addition of nickel. This method offers several advantages over labeling with fluorescently tagged antibodies. Since the product is generated by an enzymatic reaction, the signal is amplified and increases the likelihood of detecting a low-abundance protein. The label is also permanent and not susceptible to fading or photobleaching. Therefore, processed samples can be stored indefinitely and examined and reexamined as many times as desired. The results can also be visualized directly by bright-field microscopy, which makes this approach accessible to almost any laboratory. Finally, it is easy to detect landmarks in the sections and determine the precise location of labeling within the brain.

B. Methods

Proper sample fixation is essential for successful immunolabeling. The best way to ensure sufficient fixation of the brain is to perfuse the animal, which allows the fixative to fully penetrate the tissue. The animal is anesthetized or sacrificed and the chest cavity is exposed. A 26-gauge needle is inserted into the left ventricle of the heart, and phosphate-buffered saline (PBS, pH 7.6) is slowly injected by a syringe or peristaltic pump at a rate of 5 ml/min. At the same time a nick is made in the right atrium to allow the blood to flow out. The purpose of the PBS injection is to clear the tissues of blood. This is important because red blood cells can react with the enzymatic assay and lead to a high background. Effective clearing of the blood is indicated when the flow-through becomes clear. On average, we inject approximately 10–20 ml of PBS into an adult mouse. This is followed by injection of fixative, which is typically 4% (w/v) paraformaldehyde (PFA) in PBS. We have found that fixation works best with freshly made PFA. On average, we inject approximately 10–20 ml of 4% PFA into an adult mouse. The brain is then dissected from the head and further fixed in 4% PFA for 4–24 h at 4°C, followed by cryoprotection in 30% (w/v) sucrose in PBS for 16–24 h with gentle rocking. Cryoprotection is complete when the brain sinks in the sucrose.

The brain is then placed on the platform of a freezing microtome in the desired orientation, frozen with carbon dioxide, and sectioned at a thickness of 50–60 μm into PBS. The floating sections are then permeabilized with PBS containing 0.3% (v/v) Triton X-100 and 10 mg/ml BSA (PBT). This step allows antibodies to freely diffuse into the tissue and bind the epitope. Sections are incubated in primary antibody diluted in PBT for 16–24 h at 4°C with constant agitation. The sections are then rinsed in PBT and sequentially incubated in IgG (secondary antibody) diluted in PBT and peroxidase antiperoxidase (tertiary antibody) diluted in PBT for 1 h each at room temperature with constant agitation. The primary antibody and peroxidase antiperoxidase are generated in the same species, while the IgG is directed against that species and acts as a linker by binding both the primary antibody and the peroxidase antiperoxidase complex. The sections are then rinsed in PBS and processed using the glucose oxidase procedure. In this procedure, sections are incubated in a solution containing nickel ammonium sulfate, glucose, ammonium chloride, DAB, and glucose oxidase for 5–10 min. Sections are removed from the solution at intervals and observed under a light microscope to determine the extent of the reaction. When an acceptable level of staining is achieved, the sections are removed from the DAB solution, the reaction is stopped by placing the sections in 0.1 M sodium acetate (pH 6.0) solution, and then rinsed in PBS. The sections are then mounted on glass slides and allowed to air dry. They are dehydrated through a graded series of alcohol, cleared in xylene, and coverslipped using a mounting medium such as Permount.

C. Materials

Nickel ammonium sulfate solution:

1.5 g nickel ammonium sulfate in 50 ml of 0.1 M sodium acetate, pH 6.0

DAB solution:

50 Mg diaminobenzidine in 50 ml distilled water

Mix these two solutions together. Add the following in sequential order:

200 Mg beta-D-glucose
40 Mg of ammonium chloride
2 Mg glucose oxidase

Note: the amount of glucose oxidase varies per lot. Check the activity in units/mg dry weight and use a total of 550 units/100 ml solution as a standard.

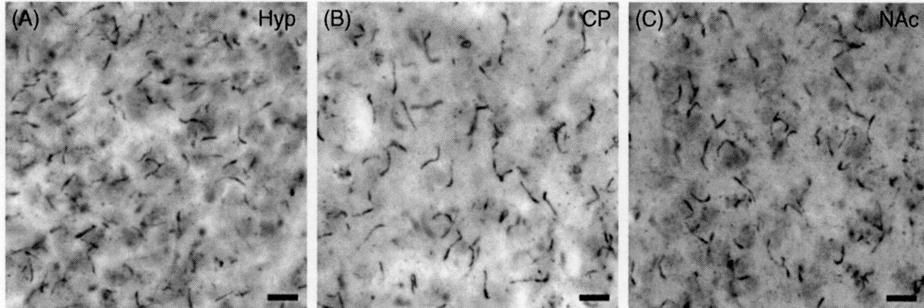

Fig. 1 Immunoenzymatic visualization of Mchrl ciliary localization in the adult mouse brain. (A) Hypothalamus (Hyp); (B) Caudate Putamen (CP); (C) Nucleus Accumbens (NAc). Scale bars are 10 μm.

D. Results

A mouse brain was fixed, processed for floating sections, and labeled with an antibody against Mchr1, as described above. Visualization of the subcellular localization of Mchr1 using bright-field microscopy reveals abundant Mchr1-positive cilia in several brain regions, including the hypothalamus, caudate putamen, and nucleus accumbens (Fig. 1). Neuronal cilia are rod-shaped structures that often appear bent or kinked and are typically around 10 μm in length (Fig. 1). Notably, the labeling of receptors on the ciliary membrane is usually significantly darker than labeling of receptors on the plasma membrane, which helps distinguish the cilia. The increased signal is presumably due to the receptors being concentrated on the relatively small surface area of the ciliary membrane.

E. Discussion

Occasionally, antibodies do not work well on tissues fixed with 4% PFA. This problem can sometimes be overcome by using HistoChoice, which is a fixative that is not formaldehyde-based. However, brains fixed solely with HistoChoice are not as firm as PFA-fixed brains and the floating sections tend to disintegrate during processing. Therefore, we perfuse the animal and postfix the brain with a 1:1 mixture of HistoChoice and 4% PFA when processing for floating sections.

III. Immunofluorescent Labeling of Neuronal Cilia in Brain Sections

A. Rationale

In this approach, neuronal cilia are labeled by a primary antibody to a particular marker and then visualized on a wide-field fluorescence or confocal microscope by the excitation of a fluorophore. The fluorophore can be conjugated to the primary antibody (direct fluorescence) or conjugated to a secondary antibody that specifically binds to the primary antibody (indirect fluorescence). The advantages of fluorescent immunolabeling include the ability to simultaneously label samples with multiple markers (i.e., multiple ciliary markers to look for colocalization or a ciliary marker and a cell-type-specific marker) and to generate higher resolution images, especially when analyzed by confocal microscopy.

B. Materials and Methods

After the brain is fixed and cryoprotected, as described above, it is placed in a 5-ml disposable plastic beaker containing Optimal Cutting Temperature (OCT) compound in the desired orientation. Note that the tissue will be sectioned from the bottom to the top. The beaker containing the submerged brain is then placed in a dry ice ethanol bath until frozen (the OCT will change from clear to white). The plastic container is then peeled away and the frozen block is mounted on a stub and sectioned in a cryostat at a thickness of 10–30 μm. Sections are collected onto positively charged slides, briefly dried on a warming plate at 37°C, and can be stored in a slide case at −20°C for several months. To label the sections, slides are first briefly rinsed in PBS and then placed in PBS containing 0.3% (v/v) Triton X-100, 2–5% (v/v) serum, 0.02% (w/v) sodium azide, and 10 mg/ml BSA for approximately 1 h to permeabilize the sections. The serum needs to be from the same species as the species in which the secondary antibody is generated and will reduce nonspecific binding of the secondary antibody and lower the background labeling. These steps can be carried out either by placing the slides into a slide chamber or by placing the slides on an immobile flat surface and applying just enough solution to cover the sections. Primary antibodies are then diluted in PBS containing 2–5% (v/v) serum, 0.02% (w/v) sodium azide and 10 mg/ml BSA (PBS+), applied to the tissue sections, and incubated for 16–24 h at 4°C. Alternatively, the primary antibody incubation can be performed at room temperature for 2–4 h. We have found that an easy way to perform antibody incubations is to place the slides in a humidified chamber (i.e., a lidded plastic container lined with a moist paper towel), add just enough diluted antibody solution to cover the sections, and then place a piece of parafilm cut to the size of the slide directly on the solution. The parafilm ensures the solution completely coats the sections and prevents the solution from evaporating or leaking from the slide. Furthermore, much less volume is required (100–200 μl) and therefore less primary antibody is needed. After primary antibody incubation, the sections are washed three times with PBS+, and then incubated with corresponding secondary antibodies (conjugated to a specific fluorophore such as those in the Alexa or CY series) diluted in PBS+ for 1 h at room temperature. If desired, a nuclear stain can be included in the diluted secondary antibody solution. The sections are then washed two times with PBS+, two times with PBS, and coverslipped with mounting medium. The slide is ready for analysis once the mounting medium has dried.

C. Results

A mouse brain was fixed, processed for frozen sections, and labeled with an antibody against ACIII, as described above. Visualization of the subcellular

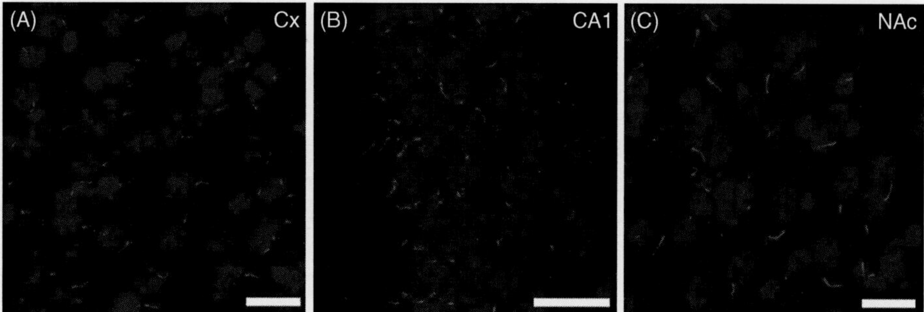

Fig. 2 Immunofluorescent visualization of ACIII ciliary localization in the adult mouse brain. (A) Cortex (Cx); (B) CA1 region of the hippocampus; (C) Nucleus Accumbens (NAc). Nuclei were stained with DRAQ5. Scale bars are 20 μm. (See Plate no. 3 in the Color Plate Section.)

localization of ACIII using confocal microscopy reveals abundant ACIII-positive cilia in several brain regions, including the cortex, hippocampus, and nucleus accumbens (Fig. 2).

D. Discussion

A potential complication arises when colabeling with primary antibodies generated in the same species. One possible solution is to use primary antibodies that have been conjugated with different fluorophores. This eliminates the need for secondary antibody detection and cross reactivity. Alternatively, a double-labeling technique can be used (Wessel and McClay, 1986).

IV. Visualization of Neuronal Cilia *In Vitro*

A. Rationale

The strength of studying neuronal cilia *in vitro* is the ability to perform functional studies, such as knock-down of ciliary proteins or rescue of neurons with ciliary defects. It is also possible to perform live-cell imaging of neuronal cilia in cultures. We have found that all of the ciliary GPCRs (i.e., Sstr3, 5-Ht6, and Mchr1) localize to cilia when heterologously expressed in ciliated cells. Therefore, it is possible to express a fusion construct of these proteins and a fluorescent protein in neurons and visualize cilia on live cells.

B. Methods

Our neuronal culture procedure is adapted from several published methods (Askwith *et al.*, 2004; Brewer *et al.*, 1993; Wemmie *et al.*, 2002). Although the procedure was originally developed for the culturing of hippocampal neurons, we have successfully cultured neurons from the hypothalamus, nucleus accumbens, cerebral cortex, and amygdala under these conditions. The neurons are isolated from the brains of mouse pups on the day of their birth (P0). The pups are anesthetized with isoflurane, decapitated with surgical scissors, and the brains are dissected out and placed in 35-mm dishes containing a sterile solution of Leibovitz's L-15 medium with 0.2 mg/ml BSA. The brain region of interest is then dissected out under a dissecting microscope and torn into small pieces. The rest of the procedure is performed in a tissue culture hood using sterile technique. The tissue is transferred into a 15-ml conical tube containing 1 ml of L-15/BSA with 0.375 mg/ml papain and incubated for 15 min at 37° C with 95%O_2/5%CO_2 blowing gently over the surface of the solution. Note that when making up the L-15/BSA/papain solution it should be incubated for at least 20 min at 37°C to ensure the papain is fully dissolved before sterile filtration. Also note that the application of oxidizing gas to the digesting neurons is critical for ensuring consistently viable neurons. After incubation, the tissue is washed three times with 1–2 ml prewarmed M5-5 medium, and then pipetted up and down three times with a Pasteur pipette. The pieces are allowed to settle and the supernatant is discarded. The purpose of this initial trituration is to remove less-viable cells along the edges of the tissue. The pieces are then triturated with a full-bore Pasteur pipette, a pipette of approximately 2/3 the size of the starting diameter, and finally with a pipette of approximately 1/3 the size of the starting diameter. The reduced diameter pipettes are generated by briefly heating the pipette tips over a Bunsen burner. Each trituration consists of pipetting the pieces up and down approximately ten times with 1–2 ml M5-5 medium. After each trituration the pieces are allowed to settle and the supernatant is transferred to a fresh conical tube. The combined supernatants are then spun at $80 \times g$ for 5 Min. The medium is removed and the neurons are resuspended in neurobasal medium and plated onto poly-D-lysine coated coverslips in a 24-well dish. The number of wells the neurons are distributed into depends on the starting tissue amount, the region of the brain that was isolated, and the desired density. In general, we plate the neurons from a hippocampus or hypothalamus from one mouse onto 2–3 coverslips in 1 ml of medium/well. To ensure the cultures remain glial "free" the DNA synthesis inhibitor cytosine arabinofuranoside (ARA-C) is added to a final concentration of 10 μM after 2–3 days.

After 3–5 days in culture we transfect the neurons with a construct expressing a ciliary protein fused to a fluorescent protein. We have successfully used Lipofectamine 2000 to transfect cultured neurons, but other transfection reagents will also work. The conditioned media are removed from each well, collected in a conical tube, and kept at 37°C. The media are replaced with 0.4 ml of fresh neurobasal medium without gentamycin. The expression construct and Lipofectamine are combined according to the manufacturer's instructions. Typically, we combine 0.8 μg DNA with 5 μl of Lipofectamine in 0.1 ml Opti-MEM for each well. The DNA/Lipofectamine complex in Opti-MEM is then added to each well and incubated for 4 h. This medium is aspirated and replaced with the saved conditioned medium. The neurons are ready to image after 24–48 h.

The cultures can also be fixed and labeled for ciliary and/or cellular markers. In most cases it is sufficient to simply aspirate the culture medium, wash the cells one time with PBS, and fix them in 4% PFA for 10–20 min. This is followed by permeabilization in

PBS+ containing 0.3% Triton X-100 for 5–10 min and blocking in PBS+ for 1 h. Primary antibody diluted in PBS+ is then applied to the culture and incubated for 16–24 h at 4°C or 2–4 h at room temperature. After primary antibody incubation, the coverslips are processed and incubated with secondary antibody as described above.

C. Materials

M5-5 Medium:

90 ml Earle's minimal essential medium without L-Glutamine
5 ml fetal bovine serum (heat-inactivated)
5 ml horse serum (heat-inactivated)
0.2 ml glutaMAX supplement *or* 200 mm L-Glutamine
1 ml penicillin/streptomycin
1 ml 30% (w/v) glucose
0.25 ml insulin/selenite/transferrin

Mix and sterile filter. Can be stored at 4°C for up to 1 week.

Neurobasal Medium:

97.4 ml neurobasal medium
2 ml B27 supplement
0.25 ml glutaMAX supplement *or* 200 mm L-Glutamine
0.25 ml insulin/selenite/transferrin
0.1 ml gentamycin

Mix and sterile filter. Can be stored at 4°C for up to 1 week.

D. Results

Hippocampal neurons were cultured, transfected, fixed, and processed, as described above. The neurons were transfected with a construct expressing Sstr3 fused to enhanced green fluorescent protein (EGFP) or 5-Ht6 fused to EGFP. The cells were also labeled with an antibody against β-tubulin III, which is a marker of neurons. Visualization of the subcellular localization of the fluorescently tagged receptors by confocal microscopy shows that both Sstr3 and 5-Ht6 robustly localize to neuronal cilia (Fig. 3). Ciliary localization of these receptors can also be visualized in live cells. It is worth noting that the orientation of cilia varies on cultured neurons. This is in contrast to cultured epithelial cells where primary cilia always project upward from the center of the apical surface. Neuronal cilia can project from the top, side, or bottom of neurons and it is necessary to look for them throughout the entire focal plane.

E. Discussion

The cultures generated by the procedures described above are considered glial "free" because glial cell proliferation is inhibited by the lack of serum and the

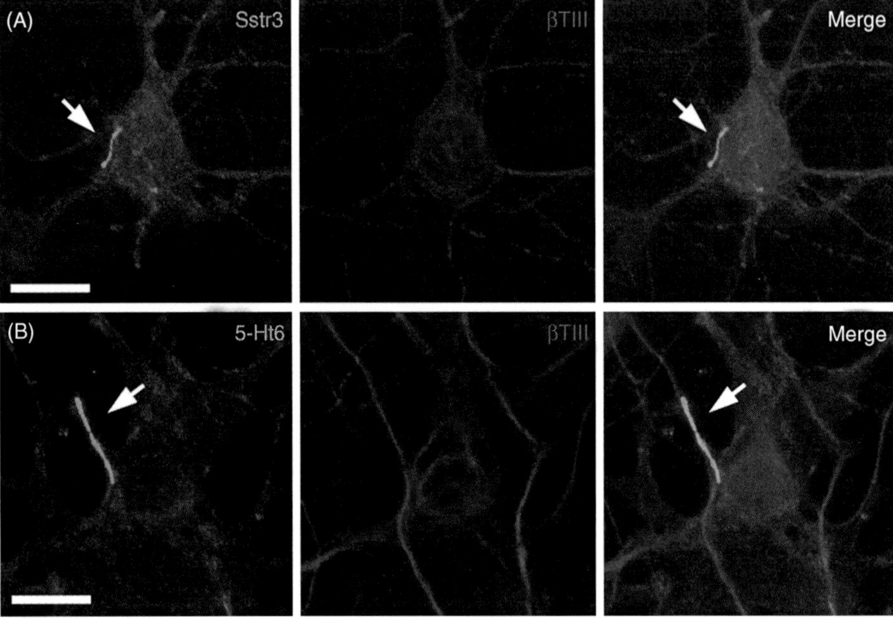

Fig. 3 Somatostatin receptor 3 (Sstr3) and serotonin receptor 6 (5-Ht6) localize to cilia when heterologously expressed in mouse neurons. (A) Image of a fixed hippocampal neuron 24 h post-transfection with a vector expressing Sstr3 fused to enhanced green fluorescent protein (EGFP) and labeled for the neuronal marker β-tubulin III (βTIII). (B) Image of a fixed hippocampal neuron 24 h post-transfection with a vector expressing 5-Ht6 fused to EGFP and labeled for βTIII. Cilia are indicated by arrows. Note the prominent ciliary localization of both Sstr3 and 5-Ht6. Nuclei were stained with DRAQ5. Scale bars are 10 μm. (See Plate no. 4 in the Color Plate Section.)

addition of ARA-C. However, it is possible to generate mass cultures that are composed of neurons and glia by plating the collected M5-5 supernatants onto collagen coated coverslips. It is still necessary to treat the cultures with ARA-C after 2–3 days to ensure that the glia do not overgrow the neurons. The advantage of mass cultures is the neurons tend to grow better, and it is less likely that the culture will fail. Neurons in mass culture do possess cilia. However, in our experience it is preferable to examine neuronal cilia in glial "free" cultures since the neurons tend to be more dispersed.

V. Summary

The techniques described here are applicable to visualizing neuronal cilia in fixed tissue as well as fixed and living cells. These techniques can be used to address the critical question of what signaling proteins reside in neuronal cilia and also provide the basis for future studies, such as live-cell imaging of protein trafficking into and out of the cilium.

Acknowledgments

We are grateful to Georgia Bishop and Candice Askwith for technical assistance and critical review of the manuscript. This work was supported by grant 5-FY05-39 from the March of Dimes Birth Defects Foundation (K.M.) and R01GM083120 from the National Institute of General Medical Sciences (K.M.).

References

Askwith, C.C., Wemmie, J.A., Price, M.P., Rokhlina, T., and Welsh, M.J. (2004). Acid-sensing ion channel 2 (ASIC2) modulates ASIC1 H+-activated currents in hippocampal neurons. *J. Biol. Chem.* **279**, 18296–18305.

Berbari, N.F., Johnson, A.D., Lewis, J.S., Askwith, C.C., and Mykytyn, K. (2008a). Identification of ciliary localization sequences within the third intracellular loop of G protein-coupled receptors. *Mol. Biol. Cell* **19**, 1540–1547.

Berbari, N.F., Lewis, J.S., Bishop, G.A., Askwith, C.C., and Mykytyn, K. (2008b). Bardet-Biedl syndrome proteins are required for the localization of G protein-coupled receptors to primary cilia. *Proc. Natl. Acad. Sci. USA* **105**, 4242–4246.

Bishop, G.A., Berbari, N.F., Lewis, J.S., and Mykytyn, K. (2007). Type III adenylyl cyclase localizes to primary cilia throughout the adult mouse brain. *J. Comp. Neurol.* **505**, 562–571.

Brailov, I., Bancila, M., Brisorgueil, M.J., Miquel, M.C., Hamon, M., and Verge, D. (2000). Localization of 5-HT(6) receptors at the plasma membrane of neuronal cilia in the rat brain. *Brain Res.* **872**, 271–275.

Brewer, G.J., Torricelli, J.R., Evege, E.K., and Price, P.J. (1993). Optimized survival of hippocampal neurons in B27-supplemented Neurobasal, a new serum-free medium combination. *J. Neurosci. Res.* **35**, 567–576.

Fuchs, J.L., and Schwark, H.D. (2004). Neuronal primary cilia: A review. *Cell Biol. Int.* **28**, 111–118.

Handel, M., Schulz, S., Stanarius, A., Schreff, M., Erdtmann-Vourliotis, M., Schmidt, H., Wolf, G., and Hollt, V. (1999). Selective targeting of somatostatin receptor 3 to neuronal cilia. *Neuroscience* **89**, 909–926.

Shu, S.Y., Ju, G., and Fan, L.Z. (1988). The glucose oxidase-DAB-nickel method in peroxidase histochemistry of the nervous system. *Neurosci. Lett.* **85**, 169–171.

Sternberger, L.A., Hardy, P.H., Jr., Cuculis, J.J., and Meyer, H.G. (1970). The unlabeled antibody enzyme method of immunohistochemistry: Preparation and properties of soluble antigen-antibody complex (horseradish peroxidase-antihorseradish peroxidase) and its use in identification of spirochetes. *J. Histochem. Cytochem.* **18**, 315–333.

Wemmie, J.A., Chen, J., Askwith, C.C., Hruska-Hageman, A.M., Price, M.P., Nolan, B.C., Yoder, P.G., Lamani, E., Hoshi, T., Freeman, J.H., Jr., and Welsh, M.J. (2002). The acid-activated ion channel ASIC contributes to synaptic plasticity, learning, and memory. *Neuron* **34**, 463–477.

Wessel, G.M., and McClay, D.R. (1986). Two embryonic, tissue-specific molecules identified by a double-label immunofluorescence technique for monoclonal antibodies. *J. Histochem. Cytochem.* **34**, 703–706.

CHAPTER 7

Immunofluorescence Staining of Ciliated Respiratory Epithelial Cells

Heymut Omran *and* Niki T. Loges

Department of Pediatrics and Adolescent Medicine, University Hospital Freiburg, Mathildenstrasse 1; 79106 Freiburg, Germany

Klinik und Poliklinik für Kinder- und Jugendmedizin - Allgemeine Pädiatrie - Universitätsklinikum Münster, Albert-Schweitzer-Strasse 33; 48149 Münster

Abstract

Respiratory epithelial cells carry multiple motile cilia. Defective ciliary motility results in reduced mucociliary airway clearance in a destructive chronic airway disease referred to as primary ciliary dyskinesia. Immunofluorescence (IF) microscopy of

978-0-12-374973-4
DOI: 10.1016/S0091-679X(08)91007-4

respiratory cells allows visualization of proteins along the length of the ciliary axoneme and distinct compartments such as the transition zone and the basal body region at the ciliary base. The advantages of the technique to localize proteins such as axonemal dynein motor components along the ciliary axoneme and their diagnostic potential are shown. Special emphasis is placed on methodological subtleties and quality controls to assure accurate IF analyses.

I. Introduction

Respiratory epithelial cells are densely covered by multiple motile 9 + 2 cilia. Transmission electron microscopy has helped to delineate the ultrastructure of this organelle. A schematic of the ciliary base and axoneme is depicted in Fig. 1 (Fliegauf *et al.*, 2007). The cilium emerges from nine triplet structures (basal bodies) surrounded by the microtubule organizing centers which convert into nine doublets (9 + 0 structure) with adjacent transition fibres also referred to as the transition zone. The transition zone also demarcates the border of the ciliary compartment and finally converts into nine doublets with attached dynein arms, which surround two single central tubules (9 + 2 structure). Immunofluorescence (IF) microscopy has emerged as a new method for analysis of ciliary subcompartments and their components (Figs. 1 and 2). Staining of distinct

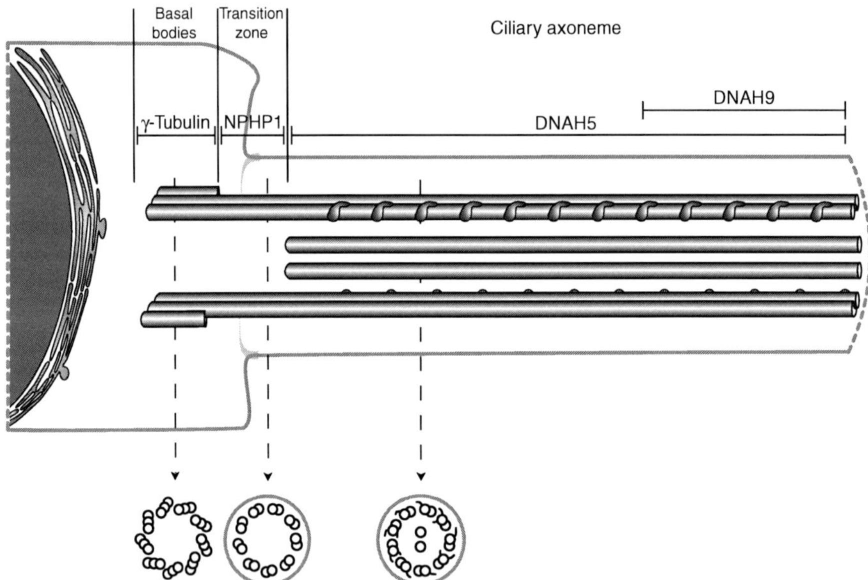

Fig. 1 Respiratory epithelial cells are densely covered by multiple motile 9 + 2 cilia. This schematic depicts subcellular compartments of a single motile respiratory cilium. Characteristic cross sections detectable by transmission electron microscopy (specified below). Specific segments can be stained by IF microscopy using site-specific marker proteins (specified above).

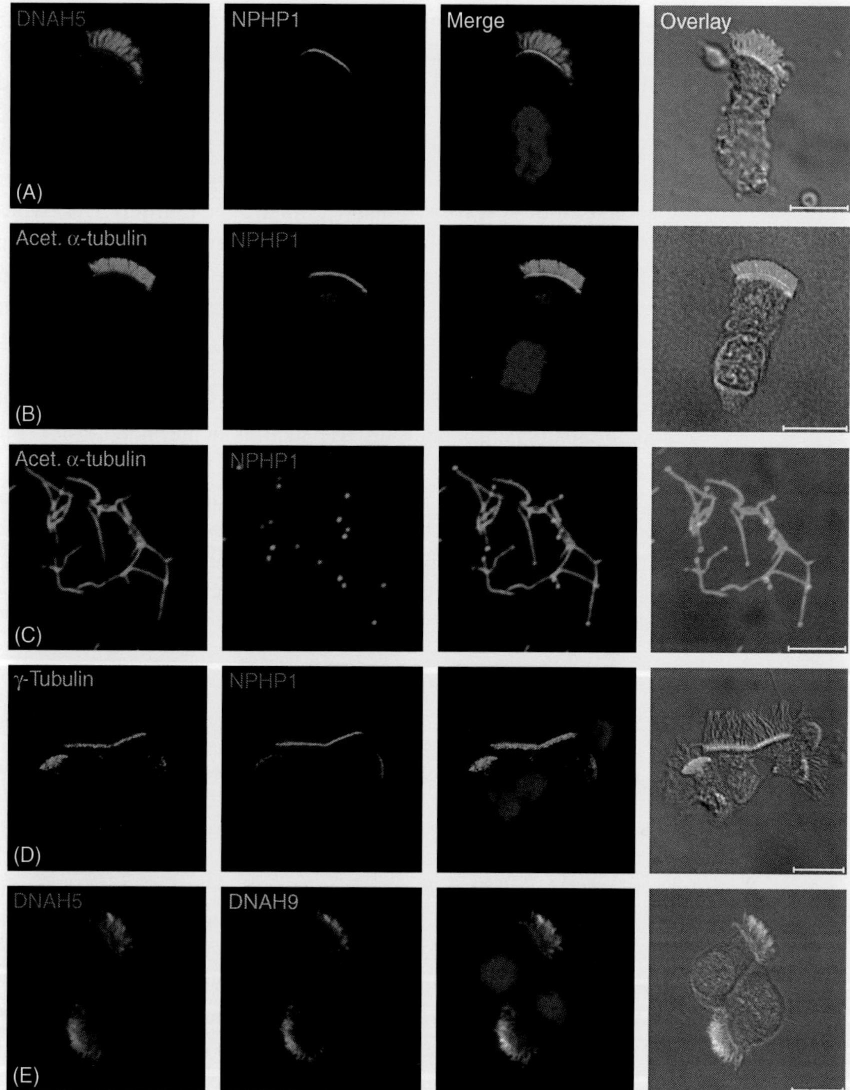

Fig. 2 Distinct proteins localize to specific ciliary subsegments. (A) Anti-DNAH5 (red) targets ODA complexes type 1 and 2 present throughout the hole ciliary axoneme, while anti-nephrocystin (NPHP1) (green) recognizes the transition zone. (B) and (C) Anti-NPHP1 (red) stains the transition zone at the ciliary base. The ciliary axoneme is stained with antiacetylated α-tubulin (green). Note in (C) single cilia detached from respiratory cells were stained. (D) Anti-γ-tubulin (green) stains the microtubule organizing centers/basal bodies region beneath the transition zone that is highlighted by anti-NPHP1 (red). (E) Anti-DNAH9 (green) recognizes solely type 2 ODA complexes present in the distal ciliary axoneme, whereas anti-DNAH5 (red) recognizes ODA complexes throughout the complete ciliary axoneme. The yellow color indicates colocalization of DNAH5 and DNAH9 in the distal ciliary axoneme. Nuclei were stained with Hoechst 33342 (blue). For double labeling monoclonal mouse antibodies (green) and polyclonal rabbit antibodies (red) were used. Scale bars in A, B, D, and E: 10 μm; in C: 5 μm. (See Plate no. 5 in the Color Plate Section.)

proteins allow recognition of the microtubule organizing centers adjacent to the basal bodies (γ-tubulin, Fig. 2D), the transition zone [nephrocystin (NPHP1), Fig. 2A–D; RPGR], and the ciliary axoneme (acetylated α-tubulin, Fig. 2B and C). In contrast to electron microscopy, IF microscopy allows us to analyze the distinct localization of ciliary or flagellar proteins (e.g., dynein heavy chains, Fig. 2E) along the length of the axoneme (Fliegauf *et al.*, 2005). These studies have shown that human respiratory cilia are composed of at least two distinct outer dynein arm (ODA) types (Fig. 2E): Type 1 (DNAH9 negative and DNAH5 positive; proximal ciliary axoneme) and type 2 (DNAH9 and DNAH5 positive; distal ciliary axoneme). This contrasts findings obtained in *Chlamydomonas* flagella, which contain only one type of ODA complex. Furthermore, the use of antibodies for the localization of ciliary components provides a novel tool to aid understanding of the molecular pathology and diagnosis of cilia-related diseases (ciliopathies, Fig. 3) caused by malfunction of motile cilia (primary ciliary dyskinesia) or immotile cilia (cystic kidney disease such as nephronophthisis) (Fliegauf *et al.*, 2005, 2006; Hornef *et al.*, 2006; Loges *et al.*, 2008; Omran *et al.*, 2008).

Here, we focus on the technique of IF staining of respiratory epithelial cells using antibodies directed against two distinct antigens and nuclear staining of DNA. We also discuss advantages and pitfalls of this technique. Emphasis will be placed on the quality controls that should be applied when polyclonal and/or monoclonal antibodies are used.

II. Antibodies

A. Polyclonal Antibodies

Usually, the immunization of rabbits or other animals (chicken, goat) is the fastest way to generate antibodies. It is in almost all cases successful and immune serum is easy to test. However, there are some disadvantages: immune serum has to be purified to avoid nonspecific labeling. In addition the animal provides a finite amount of polyclonal antibodies. More important is the fact that polyclonal rabbit serum often contains antibodies that produce an unspecific staining of the basal body region at the ciliary base. Therefore it is mandatory first to screen numerous preimmune sera of rabbits for presence of these disadvantageous antibodies in order to choose the right animal for immunization. In our experience ~80% of rabbits contain antibodies directed against the basal body region. Examples are shown in Fig. 4. This fact might also explain the high frequency of protein sublocalization to the ciliary base.

B. Monoclonal Antibodies

The value of monoclonal antibodies comes from different characteristics; their specificity of binding, their homogeneity, and their ability to be produced in unlimited quantities. In 1975, Köhler and Milstein developed a technique that allows the growth of clonal populations of cells secreting antibodies with defined specificity (Köhler and Milstein, 1975). In this method an antibody-producing cell, usually plasma cells

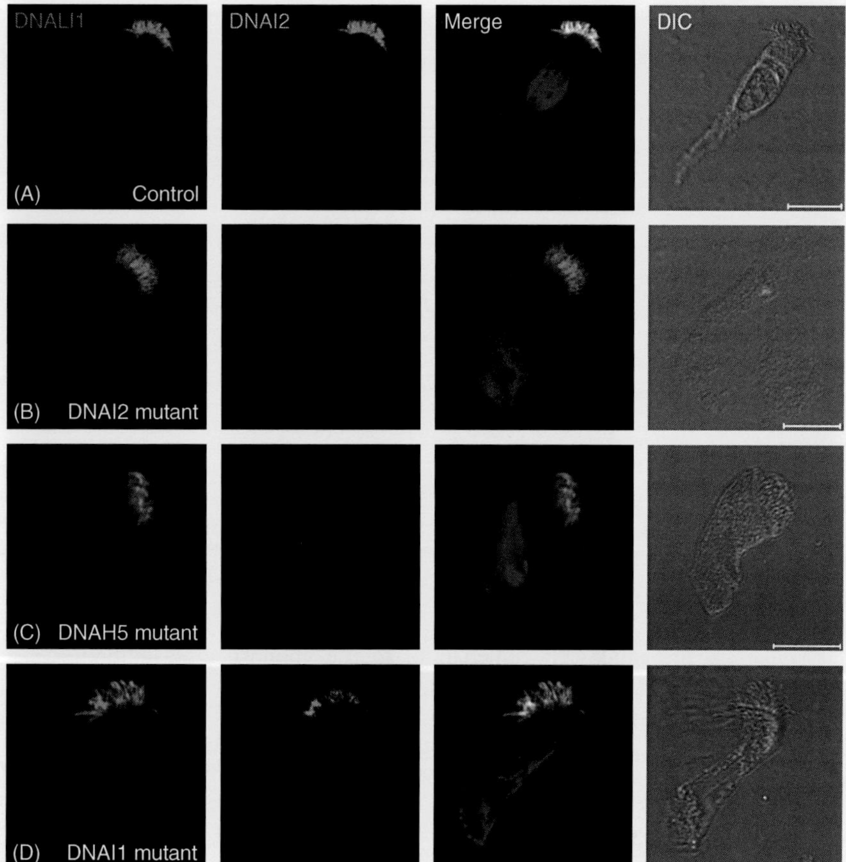

Fig. 3 Absence and/or mislocalization of ODA chain DNAI2 in respiratory epithelial cells from PCD patients carrying mutations in the genes *DNAI2*, *DNAH5*, and *DNAI1* encoding ODA components. (A) Control respiratory cell. DNAI2 is localized throughout the ciliary axoneme. (B) Mutant cell harboring a homozygous *DNAI2* mutation. Consistent with loss of function, DNAI2 is completely absent from the ciliary axoneme. (C) Mutant cell with a homozygous *DNAH5* mutation. DNAI2 is absent from the whole ciliary axoneme, because ODA assembly is disrupted. (D) Mutant respiratory cell with compound heterozygous *DNAI1* mutations. DNAI2 (green) is detectable in the proximal ciliary axoneme but absent from the distal part of the axonemes. This indicates that mutant DNAI1 inhibits assembly of DNAI2 predominantly in the distal ciliary axonemes (type 2 ODA complexes). The yellow color indicates colocalization in the proximal ciliary axoneme. Nuclei were stained with Hoechst 33342 (blue). As ciliary control, staining with anti-DNALI1 (inner dynein arm light chain; red) was performed. For double labeling monoclonal mouse antibodies and polyclonal rabbit antibodies were used. DIC: Differential Interference Contrast. Scale bars: 10 μm. (See Plate no. 6 in the Color Plate Section.)

isolated from an immunized animal (mouse), is fused with a myeloma cell, a type of B-cell tumor. These hybrid cells, so-called hybridomas, can be maintained *in vitro* and will continue to secrete antibodies with defined specificity providing an unlimited supply of antibodies. In addition, one unique advantage of hybridoma production is that impure antigens can be used to produce specific antibodies. Because hybridomas

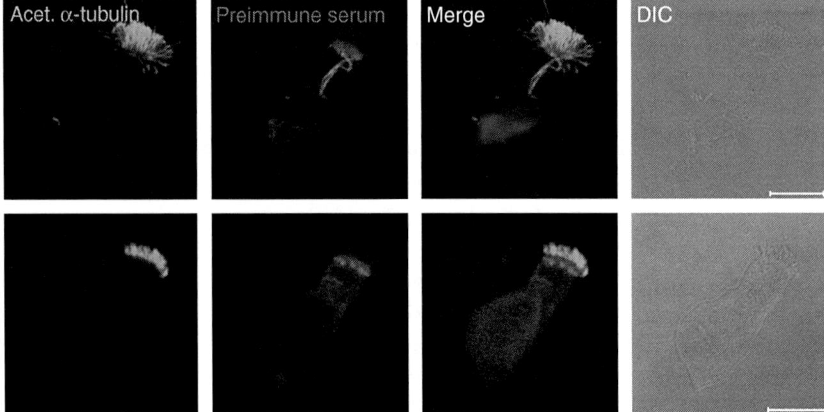

Fig. 4 Unspecific staining of basal bodies with rabbit preimmune serum. Incubation of respiratory epithelial cells with preimmune serum causes an unspecific staining in the basal bodies/microtubule organizing centers region (red). The ciliary axoneme is stained with antiacetylated α-tubulin (green). Nuclei were stained with Hoechst 33342 (blue). For double labeling monoclonal mouse antibodies and polyclonal rabbit antibodies were used. DIC: Differential Interference Contrast. Scale bars: 10 μm. (See Plate no. 7 in the Color Plate Section.)

are single-cell cloned prior to use, monospecific antibodies can be produced after immunization with complex mixtures of antigens. However, generation and testing for specific monoclonal antibodies is often more time consuming and costly to prepare than polyclonal antibodies.

C. Antibody Specificity

To prove specificity of produced antibodies we recommend using the following standard: (1) Demonstration of specificity by Western blot analyses of respiratory cell lysates and/or axonemal extracts. Ideally, the antibodies should only detect a single protein band. (2) Demonstration of an absent or truncated protein band using respiratory cell proteins from a patient harboring mutations of the gene that encodes the protein of interest. (3) IF analyses of control and mutant respiratory cells confirms antibody specificity by demonstration of mislocalization or absence of the protein of interest in the mutant cells.(4) For Western blot and IF analyses control experiments omitting each time one antibody (e.g., primary antibody) should be performed.

Ideally, identical results are obtained when polyclonal and monoclonal antibodies are used to characterize a protein of interest. If antibodies are used that have previously been reported to be specific it is mandatory to demonstrate this again, because the quality of the antibodies might have changed during transport or for other reasons. Unfortunately, many published experiments do not apply these strict criteria. Western blot analyses are either not or only incompletely shown. Others use staining experiments in the presence and absence of the antigen (used for immunization) to indicate antibody specificity. However, results of such studies are prone to result in scientific artifacts.

III. Materials and Methods

A. Materials

Glass slides (76 mm × 26 mm; R. Langenbrinck, Emmendingen, Germany)
Coverslips (24mm × 40 mm; R. Langenbrinck, Emmendingen, Germany)
Phosphate-buffered saline (PBS, Gibco, Invitrogen, Karlsruhe, Germany)
Paraformaldehyde (PFA, Merck, Ulm, Germany)
NaOH (Merck, Ulm, Germany)
Triton X-100 (Sigma, Taufkirchen, Germany)
Tween 20 (Sigma, Taufkirchen, Germany)
Skim milk (fat-free) (Roth, Karlsruhe, Germany)
Alexa Fluor 488-conjugated goat antimouse IgG (H + L) (Molecular Probes, Invitrogen, Karlsruhe, Germany)
Alexa Fluor 546-conjugated goat antirabbit IgG (H + L) (Molecular Probes, Invitrogen, Karlsruhe, Germany)
Hoechst 33342 (Sigma, Taufkirchen, Germany)
Aqueous mounting medium (Dako North America, Inc., Carpinteria, CA, USA)

B. Solutions

0.1% PBST: 500 μl Tween 20 mixed with 500 ml PBS
4% PFA: 4 g PFA dissolved in 100 ml deionized water, water should be warm (60°C), add a few drops of NaOH for dissolving of PFA, adjust pH to 7.4
0.2% Triton X: 100 μl Triton X-100 mixed with 50 ml PBS
1% (2.5%) skim milk solution: blocking solution, 0.5 mg (1.25 mg) skim milk powder dissolved in PBS

IV. Methods

A. Sampling of Respiratory Cells

Human respiratory epithelial cells are obtained by nasal-brush biopsy (Cytobrush Plus, Medscand, Malmö, Sweden) and are suspended in cell culture medium (RPMI). Note the use of isotonic saline solution or cell culture medium with serum can cause severe artifacts. Move the cytobrush gently up and down to detach respiratory cells from the brush. Suspended cell samples are then spread onto glass slides, air dried, and stored at −80°C until use. Avoid use of any coated glass slides because this will increase background signals. Alternatively, respiratory cells can first be used for *in vitro* cell cultures (e.g., *in vitro* ciliogenesis of spheroids) and then used for IF (Olbrich *et al.*, 2006). If spheroids are used, cells should be handled very gentle during the staining experiments because they detach from the slide surface more easily than freshly brushed cells. For staining of mouse respiratory cells, either sections of the nasal turbinate or trachea can be used or an inoculation loop can be employed to detach respiratory cells from the trachea.

B. Fixation and Permeabilization of Respiratory Cilia

In the literature, there are different methods and reagents for fixation described. The correct choice of method will depend on the nature of the antigen being examined and on the properties of the antibody used. There are two classes of different fixation reagents: organic solvents and cross-linking reagents. The most common organic reagents used for fixation are 100% cold methanol or acetone. The methanol or acetone fixation is an easy method; however, it frequently solubilizes and removes membrane-bound antigens while precipitating the proteins on the cellular architecture. By simple precipitation of the protein, organic solvents only provide low structural preservation. Cross-linking reagents, such as PFA, form intermolecular bridges, usually through free amino groups, thus creating a network of linked antigens. Cross-linkers conserve cell structure better than organic solvents, but may reduce the antigenicity of some cell components. Unlike the fixation with organic solvents, the usage of cross-linkers requires the addition of a permeabilization step, usually with Triton X-100, to allow access of the antibody to the specimen.

For fixation of the axoneme and axonemal structures, incubation in 4% PFA for 15 min is recommended. Afterward, incubation in 2% Triton X-100 for 15 min is necessary for permeabilization to make antigens available for the antibodies.

C. Blocking and Primary Antibody Incubation

Before the incubation of tissues or cells with primary antibodies, blocking of unspecific antigens is necessary to prevent nonspecific binding. The cheapest way to block tissue or cells is the use of fat-free skim milk (1–5%). This step can be done overnight at 4°C with 1% skim milk dissolved in PBS or for 1 h at room temperature with 5% skim milk. Another possibility to avoid nonspecific antibody binding is the use of 5% serum. The serum should originate from the same species where the secondary antibody was raised.

Incubation of samples with primary antibodies can be performed overnight at 4°C or at room temperature for a few hours. Usually, when using mouse hybridoma supernatants containing monoclonal antibodies, incubation overnight at 4°C is recommended. In case of polyclonal rabbit or goat antibodies, incubation at room temperature for 2–3 h is enough. However, for each antibody the optimal incubation time and dilution has to be individually determined.

D. Antibody Labeling

Labeling of the primary antibodies using secondary antibodies is the crucial step of the IF staining method. Secondary antibodies may be polyclonal or monoclonal, and are available with specificity for whole Ig molecules or antibody fragments such as the Fc or Fab regions. Typically, secondary antibodies are labelled with probes that make them useful for detection, like alkaline phosphatase (immunohistochemistry), horseradish peroxidase (immunoblot), or fluorescent dyes (IF). It is highly recommended to use both secondary antibodies from the same host species to avoid cross-reaction. However, before using secondary antibodies, the best dilution has to be determined as well as their specificity. For this purpose, staining of control cells or tissue without primary antibodies should be performed.

E. IF Microscopy

Standard IF microscopes can be applied for cellular sublocalization of proteins. For higher spatial resolution and colocalization of proteins, the use of high-resolution IF analysis is recommended. To obtain confocal images we use a Zeiss laser scanning microscope (Axiovert 200 LSM510 META) using a 63×1.2 numerical aperture water immersion or a 100×1.3 numerical aperture oil immersion objective. A four-channel, eight-bit multitracking scan mode is used with a 1024×1024 frame size and fourfold average line scan settings. Images are processed with the regular Zeiss LSM510 software.

V. Discussion

A. Cross-Species Specificity of Antibodies

Some antibodies can not only be exploited for use in a single species but also in a wide range of organisms such as mice. If this cross-species reactivity is intended, it is best to use evolutionary conserved protein fragments for immunization. To verify antibody specificity identical standards as detailed in Section II.C. are required. An example is shown for a mouse monoclonal antibody originally targeted against human DNAH5, which also specifically detects the orthologous murine Dnahc5 (Mdnah5) protein (Francis *et al.*, 2009; Fig. 5).

B. Immunofluorescence Analyses to Target Genetic Testing

For many years transmission electron microscopy has been the standard for diagnosis of primary ciliary dyskinesia (PCD), a hereditary disorder characterized by motility defects of respiratory cilia and sperm flagella (Fliegauf *et al.*, 2007). However, recent genetic studies have clearly demonstrated that some PCD variants (e.g., *DNAH11*-defective PCD) can exhibit normal ultrastructure but functional ciliary beating defects (Schwabe *et al.*, 2008). In addition it is known that subtle defects involving, for example, the inner dynein arms are also difficult to discern by standard transmission electron microscopy (Escudier *et al.*, 2002). Thus it is mandatory to introduce novel diagnostic techniques. IF staining has several advantages. First, this method is performed in respiratory epithelial cells obtained by noninvasive transnasal brushings, and this material can be used at the same time for functional analysis of ciliary beat frequency and pattern. Second, samples dried on glass slides can be transported easily to the performing laboratories. Third, secondary ciliary changes do not affect axonemal localization of ODA components such as DNAH5 (Olbrich *et al.*, 2006). Fourth, the method is able to detect changes along the entire ciliary

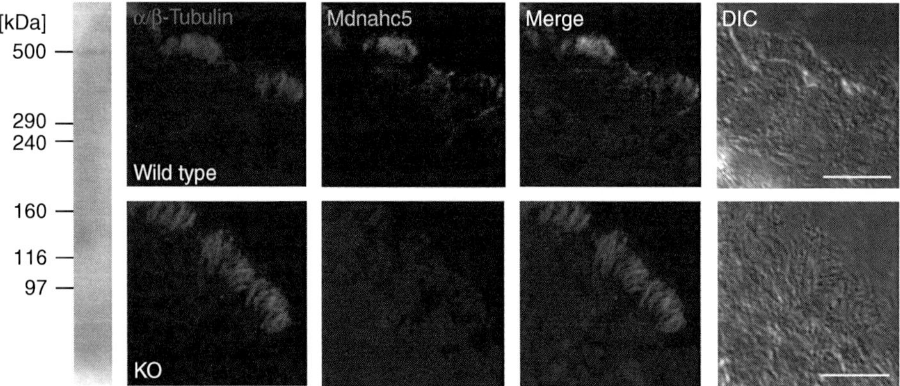

Fig. 5 Cross-species specificity of mouse monoclonal anti-DNAH5 antibodies. Western blot analysis of protein extracts from wild-type mouse tracheas (on the left). Anti-DNAH5 antibodies specifically detect a single band with the predicted size (~500 kDa). IF staining of mutant (Mdnah5-KO) and wild-type mouse respiratory epithelial cells with anti-DNAH5 antibodies (green). As observed in human cells, DNAH5 is localized throughout all analyzed mouse respiratory ciliary axonemes, but absent from the ciliary axonemes of Mdnah5-KO mice. Anti-α/β-tubulin was used as a cilia marker (red). The yellow color indicates colocalization. Nuclei were stained with Hoechst 33342 (blue). Note, the use of nonpurified mouse hybridoma supernatants resulted in higher background (wild type, upper panel). For double labeling monoclonal mouse antibodies and polyclonal rabbit antibodies were used. DIC: Differential Interference Contrast. Scale bars: 10 μm. (See Plate no. 8 in the Color Plate Section.)

axoneme. This aspect is of particular significance because with electron microscopy, localization of the examined cross sections along the ciliary axoneme is not possible. IF microscopy has been shown to be a valuable tool to identify outer and inner dynein arm defects (Fliegauf *et al.*, 2005; Hornef *et al.*, 2006; Loges *et al.*, 2008; Omran *et al.*, 2008, Fig. 3). The results of antigen mapping by IF are also very helpful to target genetic testing in PCD. With increasing numbers of antibodies available for IF analyses, this method will gain importance in the diagnostic work-up for PCD.

Interestingly, hereditary cystic kidney diseases such as nephronophthisis are also characterized by defective cilia function (Fliegauf *et al.*, 2007). IF analyses have shown that cystoproteins defective in these disorders not only localize to monocilia but also to motile respiratory cilia (Fliegauf *et al.*, 2006). This knowledge can be even exploited for diagnostic purposes, because in respiratory cells from patients with homozygous *NPHP1* deletions (most prevalent genetic defect in nephronophthisis patients) nephrocystin is absent from the transition zone.

Acknowledgments

We are grateful to the patients and their families for their participation in this study. We thank the German patient support group Kartagener Syndrom und Primaere Ciliaere Dyskinesie e.V. We thank H. Olbrich for help in preparing the figures. This work is supported by Deutsche Forschungsgemeinschaft grants DFG Om 6/4, DFG Om 6/5, GRK1104, and SFB592 (to HO).

There is no conflict in the interests of any of the authors who have contributed to this manuscript and the work associated with it.

References

Escudier, E., Couprie, M., Duriez, B., Roudot-Thoraval, F., Millepied, M.C., Prulière-Escabasse, V., Labatte, L., and Coste, A. (2002). Computer-assisted analysis helps detect inner dynein arm abnormalities. *Am. J. Respir. Crit. Care Med.* **166**, 1257–1262.

Fliegauf, M., Benzing, T., and Omran, H. (2007). Cilia: Hair-like organelles with many links to disease. *Nat. Rev. Mol. Cell Biol.* **8**, 880–893.

Fliegauf, M., Horvath, J., von Schnakenburg, C., Olbrich, H., Muller, D., Thumfart, J., Schermer, B., Pazour, G.J., Neumann, H.P., Zentgraf, H., Benzing, T., and Omran, H. (2006). Nephrocystin specifically localizes to the transition zone of renal and respiratory cilia and photoreceptor connecting cilia. *J. Am. Soc. Nephrol.* **17**, 2424–2433.

Fliegauf, M., Olbrich, H., Horvath, J., Wildhaber, J.H., Zariwala, M.A., Kennedy, M., Knowles, M.R., and Omran, H. (2005). Mislocalization of DNAH5 and DNAH9 in respiratory cells from primary ciliary dyskinesia patients. *Am. J. Respir. Crit. Care Med.* **171**, 1343–1349.

Francis, R.J., Chatterjee, B., Loges, N.T., Zentgraf, H., Omran, H., and Lo, C.W. (2009). The initiation and maturation of cilia generated flow in the newborn and postnatal mouse airway. *Am. J. Physiol. Lung Cell. Mol. Physiol.* **296**, L1067-1075.

Hornef, N., Olbrich, H., Horvath, J., Zariwala, M.A., Fliegauf, M., Loges, N.T., Wildhaber, J., Noone, P.G., Kennedy, M., Antonarakis, S.E., Blouin, J.L., Bartoloni, L., *et al.* (2006). DNAH5 mutations are a common cause of primary ciliary dyskinesia with outer dynein arm defects. *Am. J. Respir. Crit. Care Med.* **174**, 120–126.

Köhler, G., and Milstein, C. (1975). Continuous cultures of fused cells secreting antibody of predefined specificity. *Nature* **256**, 495–497.

Loges, N.T., Olbrich, H., Fenske, L., Mussaffi, H., Horvath, J., Fliegauf, M., Kuhl, H., Baktai, G., Peterffy, E., Chodhari, R., Chung, E.M., Rutman, A., *et al.* (2008). DNAI2 mutations cause primary ciliary dyskinesia with defects in the outer dynein arm. *Am. J. Hum. Genet.* **83**, 547–558.

Olbrich, H., Horváth, J., Fekete, A., Loges, N.T., Storm van's, Gravesande, K., Blum, A., Hörmann, K., and Omran, H. (2006). Axonemal localization of the dynein component DNAH5 is not altered in secondary ciliary dyskinesia. *Pediatr. Res.* **59**, 418–422.

Omran, H., Kobayashi, D., Olbrich, H., Tsukahara, T., Loges, N.T., Hagiwara, H., Zhang, Q., Leblond, G., O'Toole, E., Hara, C., Mizuno, H., Kawano, H., *et al.* (2008). Ktu/PF13 is required for cytoplasmic pre-assembly of axonemal dyneins. *Nature* **456**, 611–616.

Schwabe, G.C., Hoffmann, K., Loges, N.T., Birker, D., Rossier, C., de Santi, M.M., Olbrich, H., Fliegauf, M., Failly, M., Liebers, U., Collura, M., Gaedicke, G., *et al.* (2008). Primary ciliary dyskinesia associated with normal axoneme ultrastructure is caused by DNAH11 Mutations. *Hum. Mutat.* **29**, 289–298.

CHAPTER 8

Immunoprecipitation to Examine Protein Complexes

Gregory J. Pazour

Program in Molecular Medicine, University of Massachusetts Medical School, Worcester, Massachusetts 01605

Abstract

Immunoprecipitation approaches are widely used in the field of cell biology. This method is ideally suited to probe interactions between proteins that make up highly complex organelles like eukaryotic cilia and flagella. In this chapter, I discuss general methods for carrying out and analyzing immunoprecipitation experiments and discuss the controls that should be included in all valid immunoprecipitation experiments.

I. Introduction

Immunoprecipitation is a powerful technique to probe protein–protein interactions and is one of the most widely used techniques in modern cell biology. Unfortunately, it is also

978-0-12-374973-4
DOI: 10.1016/S0091-679X(08)91008-6

one of the most misused techniques. In my opinion, a significant portion of published immunoprecipitations are not controlled well enough to make definitive conclusions. In this chapter I will discuss the general methodology for performing and analyzing an immunoprecipitation experiment. Specific protocols for immunoprecipitations of particular ciliary components can be found in Chapter 12, by DiPetrillo and Smith, volume 92; Chapter 10 by Sakato, volume 91; and Chapter 9 by Bodlt *et al.* (this volume). I will also discuss adequate controls. To illustrate proper controls, I have chosen two examples from the literature that demonstrate well controlled immunoprecipitations.

II. General Procedure

A. Solubilization of Starting Material

The first step in an immunoprecipitation experiment is the solubilization of the protein of interest from the starting material. This is a particularly important consideration when working with ciliary proteins as components of the axoneme can be difficult to solubilize (see Chapter 13 by Kelekar *et al.*, volume 92). Conditions for the disruption of the *Chlamydomonas* cilium are well established and there are good procedures available for the solubilization of many of the macromolecular complexes (see Chapter 3 by King, volume 92). These procedures often start with HMDEK buffer, which is of relatively low ionic strength, and add additional salts to cause the release of particular components from the axoneme. For example, HMDEK plus 0.6 M KCl removes outer dynein arms from the cilium. Other structures such as radial spokes are resistant to this treatment but can be released by KI or NaBr (see Chapter 13 by Kelekar *et al.*, volume 92). Other complexes may be particularly resistant to solubilization and require more extreme approaches. For example, studies on ankyrin-G-coupled transport to the outer segment of photoreceptors showed that ankyrin-G required sodium dodecyl sulfate (SDS) for solubilization. To maintain protein–protein interactions under these denaturing conditions, the complexes were cross-linked with a reversible agent prior to extraction and immunoprecipitation (Kizhatil *et al.*, 2009).

Solubilization of mammalian cells and tissues often uses a variation on the radio immunoprecipitation assay buffer (25 mM Tris-HCl pH 7.6–8.0, 150 mM NaCl, 1% NP-40 or Triton X-100, 1% sodium deoxycholate, 0.1% SDS). The ionic detergents sodium deoxycholate and sodium lauryl sulfate are often omitted as they can be especially disruptive to protein–protein interactions. Phosphate-buffered saline (PBS)-based extraction buffers are also common. Typically, nonionic detergents such as NP-40 or Triton X-100 are added to solubilize membranes and reduce nonspecific binding. Zwitterionic detergents such as CHAPS and CHAPSO (Pierce: Rockford, IL USA) are reported to be effective at reducing nonspecific interactions while preserving legitimate complexes. Protease inhibitors can be added if degradation is an issue.

After treatment of the samples with extraction buffer, insoluble material should be removed by centrifugation and the efficiency of the solubilization determined by comparing relative amounts of target protein in the supernatant and pellet by western blotting.

B. Preclearing

A common source of background in immunoprecipitation experiments is the nonspecific binding of proteins to the affinity matrix. Often this can be significantly reduced by incubating the extract with the affinity matrix prior to the addition of antibody. Many immunoprecipitations use an IgG-binding protein (protein A, protein G, or anti-IgG immunoglobulins) conjugated to agarose beads. In this case, the preclearing step may be done with the anti-IgG-conjugated beads or with unconjugated beads. The latter is less expensive and is often equally effective. To preclear, incubate the extract with the affinity matrix for 0.5–2 h and then remove the matrix and discard it along with any nonspecifically bound proteins.

C. Formation of Antigen–Antibody Complexes

After preclearing the sample, the primary antibody is added to the unbound material and incubated to allow binding of the primary antibody to its antigen. The amount of antibody required ranges from a few micrograms up to 100 μg or more. This should be determined empirically by finding the smallest amount necessary to pull down all the antigen from the extract. Adding excess antibody may contribute to nonspecific background, can interfere with SDS-PAGE and western blotting and is expensive, so it is best to use the smallest amount necessary. The time required for binding is often 1–2 h but can be as short as 30 min to as long as overnight. Binding is typically accomplished at 4° C but can be done at room temperature if proteolysis is not a problem. During the initial binding, the anti-IgG affinity matrix should be washed to remove any impurities and equilibrated with the same buffer as the immunoprecipitation. The washed matrix is then added to the antibody plus extract mixture and incubated for an additional 0.5–2 h, usually while being inverted on a rotator. Finally, the affinity matrix with captured antibody–antigen complexes is washed and the bound proteins eluted for analysis.

A variety of different elution buffers can be used. SDS-containing denaturing solutions are often used as the eluate is ready to be loaded onto a gel for analysis. SDS-containing solutions are extremely effective at removing the protein bound to the resin but may increase background as SDS will remove proteins specifically bound to IgG and those bound nonspecifically to the resin. Low pH (0.2 M glycine, 1 mM EDTA, pH 2.16) may also be used to elute the complexes from the resin. In certain instances, the peptide that was used to generate the primary antibody can be used to compete the complex off the antibody (Zheng *et al.*, 1995). If the peptide elution approach is applicable, it is often cleaner than SDS or low pH elution as it disrupts specific antibody/protein complexes without affecting nonspecific interactions. Furthermore, peptide elution is preferred as IgG remains bound to the matrix and will not interfere with later protein analysis.

D. Analysis of the Immunoprecipitated Proteins

Eluted proteins are now ready for analysis. The most common approach is to probe for coimmunoprecipitating proteins by western blotting. This is a highly sensitive way to test hypotheses that two proteins are present in the same complex. A typical problem with this approach is that the primary antibodies used for the immunoprecipitation are detected by the secondary antibody used for the western blot analysis and this can obscure signals from proteins of about the same weight as the IgG subunits. This can be avoided if the antibodies used for western blotting are from a different species than those used for the immunoprecipitation. Another general approach is to examine precipitated proteins by stained SDS-PAGE gels. This is a powerful approach to identify novel interacting proteins because any new proteins detected may be identified by mass spectrometry. If the goal is mass spectrometry, the staining methods must be compatible with this approach. To aid the identification of proteins by mass spectrometry one may use a new fluorescent stain like Sypro Ruby (Invitrogen: Carlsbad, CA, USA) or a silver stain method that does not include glutaraldehyde treatment.

III. Controls for Immunoprecipitation Assays

Poor quality immunoprecipitations can show interactions between any two proteins in the cell. This of course is not useful, and the challenge is to carry out these experiments in a way that detects legitimate interactions without showing spurious ones. Determining whether a protein coimmunoprecipitates because it is a *bona fide* interactor or due to illegitimate interactions is the biggest challenge to this approach. The first level validation is to show that the coimmunoprecipitating protein is not brought down in control immunoprecipitations. The minimal controls required for valid immunoprecipitations must show: (1) an interaction between the specific primary antibody and its antigen is required for the coimmunoprecipitating protein to be brought down and (2) that the immunoprecipitate is not just a dilute version of the starting material (e.g., most proteins from the starting material are not found in the precipitate).

A common way to show that the coimmunoprecipitating protein requires an interaction between the specific primary antibody and its antigen is to replace the specific primary antibody with a control antibody. To be legitimate, the control antibody should be as similar as possible to the experimental antibody. For example, if you are using immune serum as the antibody source, a similar volume of preimmune serum is a good control. If you are using a monoclonal as the primary antibody, the use of a similar monoclonal antibody raised against a different protein is a good control. If you are using an affinity purified antibody, then the use of similarly purified antibody directed to a different protein is a good control. Purified IgG purchased from a commercial vendor (Sigma, St Louis, MO, USA, etc.) should not be used as the source of control antibody because this material is produced by methods that are different from the way that most antibodies are handled. It is not clear that it is an equivalent reagent and calls into question the legitimacy of immunoprecipitations, where this antibody was used as the control. Furthermore, this antibody is often added to the control samples in great excess to the amount of IgG in the experimental samples and so may be serving as additional blocking agent to reduce background in the control samples. Another suitable control is to carry out the immunoprecipitations on material derived from cells lacking the target protein. In many instances, this may be accomplished by preparing extracts from cells that are either mutated for the gene of interest or have an RNAi-induced reduction in the gene product. The failure to bring down the coimmunoprecipitating protein by the experimental IgG when its antigen is missing is compelling proof of a genuine interaction.

A common source of artifact in immunoprecipitations is caused by the final eluate being just a dilute version of the starting material. Western blotting is an extremely sensitive technique; therefore, the presence of even a small amount of the starting material carried through can appear like a real interaction. This can be caused by nonspecific binding to the matrix or by a failure to fully wash the unbound proteins away from the matrix. To test for this artifact, the precipitate should be examined for the absence of control proteins that were in the starting material but would not be expected to be in the immunoprecipitation. In addition, by keeping track of the volume of the starting material and eluate, one can determine the relative amounts. Under most circumstances, the coprecipitating proteins should be enriched in the precipitate as compared to the starting material and this is good evidence that the interaction is not an artifact. However, there are circumstances where this may not be true. For example, if the coimmunoprecipitating protein is part of multiple complexes and you are only examining one of the complexes. Analyzing the results on stained gels is another excellent way to detect this artifact. If the coimmunoprecipitation is legitimate, interacting proteins will often be much more abundant than the majority of the proteins on the gel, whereas nonspecific proteins will not enriched above the background on the gel. However, this may not be a valid control if the proteins of interest are low abundance.

As with any biological technique, confidence in the results is gained by seeing the interactions in multiple ways. For example, carrying out reverse immunoprecipitations where an antibody against the interacting protein is used to test if it coimmunoprecipitates the original protein. In addition, other approaches like sucrose gradient centrifugation or size-exclusion chromatography can be used to explore if the proteins cosediment or coelute. These are helpful complementary approaches to immunoprecipitation as they provide information on the size of the complex in which the protein of interest is located. In addition, these approaches may offer information regarding the stability of the complex in various buffers, which can influence the conditions of the immunoprecipitation.

A. Examples of Well-Controlled Immunoprecipitations

A large percentage of published immunoprecipitations are at best inconclusive, and at worst misleading. However, instead of highlighting bad examples, I have chosen two examples from the literature that are well controlled and highlight the power of this technology.

In the first example (Fig. 1), Wargo *et al.* (2005) used immunoprecipitation to characterize the PF6-containing complex from the central pair of *Chlamydomonas*

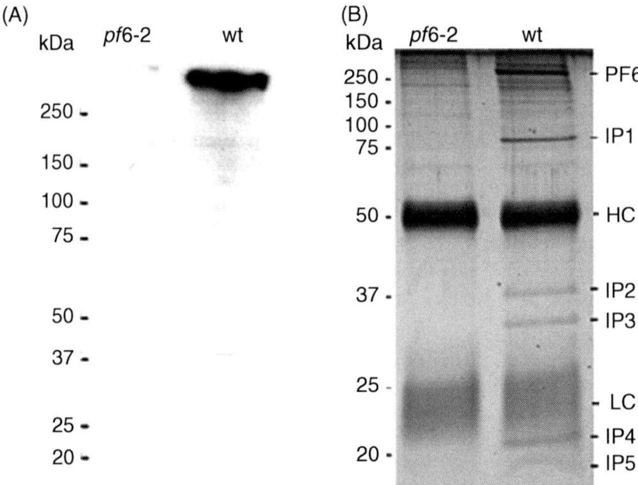

Fig. 1 (A, B) Western blots of axonemes and silver-stained gel of immunoprecipitates. (A) Western blot of axonemes isolated from wild-type (wt) and *pf6-2 Chlamydomonas reinhardtii* cells probed with anti-PF6 antibodies. A band with a molecular weight consistent with that of PF6 is present in wt axonemes and absent in *pf6-2* axonemes. (B) Silver-stained gel of immunoprecipitates obtained from wt and *pf6-2* axonemal extracts using anti-PF6 antibodies. In wt axonemal extracts, five polypeptides in addition to PF6 are specifically precipitated using anti-PF6 antibodies. These polypeptides are not precipitated in *pf6-2* axonemal extracts. The precipitating immunoglobulin heavy and light chains are labeled HC and LC, respectively. This was originally published as Fig. 1 in Wargo *et al.* (2005). Used with permission of the authors and *The Journal of Cell Science.*

flagella. To do this, they raised an antibody to PF6 that recognized a single band on western blots of axonemes purified from wild-type cells but did not detect anything in similar extracts purified from *pf6* mutant axonemes. They then used this antibody to immunoprecipitate complexes from axonemal extracts made from either wild type or *pf6* mutant axonemes. Analysis by silver stain showed that PF6 and five additional proteins were immunoprecipitated from the axonemal extracts derived from wild-type cells but not from similar extracts made from *pf6* mutant cells. While this is a relatively simple immunoprecipitation, it is well controlled. The fact that these five proteins did not immunoprecipitate from the mutant extracts that lacked the PF6 protein indicates that the interaction requires PF6. The fact that these proteins stood out above the others on a stained gel, indicates that they are not just background proteins. In the original (Wargo *et al.*, 2005) and in a subsequent manuscript (Dymek and Smith, 2007), this group further verified the specificity of the interaction between PF6 and the novel complex by a variety of means including reverse immunoprecipitations, sucrose gradients, and genetic studies.

In the second example (Fig. 2), Kizhatil and colleagues examined the role of ankyrin-G in the trafficking of cyclic nucleotide-gated channels into photoreceptor outer segments (Kizhatil *et al.*, 2009). In this example, they isolated rod outer segments

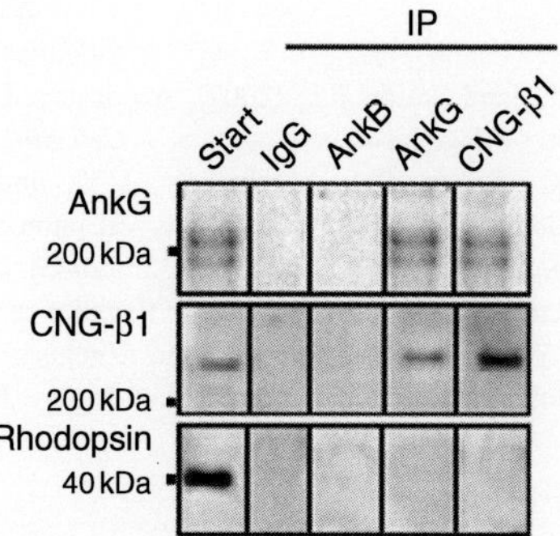

Fig. 2 Coimmunoprecipitation of ankyrin-G with CNG-β1 channels from bovine rod outer segment extracts. Isolated bovine rod outer segments were treated with the cleavable cross-linker DTSSP, solubilized with a sodium dodecyl sulfate-containing buffer and subjected to immunoprecipitation. Antibodies used for precipitations are indicated on the top (immunoglobulin G, IgG), antibodies used for protein detection are indicated on the left. This was originally published as Fig. 2 in Kizhatil *et al.* (2009). Used with permission of the authors and *Science*.

from bovine eyes and cross-linked complexes with a reversible chemical cross-linker before solubilization. The cross-linker was required because ankyrin-G was poorly solubilized with nonionic detergents and required harsh conditions for extraction. Following extraction, immunoprecipitations were carried out with antibodies to both ankyrin-G and a subunit of the cyclic nucleotide-gated channel. In each case, the antibodies were able to precipitate both their antigen and their binding partner. The ankyrin-G immunoprecipitation was well controlled by the use of a similar antibody directed against ankyrin-B, which did not precipitate any of the cyclic nucleotide-gated channel. The authors also used a “nonspecific rabbit immunoglobulin” as an additional control but since there are no details on what this is, its value as a control is limited. To control for nonspecific carry through of starting material into the final elute, the authors probed the immunoprecipitations with antibodies to rhodopsin. This protein is very abundant in their preparations and the fact that none was detected in the final elute suggests that the immunoprecipitation was very clean.

IV. Summary

Immunoprecipitation is a powerful technique for the detection of protein–protein interactions. The addition of proper controls to these experiments will strengthen the value of these approaches to scientific discovery.

Acknowledgments

I thank Jovenal SanAgustin, Brian Keady, John Follit, and Stephen King for critically reading this manuscript. My laboratory is supported by grant GM060992 from the National Institutes of Health.

References

Dymek, E.E., and Smith, E.F. (2007). A conserved CaM- and radial spoke associated complex mediates regulation of flagellar dynein activity. *J. Cell Biol.* **179**, 515–526.

Kizhatil, K., Baker, S.A., Arshavsky, V.Y., and Bennett, V. (2009). Ankyrin-G promotes cyclic nucleotide-gated channel transport to rod photoreceptor sensory cilia. *Science* **323**, 1614–1617.

Wargo, M.J., Dymek, E.E., and Smith, E.F. (2005). Calmodulin and PF6 are components of a complex that localizes to the C1 microtubule of the flagellar central apparatus. *J Cell Sci.* **118**, 4655–4665.

Zheng, Y., Wong, M.L., Alberts, B., and Mitchison, T. (1995). Nucleation of microtubule assembly by a gamma-tubulin-containing ring complex. *Nature* **378**, 578–583.

CHAPTER 9

Tandem Affinity Purification of Ciliopathy-Associated Protein Complexes

Karsten Boldt[*], Jeroen van Reeuwijk[†,‡], Christian Johannes Gloeckner[*], Marius Ueffing[*,§], *and* Ronald Roepman[†,‡]

[*]Department of Protein Science, Helmholtz Zentrum München, 85764 Neuherberg, Germany

[†]Department of Human Genetics, Radboud University Nijmegen Medical Centre, 6500 HB Nijmegen, The Netherlands

[‡]Nijmegen Centre for Molecular Life Sciences, Radboud University Nijmegen Medical Centre, 6500 HB Nijmegen, The Netherlands

[§]Institute of Human Genetics, Klinikum rechts der Isar, Technical University of Munich, 81675 Munich, Germany

METHODS IN CELL BIOLOGY, VOL. 91

978-0-12-374973-4
DOI: 10.1016/S0091-679X(08)91009-8

Abstract

Ciliary dysfunction has recently been recognized as a cause for a growing number of genetically inherited disorders termed ciliopathies. Ciliopathy-associated proteins are organized in cell/context-specific complexes and in shared regulatory circuits in cilia of affected tissues. Thus, the identification of protein interactions involved in ciliary function provides a valid starting point to molecularly dissect normal ciliary function in a context and tissue specific fashion, identify novel functional candidate genes for ciliopathies as well as uncover the molecular defects that cause ciliary disease on the cellular level. Numerous methods have been developed over the years to categorize protein–protein interactions as well as to isolate native protein complexes. This chapter presents the details of an optimized tandem affinity purification (TAP) procedure, employing a 4.6-kDa tag containing a doublet Strep-tag II and a FLAG octapeptide epitope tag. In contrast to other TAP methods, utilization of these two affinity-binding moieties eliminates the need for a proteolytic cleavage step and allows the undisturbed isolation of the native protein complex binding to the tag-fusion protein. The small size of the synthetic and hydrophilic moieties of the Strep/FLAG TAP tag greatly reduce nonspecific protein binding as well as steric hindrance. We have employed this tag successfully for the identification of the lebercilin interactome, a ciliary and ciliopathy-associated protein network. Promising developments include quantitative proteomics (stable isotope labelling with amino acids in cell culture; SILAC) and BAC (bacterial artificial chromosome) recombineering to express tagged genes in higher eukaryotes, further expanding the versatility of this procedure.

I. Introduction

A. Ciliopathies and Ciliary Proteins

Cilia are finger-like, microtubule-based projections from the cell surface that are derived from the mature centriole and perform essential motile and sensory functions. Once thought to be vestigial organelles, cilia were found in the last decade to be crucial for transduction of extracellular signals and regulation of many biological processes. The term "ciliopathy" was coined to describe the class of human genetic diseases whose etiologies lie in defective cilia. In 1976, Bjorn Afzelius was the first scientist to associate ciliary dysfunction with a clinical phenotype. About three decades ago, he described that patients with Kartagener syndrome, which is characterized by frequent infections in the respiratory system, infertility, and *situs inversus*, had immotile and structurally abnormal cilia due to the lack of outer dynein arms (Afzelius, 1976; Badano *et al.*, 2006). Primary cilia, however, were still considered as passive, non-functional remnants. It was not until recently when defects in proteins that localize to the basal body and axoneme of cilia were causally related to human disease. In 2000, the first report of a human hereditary disorder, polycystic kidney disease (PKD) appeared demonstrating its association with primary cilia defects. The relevance struck when it became clear that the mouse model for PKD, *Tg737*^{orpk} in which *polaris* is mutated, had abnormal or absent cilia (Pazour *et al.*, 2000). Since then >30 genes that encode ciliary proteins have been found to cause a diverse set of disorders, collectively termed "ciliopathies" (Badano *et al.*, 2006). These disorders are defined by overlapping clinical criteria that include retinal degeneration; renal, liver, and pancreatic cysts; polydactyly; *situs inversus*; mental retardation; and encephalocele. Some of these proteins have subsequently been revealed to be physically or functionally associated, with limited connections to other crucial biological processes, such as Wnt signaling, Shh signaling, planar cell polarity, and cell cycle control (reviewed in Berbari *et al.*, 2009 and Gerdes *et al.*, 2009). Early proteomics studies have suggested a discrete repertoire of about 1000 proteins within the organelle (i.e., <5% of the proteome). So far, it remains open for most of these proteins how they specifically relate to each other and function within cellular pathways and networks that are still in need of organization into pathways and networks (Gherman *et al.*, 2006; www.ciliaproteome.org).

B. Ciliopathy-Associated Protein–Protein Interaction Network

The perception of proteins acting as elements in cell-specific structural and regulatory networks may explain how single gene mutations produce very complex cellular aberrations or disease phenotypes, but also why disease maybe restricted to specific organs or cells in the body. We and others provided evidence, that ciliopathy-associated proteins are organized in cell/context-specific complexes and/or in shared regulatory circuits in cilia of affected tissues (Fig. 1) (Arts *et al.*, 2007; den Hollander *et al.*, 2007; Gosens *et al.*, 2007; Gorden *et al.*, 2008; Loktev *et al.*, 2008; Nachury *et al.*, 2007; Roepman and Wolfrum, 2007; Roepman *et al.*, 2000, 2005; van Wijk *et al.*, 2006). Revealing the composition of specific modules and molecular building blocks within the putatively large ciliary protein–protein interaction network has been valuable toward discovery of novel ciliopathy genes. Yet, a large gain in analytical power is needed to acquire comprehensive as well as in-depth knowledge on the composition, wiring, dynamics, and associated signaling pathways of such functional modules and associated protein networks. The resulting data can subsequently serve as a knowledge base to enable hypothesis-based studies on the discrete defects underlying specific ciliopathies.

C. Rationale

Cilia are structurally and topologically tightly organized macromolecular protein complexes that act as specialized molecular machines. The backbones of these machines are polar polymeric tubular structures, yet their function is exerted via dynamically acting protein modules. Specificity within these complexes is conferred using combinatory principles combined with specific signaling inputs and outputs. Dynamics of multifunctional protein complexes are originated by signal-dependent adaptations that are materialized as structural alterations,

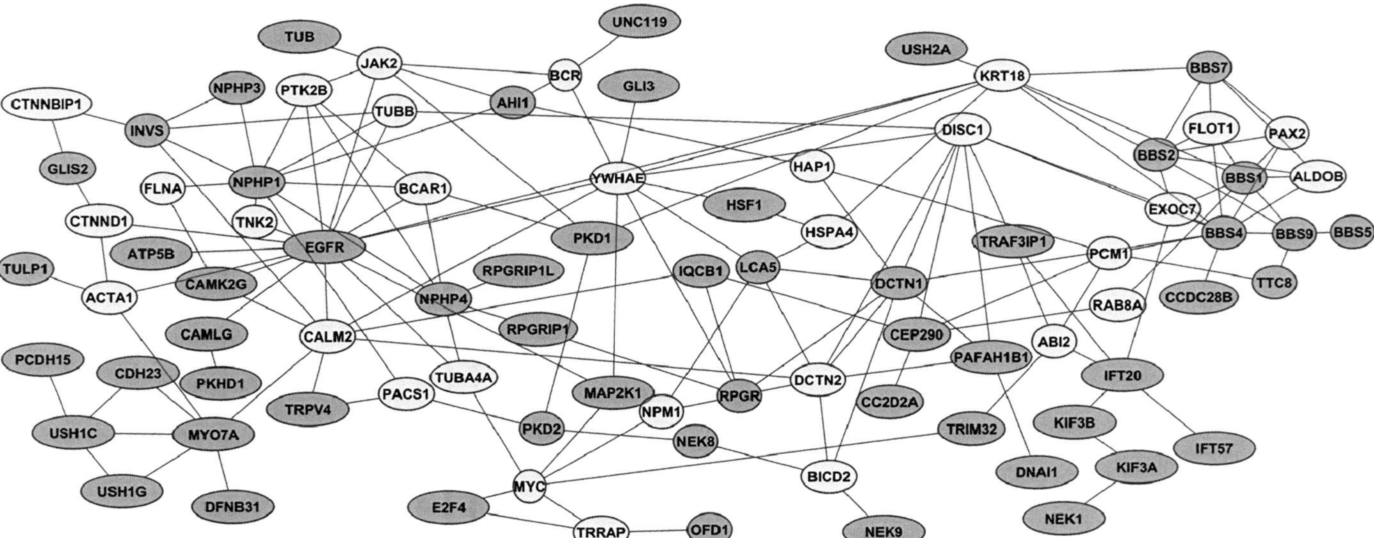

Fig. 1 Protein–protein interaction network of ciliopathy-associated proteins. The network illustrates the high degree of connectivity of proteins known to be associated with human ciliopathies or with developmental or physiological defects in animal model systems (indicated by gray spheres) (Marshall and Nonaka, 2006; Quinlan *et al.*, 2008) (www.ciliaproteome.org). Proteins that interact with at least two known ciliopathy proteins are indicated by white spheres. Protein–protein interactions were literature derived or derived from public protein interaction databases such as BioGRID (www.thebiogrid.org), HPRD (www.hprd.org), BIND (www.bind.ca), and *homo*MINT (http://mint.bio.uniroma2.it/HomoMINT). The network was visualized using Cytoscape (www.cytoscape.org).

posttranslational modifications, proteolytic cleavages, protein translocation, and interexchange in binding partners. Dynamic changes in a given protein complex are connected to protein regulatory networks and result in systemic and integrated quantitative signal outputs.

Several methods are available to define the composition of protein complexes. This methodology includes biochemical methods such as affinity binder-based protein complex purification as well as tagged affinity purification, cosegregation, and fractionation and surface plasmon resonance, just to mention a few. Established methodologies also include cell-based analysis, such as yeast two hybrid (Cagney and Uetz, 2001), split ubiquitin (Lee and Lee, 2004), FRET/BRET (fluorescence/bioluminescence resonance energy transfer) (Pfleger and Eidne, 2006), and FRAP (fluorescence recovery after photobleaching) (Reits and Neefjes, 2001). These methodologies primarily allow identifying the nature of an interaction (direct/indirect), its topology and/or, in case of live-cell imaging, its time dependency. Although the analysis of the functional relevant dynamics within macromolecular protein complexes remains to be a major challenge, tandem affinity purification (TAP) has developed into a powerful and (cost-) effective tool to rapidly categorize the protein–protein interactions of a predefined protein of interest under native conditions.

D. Strep-Flag Tandem Affinity Purification

Numerous methods have been developed over the years to categorize protein–protein interactions as well as isolate protein complexes. Affinity purification has been developed as a prime method of choice to isolate specific proteins or protein complexes (Berggard *et al.*, 2007; Collins and Choudhary, 2008; Vermeulen *et al.*, 2008). Techniques like GST (glutathione S-transferase) pull-down purification and immunoprecipitation have, however, long been hampered by low signal-to-noise ratios due to a high background of nonspecific protein binding and limited detection sensitivity or unsatisfactory yields. Recent improvements in mass spectrometry (MS) technology allow highly efficient detection of proteins from complex mixtures (Hanke *et al.*, 2008). MS in combination with isotope labeling as well as label-free quantitative methods has greatly enhanced the analytical depth as well as the ability to discriminate true from false-positive interactions. Continuous development of tandem affinity tags and complementing affinity matrices are equally important as they can efficiently reduce the background caused by nonspecific binding of proteins and at the same time ensure affinity capture of native protein complexes with a much higher yield than achived by immunoprecipitation.

Originally, TAP was developed to analyze protein interactions in yeast (Gavin *et al.*, 2002; Rigaut *et al.*, 1999). The original TAP tag has a molecular weight of 21 kDa and is composed of a Protein-A tag and a calmodulin-binding peptide (CBP) tag, separated by a TEV (tobacco etch virus) protease cleavage site. As recombinant DNA fragments can easily be inserted into the yeast genome by site-specific homologous recombination, the yeast genes were tagged with a dual tag and native protein complexes were isolated from the yeast cells by TAP, separated by SDS-PAGE, and analyzed by MS. The original TAP tag excellently suited this purpose.

Although the original TAP method has proven to be suitable for the purification of some mammalian protein complexes as well (Bouwmeester *et al.*, 2004), the development of various modified versions for this purpose show the need for alternatives (Collins and Choudhary, 2008). TEV protease cleavage is expensive, time-consuming, and in many cases incomplete. As time negatively affects any reversible biochemical interaction, any purification procedure of native protein complexes should be done in a minimum of time. Additionally, the CBP tag and the elution by the calcium-chelating agent EDTA interfere with calcium-dependent protein complex assembly.

To address these concerns, we have developed the Strep/FLAG tandem affinity purification (SF-TAP) tag (Gloeckner *et al.*, 2007), which combines a tandem Strep-tag II (Junttila *et al.*, 2005; Skerra and Schmidt, 2000) and a FLAG tag resulting in a very small (4.6 kDa) tag. An overview of the tag sequence is shown in Fig. 2A. Given its design, it omits the aforementioned hindrances, without compromising its efficiency. The first step uses desthiobiotin for elution of the SF-TAP fusion protein from the Strep-Tactin matrix. Subsequently, the FLAG octapeptide is used in the second step for elution of the SF-TAP fusion protein from the anti-FLAG M2 affinity matrix. Thereby, elution for both steps is possible without the necessity of time-consuming proteolytic cleavage. Thus, the optimized SF-TAP protocol allows efficient and fast purification of protein complexes from mammalian cells within 2–3 h. The result of a TAP experiment is shown in Fig. 2B, and a flow chart of the SF-TAP procedure is shown in Fig. 2C.

We have previously employed this approach to identify the interactome of the ciliary and ciliopathy-associated protein lebercilin. This showed its efficacy in dissecting the architecture of the ciliopathy-associated protein networks by unveiling cellular signaling as well as axonemal transport-associated protein complex members (den Hollander *et al.*, 2007).

In this chapter, we describe the detailed workflow starting with the cell culture work needed for recombinant expression of the SF-TAP fusion proteins, followed by the SF-TAP protocol and ending with preparation of the samples for mass spectrometric analysis.

Since the sample preparation for MS is a crucial step in the whole process, a special focus has been laid on this part. For the identification of associated proteins following SF-TAP, the volume of the eluates is reduced by ultrafiltration using centrifugal units with a low-molecular-weight cutoff or by chloroform/methanol precipitation. The samples are then directly subjected to proteolytic cleavage for analysis on a nano liquid chromatography (LC)-coupled electrospray ionization (ESI) tandem mass spectrometer. For complex samples, which contain many proteins, an alternative protocol for SDS-PAGE prefractionation, including a method for sensitive MS-compatible Coomassie protein staining followed by in-gel proteolytic cleavage is provided. By reducing sample complexity, prefractionation helps to increase the number of protein identifications on recommended state-of-the-art LC-coupled tandem mass spectrometers.

E. General Considerations

Some important points should be considered with the TAP of ciliary proteins, specifically regarding the cloning of the expression constructs and the choice of the

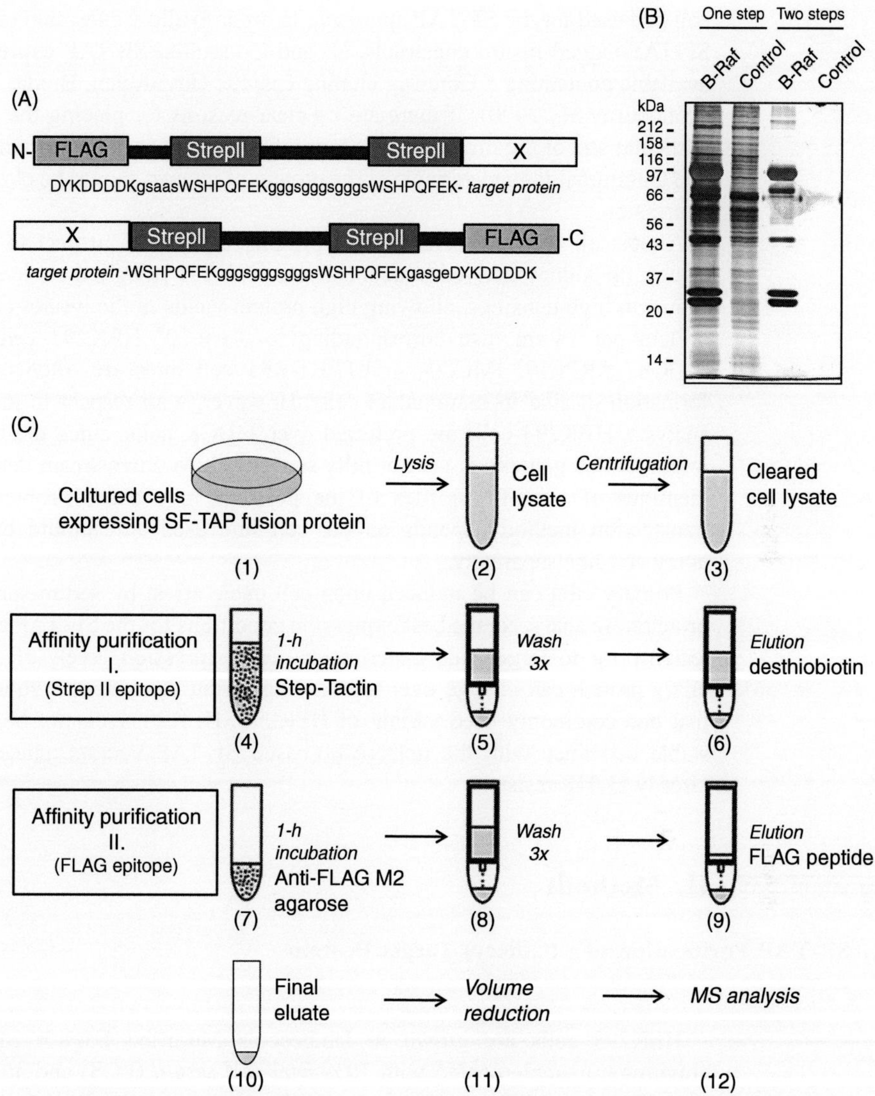

Fig. 2 The SF-TAP procedure. (A) Diagram of the N-terminal (top) and C-terminal (bottom) SF-TAP tag. (B) Comparison of a one-step purification (Strep-tag II purification: lanes 1 and 2) with a two-step purification (Strep-tag II/FLAG tag purification: lanes 2 and 3). HEK293 cells served as negative control to determine unspecific background (lanes 2 and 4). Each purification was performed with 26108 HEK293 cells stably expressing SF-tagged B-Raf. Half the amount of each eluate was separated on a 10% SDS-PAGE gel. Proteins were stained with silver. Copyright Wiley-VCH Verlag GmbH & Co. KGaA. Reproduced with permission from Gloeckner *et al.* (2007). (C) Flow chart of the SF-TAP procedure.

cell line used for the SF-TAP approach. In order to allow a fast and efficient cloning of SF-TAP-tagged fusion constructs, N- and C-terminal SF-TAP expression vectors are available containing a Gateway cloning cassette (Invitrogen, Breda, The Netherlands) (Walhout *et al.*, 2000). If there are no clear reasons for placing the SF-TAP tag on a particular site of the protein (i.e., a terminal targeting or modification signal), both, N- and C-terminal-tagged variants of the protein of interest should be cloned and tested for expression.

Almost all cells are able to form primary cilia. With respect to transfection efficiency, the kidney-derived human cell line HEK293 is a good choice. HEK293 can be grown to high densities, allowing high protein yields in the lysates (10–15 mg of total protein per 14-cm dish corresponding to $\sim 1 \times 10^8$ HEK293 cells). Alternatively, MDCK, ARPE19, IMCD3, or hTERT-RP1 cell lines are often used to study the formation of cilia in mammalian cells. However, with respect to mass spectrometric analysis, HEK293 cells are preferred over MDCK cells, since the latter is of canine origin, which genome is yet not fully sequenced, so downstream determination of the identities of the MS profiles of the peptides may cause problems. The optimal transfection method depends on the cell line used and should be established for every cell line separately.

Primary cilia can be induced upon cell cycle arrest by serum starvation. If ciliary proteins are analyzed, the best expression conditions for the SF-TAP analysis should be tested prior to large-scale experiments. The expression levels of an overexpressed cilary protein can change over time during serum starvation. It should also be noted that one commonly used variant of HEK293, HEK293T cannot be used to generate stable cell lines with the pcDNA3.0-based SF-TAP vectors, since this cell line is already G-418-resistant.

II. Methods

A. SF-TAP Purification of a (Ciliary) Target Protein

1. Cell Culture

HEK293 cells are grown in Dulbecco's Modified Eagle's Medium (DMEM) (Invitrogen) supplemented with 10% fetal calf serum (FCS) and appropriate antibiotics. Growth conditions for many cell lines are given in the American type culture collection (ATCC) human and animal cell lines database (http://www.atcc.org/).

2. SF-TAP Vectors

The SF-TAP tag is available as N- or C-terminal tag. The mammalian SF-TAP expression vectors are based on pcDNA3.0 (Invitrogen). In order to achieve rapid cloning of different SF-TAP fusion proteins, Gateway cloning cassettes (Invitrogen) have been inserted into the vectors allowing the generation of expression vectors by site-directed recombination. Detailed protocols are supplied by the manufacturer (Invitrogen).

3. Negative Control

Although SF-TAP purifications typically exhibit low background caused by non-specific binding of proteins to the affinity matrix, a suitable negative control should be used in every experiment. Cells transfected with the empty expression vectors may be used in the same amount as for the SF-TAP-tagged bait protein. However, the tag is quite small and only expressed at low levels if not fused to a protein. Thus, the untransfected cell line or an unrelated, SF-TAP-tagged protein are acceptable alternatives for a negative control.

4. Materials (for Transfection, Cell Lysis, and SF-TAP Purification)

Transfection reagent of choice
Cell scraper
Millex GP 0.22-μm syringe-driven filter units (Millipore, Amsterdam, The Netherlands)
Microspin columns (GE-Healthcare, Diegem, Belgium)
Microcon YM-3 centrifugal filter devices (Millipore)
TBS buffer: 30 mM Tris-HCl, pH 7.4, 150 mM NaCl
Lysis buffer: TBS buffer supplemented with Protease Inhibitor Cocktail (Roche, Almere, The Netherlands), Phosphatase Inhibitor Cocktail I and II (Sigma, Zwijndrecht, The Netherlands) and 0.5% NP40 (Roche).

Note: Phosphatase inhibitors are necessary if phosphorylation is to be studied or if phosphorylation is important for protein complex formation. They can be left out if the analysis of protein phosphorylation is not of interest.

Wash buffer: TBS buffer supplemented with Phosphatase Inhibitor Cocktail I and II and 0.1% NP40.
Desthiobiotin elution buffer: dilute $10\times$ buffer E (IBA; Westburg, Leusden, The Netherlands) 1 in 10 with water (final concentration: 2 mM desthiobiotin).
Strep-Tactin Superflow (IBA)
Anti-FLAG M2 agarose (Sigma-Aldrich, Zwijndrecht, The Netherlands)
FLAG peptide stock solution: dissolve 1 mg of FLAG peptide (Sigma-Aldrich) in 800 μl TBS buffer (store stock solution at −80°C).
FLAG elution buffer: dilute FLAG peptide stock solution 1 in 25 in TBS buffer (final concentration: 200 μg/ml FLAG peptide).

Procedures

Tissue culturing

1. Seed HEK293 cells on 14-cm plates (density: $\sim 1\text{–}2\times 10^7$ cells per dish).
   Note: The amount of cells used for SF-TAP purification can be varied depending on the expression levels of the bait protein. Usually, four 14-cm dishes, corresponding to a final amount of $\sim 4\times 10^8$ HEK293 cells, provide a good starting point.
2. Grow cells overnight.

3. Transfect cells with the SF-TAP/target protein expressing plasmids using a transfection reagent of choice (according to manufacturer's protocols).

   Note: HEK293 cells can be easily transfected with lipophilic transfection reagents. The transfection efficiency is usually >80%. For a typical SF-TAP experiment, 1–4 μg plasmid per 14-cm dish is used. Depending on the cell type, other transfection reagents may be favorable.
4. Let cells grow for 48 h.

   Note: To induce cilia, cells can be starved in DMEM without FCS for 12 h prior to harvesting. Comparison of the identified protein complex with or without serum starvation could provide information about the requirement of cilia for protein complex recruitment.

Cell lysis

5. Remove medium from the plates.
6. Optional: rinse cells in warm phosphate-buffered saline (PBS).

   Note: Rinsing the cells with PBS is necessary if the cells were not serum starved to remove the serum and enable determination of protein concentrations.
7. Scrape of cells in 1 ml lysis buffer per 14-cm plate on ice using a cell scraper and combine lysates of each condition.
8. Lyse cells for 15 min on ice, mix the lysates during incubation.
9. Pellet cell debris including nuclei by centrifuging 10 min, 10,000 × *g*, 4°C.
10. Clear lysate supernatant by filtration through 0.22-μm syringe filters.

SF-TAP purification

11. Prepare Strep-Tactin Superflow resin: wash resin twice with TBS and once with lysis buffer.
12. Incubate lysates with 50-μl/plate Strep-Tactin Superflow resin for 1 h at 4°C (use a tumbler to keep the resin evenly distributed).

    Note: A maximum of 200 μl settled resin per spin column should not be exceeded. If more than four 14-cm dishes (~4 × 10^8 HEK293 cells) are used, reduce the volume per plate or use additional spin columns.
13. Centrifuge for 30 s at 7000 × *g*, remove most of the supernatant and transfer resin to microspin columns.

    Note: Snap of bottom closure of the spin columns prior to use. The maximum volume of the spin columns is 650 μl. The maximum amount of settled resin should not exceed 200 μl. Using higher amounts of resin would increase the background and lower the efficiency of elution. Thus, the spin columns are suitable for small and medium scale purifications. If larger scales are needed, 10-ml gravity flow columns (Bio-Rad or similar) can be used instead.
14. Remove remaining supernatant by centrifugation (5 s at 100 × *g*), wash 3 × with 500 μl wash buffer (centrifuge 5 s at 100 × *g* each time to remove the supernatant).

    Note: Replug spin columns with inverted bottom closure prior to adding the elution buffer.

Note: Avoid allowing the resin to run dry. Depending on the bait protein, this markedly reduces the yield!

15. Add 500 µl desthiobiotin elution buffer and gently mix the resin for 10 min at 4°C.
16. Remove the plug of the spin column, transfer the column to a new collection tube, and harvest the eluate by centrifugation (10 s, 2000 × *g*).

    Note: If spin columns were closed by the top screw cap during incubation with elution buffer, they need to be removed prior to centrifugation. Spin columns must be left open (without screw cap) during centrifugation to allow pressure balance.
17. Prepare FLAG M2 agarose resin: wash resin 3 × in TBS buffer (25-µl resin per plate are needed).
18. Transfer eluate to 25 µl per 14-cm plate anti FLAG M2 agarose in microspin columns.
19. Incubate for 1 h at 4°C (on an end-over-end tumbler).
20. Wash once with 500 µl wash buffer and twice with 500 µl TBS buffer (centrifuge 5 s at 100 × *g* each time to remove the supernatant).
21. For elution, incubate with 4 × bead volume (at least 200 µl) FLAG elution buffer for 10 min, gently mix the resin several times.

    Note: To ensure efficient elution of SF-TAP proteins from the anti-FLAG M2 resin, the volume of FLAG elution buffer should be at least four fold the volume of the resin. The samples should be frequently mixed during elution. A second elution step can be used to increase elution efficiency.
22. After incubation, remove the plug of the spin column, transfer it to a new collection tube and harvest the eluate by centrifugation (10 s at 2000 × *g*).
23. Take 10–20 µl of the eluate for an SDS-PAGE analysis in order to determine the yield prior to the mass spectrometric analysis.

    Note: SF-TAP proteins can be detected using the anti-FLAG M2 antibody (Sigma-Aldrich); dilution: 1:1000 to 1:5000 in 5% nonfat milk powder in TBS buffer, 0.1% Tween 20).

    Note: In principle, the purification steps can be done in any order. However, if the eluates are directly subjected to LC-MS/MS analysis, the Strep-tag/Strep-Tactin system should be used first and the FLAG-tag/anti-FLAG M2 affinity resin purification performed second. The desthiobiotin used for elution of Strep-tagged proteins binds to the C18 matrix with high affinity, outcompeting the peptides. High amounts of biotin or desthiobiotin bind almost irreversibly to the C18 matrix under the conditions used for reversed phase chromatography.

B. Sample Preparation for Mass Spectrometry

The direct mass spectrometric analysis of the SF-TAP eluate is a straightforward analysis strategy. For this purpose, the eluates need to be concentrated, preferentially by protein precipitation. The pellets can be directly subjected to tryptic proteolysis prior to LC–MS/MS analysis. A surfactant (RapiGest) is used to increase the solubility of the precipitated proteins (Yu *et al.*, 2003). Depending on the complexitiy of the eluates and the speed of the mass spectrometer, a preseparation of the samples by SDS-PAGE combined with tryptic in-gel proteolyis might help to increase the number of identified proteins. Alternatively, the number of identified proteins can be increased by applying a two-dimensional LC separation of the peptides directly coupled to the ESI mass spectrometer like the multidimensional protein identification technology (MudPIT) method (Wolters *et al.*, 2001).

1. Chloroform Methanol Precipitation According to Wessel and Flügge (1984)

Materials

SF-TAP eluate (from A. SF-TAP purification, step 22)
Chloroform (AR grade)
Methanol (AR grade)
2-ml polypropylene sample tubes
Deionized water

Procedure

1. Transfer 200 μl SF-TAP eluate to a 2-ml sample tube.
   Note: All following steps are done at ambient temperature.
2. Add 0.8 ml of methanol, vortex, and centrifuge for 20 s.
   Note: Use 9000 × *g* for all centrifugation steps.
3. Add 0.2 ml chloroform, vortex, and centrifuge for 20 s.
4. Add 0.6 ml of water, vortex for 5 s, and centrifuge for 1 min.
5. Carefully remove and discard most of the upper layer (aqueous phase).
   Note: The protein precipitate (visible as white flakes) is in the interphase. Do not remove the complete upper phase because this would disturb the protein precipitate.
6. Add 0.6 ml of methanol, vortex, and centrifuge for 2 min at 16,000 × *g*.
7. Carefully remove the supernatant and air-dry the pellet.

2. In-Solution Digest

The in-solution digest is a quick an efficient method to digest the whole SF-TAP eluate after protein precipitation. The usage of a MS-compatible surfactant helps to solubilize the precipitated proteins. In order to allow the identification of cysteine-containing peptides, random oxidation is prevented by reduction/alkylation applying dithiothreitol (DTT)/iodoacetamide treatment prior to digestion, leading to a defined mass-adduct. The digested protein sample can be directly subjected to the analysis by a LC-coupled tandem mass spectrometer.

Materials

Protein pellet (from B.1. Chloroform/methanol precipitation, step 7)
50 mM ammonium bicarbonate (freshly prepared prior to use)
RapiGest SF (Waters, Milford, MA); prepare a 2% stock solution in deionized water (10 × stock)

Note: RapiGest (sodium 3-[(2-methyl-2-undecyl-1,3-dioxolan-4-yl)methoxyl]-1-propanesulfonate) is an acid-labile surfactant which helps to solubilize and denature proteins in order to make them accessible to proteolytic cleavage (Yu et al., 2003).
100 mM DTT solution (prepared from 500 mM stock solution)
300 mM iodoacetamide solution (prepared fresh)
50 × Trypsin stock solution (0.5 μg/μl, sequencing grade, Promega, Leiden, The Netherlands), stored at −20°C
HCl (37%)
Polypropylene inserts (Supelco (Zwijndrecht, The Netherlands), #24722)
1–200 μl gel-loader tips (Sorenson Bioscience, Salt Lake City, UT)

Procedure

1. Dissolve the protein pellet in 30 μl of 50 mM ammonium bicarbonate by extensive vortexing.
2. Add 3 μl of a RapiGest stock solution (final concentration 0.2%).
3. Add 1 μl of 100 mM DTT and vortex.
4. Incubate 10 min at 60°C.
5. Cool down the samples to room temperature.
6. Add 1 μl of 300 mM iodoacetamide and vortex.
7. Incubate for 30 min at room temperature in the dark.
   Note: Samples should be protected from light since iodoacetamide is light sensitive.
8. Add 2 μl trypsin stock solution and vortex.
9. Incubate at 37°C overnight.
10. For hydrolysis of RapiGest add 2 μl HCl (37%).
    Note: For hydrolysis of the RapiGest reagent the pH must be <2.
11. Transfer samples to polypropylene inserts (remove spring).
12. Incubate for 30 min at RT
13. Place inserts in 1.5-ml reaction tubes and centrifuge for 10 min (16,100 × *g*, RT).
    Note: One hydrolysis product of the RapiGest reagent is water immiscible and can be removed by centrifugation. After centrifugation, it is visible as faint film (oleic phase) on top of the aqueous sample phase. The other hydrolysis product is an ionic water-soluble component which does not interfere with reversed-phase LC or MS analysis.
14. Carefully recover the solution between the upper oleic phase and the pellet using gel-loader tips.
15. The sample can be directly subjected to C18 HPLC separation prior to MS/MS analysis (LC-MS/MS).

III. Data Evaluation

A. Representation and Architecture of Ciliary Protein Complexes

Excellent public protein interaction databases such as *BioGRID* (www.thebiogrid.org), *HPRD* (www.hprd.org), *BIND* (www.bind.ca), and *homo*MINT (http://mint.bio.

uniroma2.it/HomoMINT) are available for analyzing protein–protein interaction data that can be used to complement data sets derived from SF-TAP approaches. In addition, a queryable catalog of the published ciliary and basal body proteins and ciliopathy animal models is maintained at www.ciliaproteome.org (Gherman *et al.*, 2006), providing an initial cross-reference option for ciliary involvement. As the SF-TAP approach generates data from entire protein complexes rather than just binary interactions, preexisting protein–protein interaction data from such databases provide valuable additions. For example, binary protein–protein interaction data can be used to specify the architecture of the identified protein complex, and annotated links to signaling pathways, such as Hedgehog signaling or Wnt signaling, but also planar cell polarity or cell cycle control, may provide clues for the functional ciliary modules in which the target proteins participate. This strategy can be also used to map functional networks of a disease-associated protein and potentially reveal new candidate genes/proteins for the disease.

Several tools have been recently developed for presentation of protein networks. One of them is Cytoscape (www.cytoscape.org) which is available as open source software and for which many network analysis plug-ins are available (Cline *et al.*, 2007).

B. Nonspecific Interactions

If the expression level of the recombinant bait proteins is very high, which is often the case under control of potent promoters like the commonly used cytomegalovirus (CMV) promoter, especially when transient expression conditions are used with high amounts of plasmid DNA, the purified protein complexes may contain high amounts of heat shock proteins such as HSP70. A decrease in the amount of DNA used for transfection may in part alleviate this problem, as well as the generation of cell lines stably expressing the SF-TAP/bait fusion protein. In addition, if less bait protein is expressed, the ratio between bait protein and native interaction partners is closer to native stoichiometry. This can also be achieved when using a different cell line for the experiments, placing the expression of the SF-TAP/target fusion protein under control of a different promoter, or employing inducible expression conditions.

However, some proteins are found in many TAP purifications and therefore might be associated nonspecifically, even if they apparently do not bind to the affinity matrix, which is used as a negative control. Thus, the best way to exclude these proteins is to compare TAP purifications of different, functionally unrelated bait proteins. Given the extensive protein network of cilia and basal bodies, this means a comparison with the interactome of proteins that are active at a different subcellular site.

C. Transient Interactions

TAPs in general enable the identification of rather stable protein complexes. This is mostly due to the dilution of complex components in the second purification step, shifting the equilibrium between association and dissociation of proteins toward dissociation from the complex, especially for weak and transient interactions. The procedure may therefore exclude or limit detection of transient protein interactions that nevertheless may play a very significant role in the function of the bait protein. For detection of transient interactions, single-step affinity purification methods should then be favored. The SF-TAP tag provides two options for that: the Strep-tag II/desthiobiotin combination and the FLAG/FLAG M2 agarose combination. As single-step purifications usually result in a much higher degree of false-positive binders, a different approach is required to discriminate specific complex components from the nonspecific binders. These false-positive binders can be sorted out by means of quantitative MS. Stable isotopes can be used to identify those proteins that bind to tag and affinity matrix by comparing an isolate pulled from cells expressing the tag alone versus one pulled from cells expressing a tag-fusion protein of interest. Labeling can be performed by stable isotope labeling of amino acids in cell culture, SILAC (Mann, 2006; Ong and Mann, 2006; Selbach and Mann, 2006) as well as through peptide labeling methods on peptide level such as isotope-coded protein labeling (ICPL) (Kellermann, 2008).) and isobaric tags for relative and absolute quantitation (iTRAQ) (Latterich *et al.*, 2008). This, however, further increases the need for proper computer assisted data analysis, which requires specialized software enabling the comparative analysis and interpretation of mass spectrometric data (Cox *et al.*, 2009).

IV. Outlook

An old saying in many languages is "show me your friends, and I'll tell you who you are." TAP has developed into a versatile and efficient approach to identify the interacting "friends" of many proteins in many different contexts, including the ciliary proteins involved in ciliopathies. This opened up novel ways to identify who these proteins are, by defining their connections in the protein–protein interaction network to modules with a known function, for example, in cellular signaling, polarity, cell cycle control, vesicle transport, or intraflagellar transport. Only with this knowledge in hand can we begin to unveil the complex molecular disease mechanisms that lead to the wide and expanding but still defined and largely typical set of clinical features that mark the ciliopathies as a group. To increase the versatility of the tag and its use even further, we and others are generating novel reagents that carry this tag. These reagents include lentiviral vectors to increase the transfection efficiency and allow targeting of primary cells, inducible vectors with different promoters to manipulate the expression levels of the tagged bait proteins, and BAC recombineering constructs to express tagged genes in higher eukaryotes under native expression levels (Poser *et al.*, 2008). The increased versatility is required to move from the model of cultured ciliated cells to animal models of the ciliopathies in order to pinpoint the molecular disruptions that cause the pathogenesis of these disorders. This knowledge may open new avenues in the development of therapeutic strategies.

Acknowledgments

This work was supported by the German Federal Ministry for Education and Research BMBF grant: 0316865A (QuantPro) to M.Ue, Helmholtz-Alliance HelMa to M.Ue, EU-grant: ProteomeBinders (FP6-026008) to M.Ue, and The Netherlands Organisation for Scientific Research (NWO) Vidi grant (91786396) to R.R.

References

Afzelius, B.A. (1976). A human syndrome caused by immotile cilia. *Science* **193**, 317–319.

Arts, H.H., Doherty, D., van Beersum, S.E., Parisi, M.A., Letteboer, S.J., Gorden, N.T., Peters, T.A., Marker, T., Voesenek, K., Kartono, A., Ozyurek, H., Farin, F.M., *et al.* (2007). Mutations in the gene encoding the basal body protein RPGRIP1L, a nephrocystin-4 interactor, cause Joubert syndrome. *Nat. Genet.* **39**, 882–888.

Badano, J.L., Mitsuma, N., Beales, P.L., and Katsanis, N. (2006). The ciliopathies: An emerging class of human genetic disorders. *Annu. Rev. Genomics Hum. Genet.* **7**, 125–148.

Berbari, N.F., O'Connor, A.K., Haycraft, C.J., and Yoder, B.K. (2009). The primary cilium as a complex signaling center. *Curr. Biol.* **19**, R526–R535.

Berggard, T., Linse, S., and James, P. (2007). Methods for the detection and analysis of protein-protein interactions. *Proteomics* **7**, 2833–2842.

Bouwmeester, T., Bauch, A., Ruffner, H., Angrand, P.O., Bergamini, G., Croughton, K., Cruciat, C., Eberhard, D., Gagneur, J., Ghidelli, S., Hopf, C., Huhse, B., *et al.* (2004). A physical and functional map of the human TNF-alpha/NF-kappa B signal transduction pathway. *Nat. Cell Biol.* **6**, 97–105.

Cagney, G., and Uetz, P. (2001). High-throughput screening for protein–protein interactions using yeast two-hybrid arrays. *Curr. Protoc. Protein Sci.* Chapter 19, Unit 19.6.

Cline, M.S., Smoot, M., Cerami, E., Kuchinsky, A., Landys, N., Workman, C., Christmas, R., Avila-Campilo, I., Creech, M., Gross, B., Hanspers, K., Isserlin, R. *et al.* (2007). Integration of biological networks and gene expression data using Cytoscape. *Nat. Protoc.* **2**, 2366–2382.

Collins, M.O., and Choudhary, J.S. (2008). Mapping multiprotein complexes by affinity purification and mass spectrometry. *Curr. Opin. Biotechnol.* **19**, 324–330.

Cox, J., Matic, I., Hilger, M., Nagaraj, N., Selbach, M., Olsen, J.V., and Mann, M. (2009). A practical guide to the MaxQuant computational platform for SILAC-based quantitative proteomics. *Nat. Protoc.* **4**, 698–705.

den Hollander, A.I., Koenekoop, R.K., Mohamed, M.D., Arts, H.H., Boldt, K., Towns, K.V., Sedmak, T., Beer, M., Nagel-Wolfrum, K., McKibbin, M., Dharmaraj, S., Lopez, I., *et al.* (2007). Mutations in LCA5, encoding the ciliary protein lebercilin, cause Leber congenital amaurosis. *Nat. Genet.* **39**, 889–895.

Gavin, A.C., Bosche, M., Krause, R., Grandi, P., Marzioch, M., Bauer, A., Schultz, J., Rick, J.M., Michon, A.M., Cruciat, C.M., Remor, M., Hofert, C., *et al.* (2002). Functional organization of the yeast proteome by systematic analysis of protein complexes. *Nature* **415**, 141–147.

Gerdes, J.M., Davis, E.E., and Katsanis, N. (2009). The vertebrate primary cilium in development, homeostasis, and disease. *Cell* **137**, 32–45.

Gherman, A., Davis, E.E., and Katsanis, N. (2006). The ciliary proteome database: An integrated community resource for the genetic and functional dissection of cilia. *Nat. Genet.* **38**, 961–962.

Gloeckner, C.J., Boldt, K., Schumacher, A., Roepman, R., and Ueffing, M. (2007). A novel tandem affinity purification strategy for the efficient isolation and characterisation of native protein complexes. *Proteomics* **7**, 4228–4234.

Gorden, N.T., Arts, H.H., Parisi, M.A., Coene, K.L., Letteboer, S.J., van Beersum, S.E., Mans, D.A., Hikida, A., Eckert, M., Knutzen, D., Alswaid, A.F., Ozyurek, H., *et al.* (2008). CC2D2A is mutated in Joubert syndrome and interacts with the ciliopathy-associated basal body protein CEP290. *Am. J. Hum. Genet.* **83**, 559–571.

Gosens, I., van Wijk, E., Kersten, F.F., Krieger, E., van der, Z.B., Marker, T., Letteboer, S.J., Dusseljee, S., Peters, T., Spierenburg, H.A., Punte, I.M., Wolfrum, U., *et al.* (2007). MPP1 links the Usher protein network and the Crumbs protein complex in the retina. *Hum. Mol. Genet.* **16**, 1993–2003.

Hanke, S., Besir, H., Oesterhelt, D., and Mann, M. (2008). Absolute SILAC for accurate quantitation of proteins in complex mixtures down to the attomole level. *J. Proteome. Res.* **7**, 1118–1130.

Junttila, M.R., Saarinen, S., Schmidt, T., Kast, J., and Westermarck, J. (2005). Single-step Strep-tag purification for the isolation and identification of protein complexes from mammalian cells. *Proteomics* **5**, 1199–1203.

Kellermann, J. (2008). ICPL—isotope-coded protein label. *Methods Mol. Biol.* **424**, 113–123.

Latterich, M., Abramovitz, M., and Leyland-Jones, B. (2008). Proteomics: New technologies and clinical applications. *Eur. J. Cancer* **44**, 2737–2741.

Lee, J.W., and Lee, S.K. (2004). Mammalian two-hybrid assay for detecting protein–protein interactions in vivo. *Methods Mol. Biol.* **261**, 327–336.

Loktev, A.V., Zhang, Q., Beck, J.S., Searby, C.C., Scheetz, T.E., Bazan, J.F., Slusarski, D.C., Sheffield, V.C., Jackson, P.K., and Nachury, M.V. (2008). A BBSome subunit links ciliogenesis, microtubule stability, and acetylation. *Dev. Cell* **15**, 854–865.

Mann, M. (2006). Functional and quantitative proteomics using SILAC. *Nat. Rev. Mol. Cell Biol.* **7**, 952–958.

Marshall, W.F., and Nonaka, S. (2006). Cilia: Tuning in to the cell's antenna. *Curr. Biol.* **16**, R604–R614.

Nachury, M.V., Loktev, A.V., Zhang, Q., Westlake, C.J., Peranen, J., Merdes, A., Slusarski, D.C., Scheller, R.H., Bazan, J.F., Sheffield, V.C., and Jackson, P.K. (2007). A core complex of BBS proteins cooperates with the GTPase Rab8 to promote ciliary membrane biogenesis. *Cell* **129**, 1201–1213.

Ong, S.E., and Mann, M. (2006). A practical recipe for stable isotope labeling by amino acids in cell culture (SILAC). *Nat. Protoc.* **1**, 2650–2660.

Pazour, G.J., Dickert, B.L., Vucica, Y., Seeley, E.S., Rosenbaum, J.L., Witman, G.B., and Cole, D.G. (2000). Chlamydomonas IFT88 and its mouse homologue, polycystic kidney disease gene tg737, are required for assembly of cilia and flagella. *J. Cell Biol.* **151**, 709–718.

Pfleger, K.D., and Eidne, K.A. (2006). Illuminating insights into protein-protein interactions using bioluminescence resonance energy transfer (BRET). *Nat. Methods* **3**, 165–174.

Poser, I., Sarov, M., Hutchins, J.R., Heriche, J.K., Toyoda, Y., Pozniakovsky, A., Weigl, D., Nitzsche, A., Hegemann, B., Bird, A.W., Pelletier, L., Kittler, R., *et al.* (2008). BAC TransgeneOmics: A high-throughput method for exploration of protein function in mammals. *Nat. Methods* **5**, 409–415.

Quinlan, R.J., Tobin, J.L., and Beales, P.L. (2008). Modeling ciliopathies: Primary cilia in development and disease. *Curr. Top. Dev. Biol.* **84**, 249–310.

Reits, E.A., and Neefjes, J.J. (2001). From fixed to FRAP: Measuring protein mobility and activity in living cells. *Nat. Cell Biol.* **3**, E145–E147.

Rigaut, G., Shevchenko, A., Rutz, B., Wilm, M., Mann, M., and Seraphin, B. (1999). A generic protein purification method for protein complex characterization and proteome exploration. *Nat. Biotechnol.* **17**, 1030–1032.

Roepman, R., Bernoud-Hubac, N., Schick, D.E., Maugeri, A., Berger, W., Ropers, H.H., Cremers, F.P., and Ferreira, P.A. (2000). The retinitis pigmentosa GTPase regulator (RPGR) interacts with novel transport-like proteins in the outer segments of rod photoreceptors. *Hum. Mol. Genet.* **9**, 2095–2105.

Roepman, R., Letteboer, S.J., Arts, H.H., van Beersum, S.E., Lu, X., Krieger, E., Ferreira, P.A., and Cremers, F.P. (2005). Interaction of nephrocystin-4 and RPGRIP1 is disrupted by nephronophthisis or Leber congenital amaurosis-associated mutations. *Proc. Natl. Acad. Sci. USA* **102**, 18520–18525.

Roepman, R., and Wolfrum, U. (2007). Protein networks and complexes in photoreceptor cilia. *Subcell. Biochem.* **43**, 209–235.

Selbach, M., and Mann, M. (2006). Protein interaction screening by quantitative immunoprecipitation combined with knockdown (QUICK). *Nat. Methods* **3**, 981–983.

Skerra, A., and Schmidt, T.G. (2000). Use of the Strep-Tag and streptavidin for detection and purification of recombinant proteins. *Methods Enzymol.* **326**, 271–304.

van Wijk, E., van der, Z.B., Peters, T., Zimmermann, U., Te, B.H., Kersten, F.F., Marker, T., Aller, E., Hoefsloot, L.H., Cremers, C.W., Cremers, F.P., Wolfrum, U., *et al.* (2006). The DFNB31 gene product whirlin connects to the Usher protein network in the cochlea and retina by direct association with USH2A and VLGR1. *Hum. Mol. Genet.* **15**, 751–765.

Vermeulen, M., Hubner, N.C., and Mann, M. (2008). High confidence determination of specific protein-protein interactions using quantitative mass spectrometry. *Curr. Opin. Biotechnol.* **19**, 331–337.

Walhout, A.J., Temple, G.F., Brasch, M.A., Hartley, J.L., Lorson, M.A., van den, H.S., and Vidal, M. (2000). GATEWAY recombinational cloning: Application to the cloning of large numbers of open reading frames or ORFeomes. *Methods Enzymol.* **328**, 575–592.

Wessel, D., and Flugge, U.I. (1984). A method for the quantitative recovery of protein in dilute solution in the presence of detergents and lipids. *Anal. Biochem.* **138**, 141–143.

Wolters, D.A., Washburn, M.P., and Yates, J.R., III (2001). An automated multidimensional protein identification technology for shotgun proteomics. *Anal. Chem.* **73**, 5683–5690.

Yu, Y.Q., Gilar, M., Lee, P.J., Bouvier, E.S., and Gebler, J.C. (2003). Enzyme-friendly, mass spectrometry-compatible surfactant for in-solution enzymatic digestion of proteins. *Anal. Chem.* **75**, 6023–6028.

CHAPTER 10

Crosslinking Methods: Purification and Analysis of Crosslinked Dynein Products

Miho Sakato

Department of Molecular Biology and Biochemistry, Wesleyan University, Middletown, Connecticut 06459

Abstract

Axonemal dyneins are multi-megadalton complexes which consist of heavy chains (HCs), intermediate chains (ICs), and light chains (LCs). The configuration and interactions among the many components within the dynein complex are not fully understood. For initial investigation of protein–protein interactions, chemical cross-linking can be easily applied to either flagellar axonemes or purified dyneins. Careful selection of crosslinker enables one to target protein–protein interactions that are constitutive and also to identify alterations in the configuration of the complex. For example, when performed in the presence of nucleotide or ligands such as Ca^{2+}, it is possible to trap transient interactions under specific physiological condition. Here I first describe the preparation of a crosslinked sample and its analysis by electrophoresis and immunoblotting using antibodies raised against a target and candidate interaction

978-0-12-374973-4
DOI: 10.1016/S0091-679X(08)91010-4

proteins. Next, when an interaction partner cannot be simply identified by immunoblotting, a crosslinked product may be isolated by immunoprecipitation, and its composition determined by mass spectrometry. These general approaches have great potential to define protein–protein interactions within any macromolecular complex of interest.

I. Chemical Crosslinking

A. Introduction

Many crosslinking reagents are commercially available and possess differing chemical characteristics including: (1) spacer arm length; (2) reactive group; and (3) water solubility. In this chapter, I introduce four crosslinkers 1-ethyl-3-(3-dimethylaminopropyl) carbodiimide hydrochloride (EDC), 1-5-difluoro-2,4-dinitrobenzene (DFDNB), dimethyl pimelimidate·2HCl (DMP), and disuccinimidyl suberate (DSS) (Table I) which have been successfully used in the analysis of *Chlamydomonas* flagellar axonemes and axonemal dynein samples [for purification methods, see Wakabayashi *et al.* (2007) and Chapter 3 by King (volume 92)]. EDC is a zero-length heterobifunctional carbodiimide which couples carboxyl groups to primary amines. The other three reagents are homobifunctional amine-reactive reagents with different spacer lengths and reaction chemistries [details are described in Benashski and King (2000) and Pierce Biotechnology (2005)]. Primary amines in the side chains of lysine residues and the N-termini of polypeptides can be targeted with these linkers. All four linkers work reasonably efficiently at physiologycal pH of ~7.5 so that HMEK/HMEA buffers can be routinely used. Note that primary amine-containing buffer systems (e.g., Tris) must be avoided as they will immediately quench the reaction. Although EDC can also be quenched by reducing reagents, the trace amounts carried over from the axoneme/dynein purification do not normally affect the crosslinking reaction.

In general, the crosslinking reaction contains dynein at 0.5–1 mg/ml, or flagellar axonemes at higher concentrations (1–5 mg/ml). Crosslinker concentrations should be titrated to determine the optimal experimental conditions where efficient crosslinking of intermolecular interactions with a target protein occurs but the generation of unfavorable and/or higher order products is minimized. I normally employ a series of concentrations ranging from 0.01 to 50 mM. Furthermore, in combination with UV

Table I
Chemical Crosslinkers

Abbreviation	Chemical Name	Spacer Arm	Reactive Group
EDC	1-Ethyl-3-(3-dimethylaminopropyl) carbodiimide hydrochloride	0 Å	$—NH_2$ + —COOH
DFDNB	1-5-Difluoro-2,4-dinitrobenzene	3 Å	$—NH_2$ + $—NH_2$
DMP	Dimethyl pimelimidate·2HCl	9.2 Å	$—NH_2$ + $—NH_2$
DSS	Disuccinimidyl suberate	11.4 Å	$—NH_2$ + $—NH_2$

All informations on this table are based on Pierce Biotechnology, Inc. (2005).

photolysis in the presence of Mg^{2+}, ATP, and vanadate (King, 1995), which cleaves the dynein heavy chain at the V1 ATPase site within the first AAA+ domain of the motor unit, crosslinking provides an effective way to determine with which heavy chain (HC) domain (stem or motor) a target LC associates (Fig. 2).

B. Materials

- HMEK buffer: 30 mM HEPES-NaOH, pH 7.5, 5 mM $MgSO_4$, 1 mM EGTA, 25 mM KCl
- HMEA buffer: 30 mM HEPES-NaOH, pH 7.5, 5 mM $MgSO_4$, 0.5 mM EGTA, 25 mM K acetate
- Flagellar axonemes from *Chlamydomonas reinharditii* wild-type cc124 strain in HMEK/HMEA buffer
- Outer arm dynein purified from flagellar axonemes of *C. reinharditii ida1* strain in HMEA buffer (note that sucrose from the purification process does not disturb the crosslinking reaction)
- 0.1 M ATP
- 0.1 M Na metavanadate [for preparation, see Shimizu (1995)]

Crosslinkers (freshly prepared immediately before use!):

- 200 mM EDC in H_2O
- 10 mM DFDNB in methanol
- 100 mM DMP in methanol
- 50 mM DSS in DMSO

Quenching reagents:

- 5× SDS-PAGE sample buffer: 0.25 M Tris-Cl, pH 6.8, 0.5 M DTT, 10% (w/v) SDS, 0.5% (w/v) bromophenol blue, 50% (v/v) glycerol
- 1 M Tris-Cl, pH 7.5

C. Methods

1. Make a series of crosslinker dilutions with the appropriate solvent to 10× the final desired concentration.
2. If performed in the presence of ligands (such as Ca^{2+}) or nucleotide, these should be added prior to crosslinking and the sample incubated for 1 h on ice. For V1 photolysis, see section below.
3. Transfer the reaction tubes to the bench at room temperature.
4. Add the 1/10 volume of crosslinker, rapidly vortex, and spin down briefly.
5. Incubate for 1 h at room temperature.
6. If the samples are to be separated by SDS-PAGE, add 1/5 volume of 5× SDS-PAGE sample buffer. Otherwise quench the reactions by adding 1/10 volume of 1 M Tris-Cl, pH 7.5 (a final concentration above 50 mM is needed).

D. V1 Photolysis

1. Warm up a hand-held UV lamp set at 365 nm for >15 min before starting.
2. Mix 1 mM ATP and 0.1 mM vanadate (final concentrations) with the protein samples in the presence of Mg^{2+}.
3. Fill an ice bucket and make the surface flat.
4. Place the reaction tubes in the ice with the caps open.
5. Illuminate the tubes by placing the UV lamp directly on top of the tubes for 1 h.
6. Proceed to step 3 for crosslinking.

II. Identification of Interaction Partners

A. Immunoblotting

To examine whether crosslinking has been successful, a monospecific antibody against the target protein is used. Initially, several crosslinkers may be examined and their concentrations titrated (Fig. 1). The crosslinked samples are separated by SDS-PAGE, transferred to nitrocellulose membrane and immunoblotted with the antibody. When a crosslinked product which migrates significantly more slowly than the target protein is generated, it is most likely to have resulted from intermolecular interaction(s) between the target and one or more proteins. The optimal crosslinker and concentration that generate high yield of the product and minimize unwanted side reactions need to be determined empirically. Next, based on the total mass of the crosslinked product, candidate interaction partners can sometimes be identified; immunoblotting should be performed if antibodies are available.

For easy comparison, the crosslinked sample can be loaded into one wide well in a gel and, after protein transfer, the nitrocellurose membrane is cut into strips (Fig. 2). Reassembly of adjacent blot strips probed with different antibodies allows one to determine whether the immunoreactive bands indeed coelectrophorese.

When one-dimensional electrophoresis does not give good enough separation for screening of candidate interaction proteins, two-dimensional electrophoresis should be tried. For isoelectric focusing (IEF) in the first dimension, an acrylamide-based gel system (e.g., Bio-rad IPG strip) is commonly employed as they are easy to use. However, this system does not focus well if the crosslinked product is larger than ~100 kDa and so in this situation an agarose-based gel is recommended (Fig. 3; Fujinoki, 2001).

When immunoblotting, some caution needs to be taken in considering which antibody to use. Either monoclonal or polyclonal antibodies are useful provided that they are monospecific. However, it sometimes occurs (especially with monoclonals) that the recognition site is masked or altered by chemical crosslinking; this can result in reduced affinity of the antibody for the target and sometimes in the complete destruction of the epitope (see right panels in Fig. 3). If an interaction cannot be identified by immunoblotting, it is sometimes possible to isolate the crosslinked product by immunoprecipitation and then determine its composition by mass spectrometry (Fig. 4).

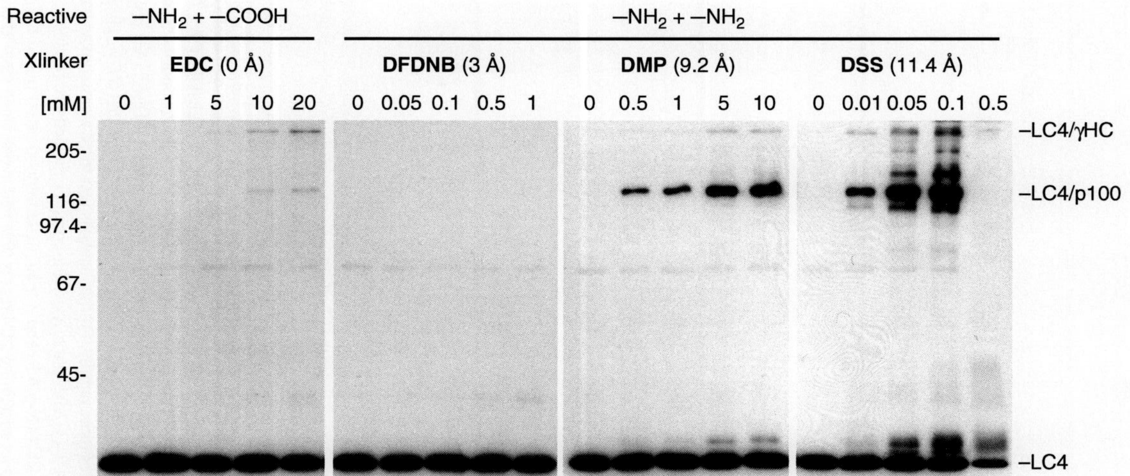

Fig. 1 Chemical crosslinking of LC4 in *Chlamydomonas* outer arm dynein. Purified outer arm dynein was treated with the carbodiimide EDC or with the amine-selective reagents DFDNB, DMP, and DSS in the presence of 1 mM Ca^{2+}. After electrophoresis in 8% acrylamide SDS gels, samples were probed for the presence of LC4. Crosslinked products containing LC4 (18 kDa) and either the γHC (~500 kDa) or p100 are evident in the EDC, DMP, and DSS samples. The p100 protein was later identified as IC1 afterwards (see Fig. 3). Reprinted from Sakato *et al.* (2007). Copyright © 2007 by The American Society for Cell Biology.

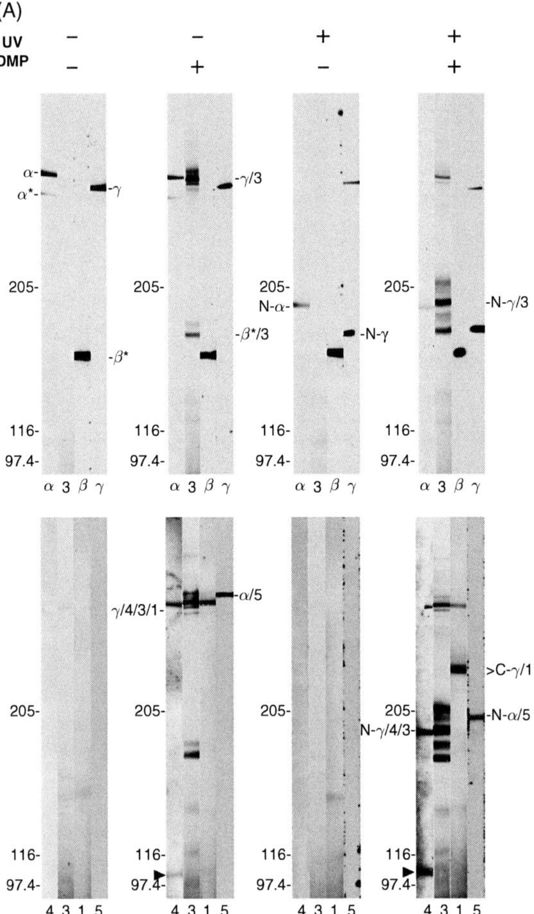

Fig. 2 Chemical crosslinking defines intradynein interactions. (A) Outer arm dynein containing an ~160-kDa truncated form of the β HC motor domain (β*) from *Chlamydomonas oda4-s7* flagellar axonemes was incubated with 1 mM ATP plus 100 μM vanadate, and half the samples were irradiated with UV light to cleave the α and γ HCs at their V1 sites. After the photolysis reaction, proteins were then subject to crosslinking with 10 mM DMP or were treated with solvent alone. Samples were electrophoresed in 4% acrylamide 4 M urea gels and probed with the 18αA, 18βC, and 12γB antibodies to detect the N-terminal regions of the three HCs, and with the R5932, R4930, CT61, and R4924 antibodies, which recognize the HC-associated LC1, LC3, LC4, and LC5 proteins, respectively. The top series of blots indicate the location of HC bands and LC3, whereas the other LCs were analyzed with respect to LC3 in the bottom series. In the presence of DMP, LC3 is crosslinked to both the β and the γ HCs (labeled β*/3 and γ/3); after photolysis the γ HC/LC3 product (N-γ/3) lacks the C-terminal motor unit and consequently migrates more rapidly. Further analysis identified a crosslinked band containing the γ HC and LC1, LC3, and LC4 (γ/4/3/1). After photolysis, this complex yielded two products: the γ HC N-terminal region crosslinked to LC3 and LC4 (N-γ/4/3) and a γ HC C-terminal domain linked to LC1 (C-γ/1). The arrowhead marks an additional product containing LC4 that is obtained in enhanced yield after UV irradiation to cleave the HCs at the V1 site; this product (LC4/p100 in Fig. 1) is further analyzed in Fig. 3. (B) Diagram illustrating the crosslinked products generated by DMP treatment and subsequent V1 photocleavage of *oda4-s7* dynein, which contains a truncated form of the β HC (β*). The relative size of the HC fragments corresponds to their apparent size after electrophoresis; it does not directly relate to their actual mass based on sequence. Reprinted from Sakato *et al.* (2007).

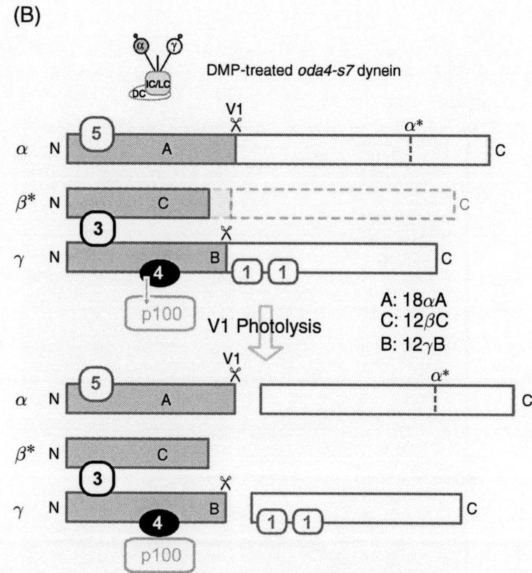

Fig. 2 (*Continued*)

B. Isolation of Crosslinked Products by Immunoprecipitation

1. Materials

- Appropriate antibody for the target protein
- TBS (50 mM Tris-Cl, pH 7.4, 150 mM NaCl)
- Centriplus50 ultrafiltration unit (Millipore, Billerica, MA, USA) presoaked with TBS + 0.1% (v/v) Tween 20 and then washed with H_2O immediately before use
- Protein G Plus beads (Pierce Biotechnology, Rockford, IL, USA), binding capacity: >20 mg human IgG per ml of settled resin (depending on the antibody to be used, Protein A-immobilized beads may be substituted)
- 20% (w/v) SDS
- 1% (v/v) Triton X-100 in TBS
- 2× SDS-PAGE sample buffer: 2/5× dilution of 5× SDS-PAGE sample buffer with H_2O

2. Method

1. Preparation of antibody-bound beads
   - Concentrate the antibody with a Centriplus50 to a final volume of ~1 ml.
   - Transfer 40 μl settled volume of Protein G Plus beads to a 1.5-ml microfuge tube.
   - Prewash the beads three times with 1 ml TBS.

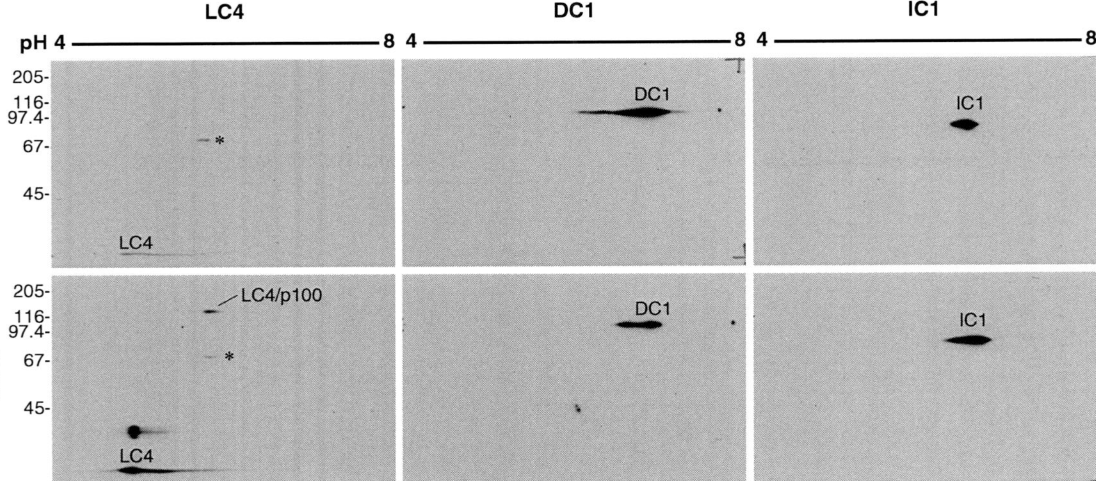

Fig. 3 Analysis of crosslinked outer arm dynein by 2D electrophoresis. Outer arm dynein purified from *Chlamydomonas oda11ida1* flagellar axonemes was crosslinked with 10 mM DMP in the presence of Ca^{2+}. The noncrosslinked (top panels) and crosslinked (bottom panels) samples were separated by IEF (first dimension) and SDS-PAGE (second dimension), transferred and probed with the CT61, anti-DC1, and 1878A antibodies to detect LC4, DC1, and IC1, respectively. The LC4-p100 crosslinked product was visible in the DMP-crosslinked sample. No corresponding spot was detected with either the anti-DC1 or the anti-IC1 antibodies. Asterisks indicate non-LC4 spots.

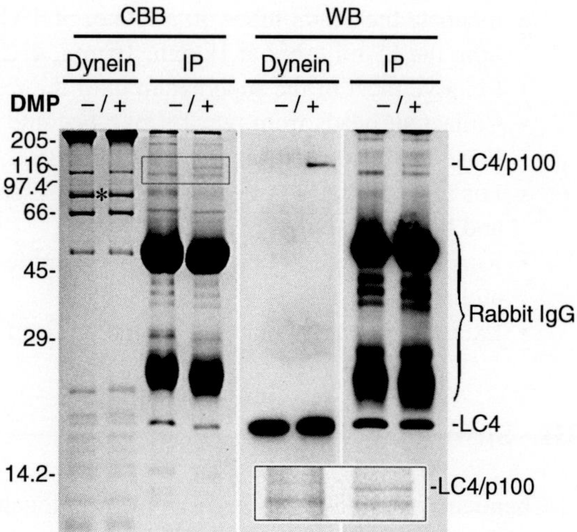

Fig. 4 Immunoprecipitation of crosslinked LC4-p100 using anti-LC4 antibody. Purified outer arm dynein was crosslinked with 10 mM DMP in the presence of Ca^{2+} (dynein), denatured, refolded, and immunoprecipitated (IP). Samples were electrophoresed in 10% tricine SDS gels and either stained with Coomassie blue (CBB) or transferred and probed with the CT61 antibody (WB). A LC4-p100 crosslinked product was immunoprecipitated in addition to LC4 as indicated at right. An inset at the bottom right shows an enlarged image of the boxed region of the Coomassie blue-stained gel. Mass spectrometry identified p100 as IC1. The asterisk indicates noncrosslinked IC1 that migrates with M_r78,000. Reprinted from Sakato *et al.* (2007). Copyright © 2007 by The American Society for Cell Biology.

- Add ~1 ml antibody into the tube containing the beads and incubate on a rotator for 1 h at room temperature.
- Briefly spin the tube and remove the supernatant.
- Wash the antibody-bound beads four times with 1 ml TBS. *If the crosslinked product migrates close to IgG bands upon electrophoresis, consider chemically immobilizing the antibody on the beads, for example, with DSS, and then eluting product with 0.2 M glycine, pH 2.8, followed by SDS-PAGE.

2. Crosslinking, denaturation, and refolding
   - Prepare a 1-ml reaction containing ~800 μg dynein which has been crosslinked and quenched with Tris buffer.
   - Split the crosslinked sample into 2×15-ml tube (each ~0.5 ml sample).
   - Add 50 μl of 20% SDS into each 15-ml tube (final concentration to 2% SDS), and gently but rapidly mix.
   - Incubate for 5 min at room temperature.
   - Add 11 ml of 1% Triton X-100 in TBS (>20× the volume of SDS), and gently mix well.
3. Immunoprecipitation
   - Transfer half the volume of the antibody-bound beads into a 15-ml tube and finally fill up to 15-ml mark with 1% Triton X-100 in TBS.

- Incubate the 15-ml tubes on a rotator at 4°C overnight.
- Spin the 15-ml tubes at 1k rpm, 10 min, 4°C.
- Remove most of the supernatant until the residual volume is ~0.5 ml.
- Collect all beads from two 15-ml tubes into one 1.5-ml microfuge tube.
- Wash the beads five times with 1 ml TBS.
- For elution, add 40 μl of 2× SDS-PAGE sample buffer, resuspend the beads, and boil for 5 min.
- Run 25 μl of the eluate in an acrylamide gel of appropriate concentration and stain the gel.
- Excise the corresponding band and perform mass spectrometry.

III. Summary

Chemical crosslinking is a powerful approach for investigation of intradynein interactions and interactions between dynein and nondynein axonemal components. Furthermore, it has the potential to identify interaction partners that had not previously been considered. In addition, it is a relatively simple approach that does not require any specialized equipment for initial study. Thus far, these methods have resulted in several interesting findings such as a physical link between outer and inner arm dyneins (DiBella *et al.*, 2004) and defining the complicated interaction networks within the intermeditate chain–light chain (IC–LC) complex of outer arm dynein (DiBella *et al.*, 2004, 2005) and the LC–LC complex of inner arm dynein (Yanagisawa and Kamiya, 2001). Although here I have described only applications involving native proteins, chemical crosslinking also can be applied to bacterially overexpressed proteins and to radioactive labeled *in vitro* translation products (King *et al.*, 1995; Sakato *et al.*, 2007).

Acknowledgments

I am greatly thankful to the former and current members of Steve King's laboratory at the University of Connecticut Health Center for all their help. I thank Drs. Ritsu Kamiya and Ken-ichi Wakabayashi (University of Tokyo) for the anti-DC1 antibody.

References

Benashski, S.E., and King, S.M. (2000). Investigation of protein–protein interactions within flagellar dynein using homobifunctional and zero-length crosslinking reagents. *Methods* **22**, 365–371.

DiBella, L.M., Gorbatyuk, O., Sakato, M., Wakabayashi, K.-I., Patel-King, R.S., Pazour, C.J. Witman, G.B., and King, S.M. (2005). Differential light chain assembly influences outer arm dynein motor function. *Mol. Biol. Cell* **16**, 5661–5674.

DiBella, L.M., Sakato, M., Patel-King, R.S., Pazour, G.J., and King, S.M. (2004). The LC7 light chains of *Chlamydomonas* flagellar dyneins interact with components required for both motor assembly and regulation. *Mol. Biol. Cell* **15**, 4633–4646.

Fujinoki, M., Ohtake, H., and Okuno, M. (2001). Serine phosphorylation of flagellar proteins associated with the motility activation of hamster spermatozoa. *Biomed. Res.* **22**, 45–58.

King, S.M., (1995). Vanadate-mediated photolysis of dynein heavy chains. *Methods Cell Biol.* **47**, 503–506.

King, S.M., Patel-King, R.S., Wilkerson, C.G., and Witman, G.B. (1995). The 78,000-M(r) intermediate chain of *Chlamydomonas* outer arm dynein is a microtubule-binding protein. *J. Cell Biol.* **131**, 399–409.

Pierce Biotechnology, Inc. (2005). "Technical Handbook: Cross-Linking Reagents."

Sakato, M., Sakakibara, H., and King, S.M. (2007). *Chlamydomonas* outer arm dynein alters conformation in response to Ca^{2+}. *Mol. Biol. Cell* **18**, 3620–3634.

Shimizu, T. (1995). Inhibitors of the dynein ATPase and ciliary or flagellar motility. *Methods Cell Biol.* **47**, 497–501.

Wakabayashi, K., Sakato, M., and King, S.M. (2007). Protein modification to probe intradynein interactions and *in vivo* redox state. *Methods Mol. Biol.* **392**, 71–83.

Yanagisawa, H.A., and Kamiya, R. (2001). Association between actin and light chains in *Chlamydomonas* flagellar inner-arm dyneins. *Biochem. Biophys. Res. Commun.* **288**, 443–447.

CHAPTER 11

Analysis of the Ciliary/Flagellar Beating of *Chlamydomonas*

Kenneth W. Foster
Department of Physics, Syracuse University, Syracuse, New York 13244-1130

Abstract

Eukaryotic flagella and cilia are alternative names, for the slender cylindrical protrusions of a cell (240 nm diameter, ~12,800 nm-long in *Chlamydomonas*

978-0-12-374973-4
DOI: 10.1016/S0091-679X(08)91011-6

reinhardtii) that propel a cell or move fluid. Cilia are extraordinarily successful complex organelles abundantly found in animals performing many tasks. They play a direct or developmental role in the sensors of fluid flow, light, sound, gravity, smells, touch, temperature, and taste in mammals. The failure of cilia can lead to hydrocephalus, infertility, and blindness. However, in spite of their large role in human function and pathology, there is as yet no consensus on how cilia beat and perform their many functions, such as moving fluids in brain ventricles and lungs and propelling and steering sperm, larvae, and many microorganisms. One needs to understand and analyze ciliary beating and its hydrodynamic interactions. This chapter provides a guide for measuring, analyzing, and interpreting ciliary behavior in various contexts studied in the model system of *Chlamydomonas*. It describes: (1) how cilia work as self-organized beating structures (SOBSs), (2) the overlaid control in the cilia that optimizes the SOBS to achieve cell dispersal, phototaxis steering, and avoidance of obstacles, (3) the assay of a model intracellular signal processing system that responds to multiple external and internal inputs, choosing mode of behavior and then controlling the cilia, (4) how cilia sense their environment, and (5) potentially an assay of ciliary performance for toxicology or medical assessment.

I. Introduction

This technical chapter discusses the analysis of ciliary beating in the context of the biciliated green alga, *Chlamydomonas*, which has rhodopsin light sensors (Foster *et al.*, 1984) and a chloroplast that can be selectively stimulated by light to perturb the normal beating pattern. Among ciliary model systems, *Chlamydomonas* cilia have the unique advantage of being dynamically modulated with control parameters due to light stimulation, revealing clearly its dynamic control. This well-studied eukaryotic organism steers with differential (asymmetrical) "planar" beating of two cilia (Fig. 1) relative to light sources and is responsive to mechanical, chemical, and light stimuli. To understand current applications and the motivation of certain analyses the relevant basics of cilia and applications are briefly reviewed. Virtually, all of the analysis applied here to the beating of *Chlamydomonas* cilia can also be applied to cilia of any organism, to sperm, or various human cells.

A. Basics of Ciliary Geometry and the Beat Cycle

Considerable effort has been aimed at understanding the cilium structure (Nicastro *et al.*, 2006; Oda *et al.*, 2007; Sui and Downing, 2006), its functioning, and the relevant

Fig. 1 The ciliary beat of *Chlamydomonas* cilia held on a micropipette, the cell body is an animation of Jyothish Vidyadharan, but the beating shape is from real data.

hydrodynamics (Blake, 2001; Brennen and Winet, 1977; Cortez *et al.*, 2004; Cosson, 1996; Dillon and Fauci, 2000; Dillon *et al.*, 2003, 2007; Gray and Hancock, 1955; Gueron and Levit-Gurevich, 1998, 1999, 2001a,b; Gueron and Liron, 1992, 1993; Johnson and Brokaw, 1979; Kinukawa *et al.*, 2005).

The core structure of each cilium is known as the axoneme (cross section shown in Fig. 2). It consists of nine doublet microtubules (db) arranged around the central pair doublet (the dynein "arms" point to the next higher numbered doublet, if numbered clockwise you are looking toward the tip, if counterclockwise you are looking toward the base). Figure 3 shows one of the doublets with its dynein motors, which drive the sliding between the doublets and spokes that connect the doublets to the central pair. The observed bending implies that the motor activity periodically varies from being higher on one side of the axoneme to being higher on the other side. During a P

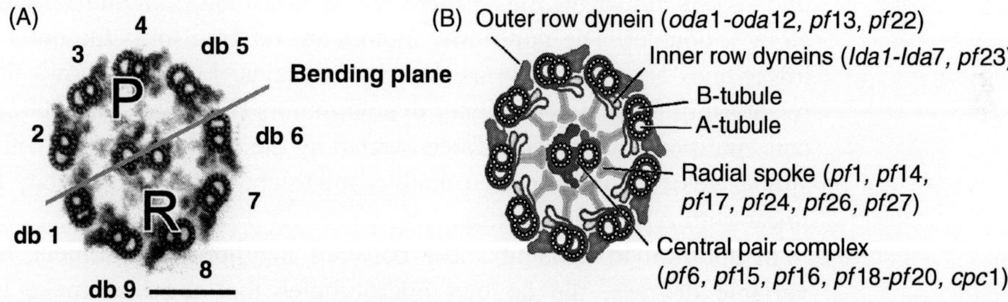

Fig. 2 (A) EM micrograph through a demembranated ciliary axoneme of *Chlamydomonas* (scale bar = 100 nm). Note the P-side consisting of db 1–5 facilitates the principal bend and the R-side consisting of db 6–9 facilitates the recovery bend. (B) Structural components of the axoneme and assembly mutations. Viewed from the cell body looking outward (Fig. 2 from Mitchell, 2000, with permission of Wiley-Blackwell). (See Plate no. 9 in the Color Plate Section.)

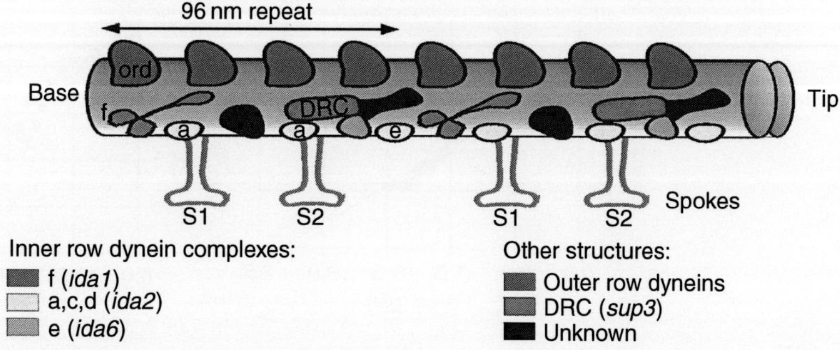

Fig. 3 One doublet showing dyneins and related structures along A tubule (Fig. 3 from Mitchell, 2000, with permission from Wiley-Blackwell). (See Plate no. 10 in the Color Plate Section.)

(principal) bend, doublets 1–4 walk (by attaching/detaching) on the adjacent higher numbered doublet toward the base of the cilium making the cilium bend with doublets 5 and 6 on the inside of the curve (negative curvature according to the convention of Brokaw, 1979, Brokaw *et al.*, 1982 and adopted here). Similarly, during an R (recovery) bend, doublets 6–9 walk on the adjacent higher numbered doublet toward the base making the cilium bend in the opposite direction (positive curvature). Since the cilium is thin and the distance between the doublets is short, even a short walk can produce a significant bend in the cilium (e.g., a 100 nm relative sliding of a doublet may induce a bend as large as 50°). The full sliding of doublets at the end of a cilium measured by *in vitro* experiments is no more than ±200 nm (Satir, 1985). A simplified way of thinking about it is to consider the axoneme as two elastic filaments that slide relative to each other resulting in the bending. The dynein motors provide the shear forces to produce the relative sliding.

The dynein motors are similar to piezoelectric motors used in cameras and cell phones. As shown in Fig. 3 there are several motors periodically arranged in rows along a doublet. The upper row motors are referred to as the outer dyneins and the lower ones are referred to as the inner dyneins. Figure 4 shows the force–velocity relationship for one of the inner dynein motors (Kojima *et al.*, 2002) at two different concentrations of ATP. The force exerted by the motor decreases with the increases in sliding velocity of the driven doublet microtubule and the velocity is approximately linear with ATP concentration.

In addition to causing shear between neighboring doublets, dyneins bind, to variable degrees, the doublet microtubules that together make up the axoneme (interior structure) of a cilium. These doublet attachments remarkably account for most of the bending or flexure rigidity of a cilium. The all dynein attached flexure rigidity, $\kappa \approx 11{,}000\,\text{pN}\,\mu\text{m}^2$, is 14 times stiffer than when the dyneins are unattached, $\kappa \approx 800\,\text{pN}\,\mu\text{m}^2$ (Okuno and Hiramoto, 1979). The potential role of dynamic stiffness in response has not yet received experimental or theoretical attention.

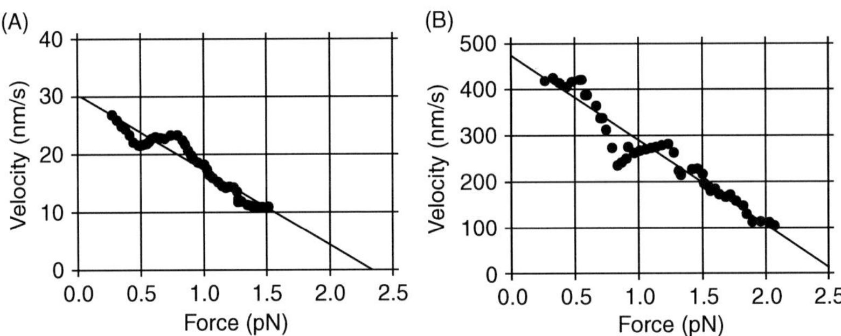

Fig. 4 Velocity–force curve for inner dynein c in the presence of (A) 5 μM ATP and (B) 100 μM ATP. The solid lines represent linear fits to the data. (Fig. 5 from Kojima *et al.*, with permission from Springer Science + Business Media).

B. Attachment of the Cilia to the Cell Body

The cilia have a very specific orientation with respect to the cell body. Looking down from above near the base the one doublet microtubule faces the other cilium, in twofold rotation or C2 symmetry (see Fig. 5), which leads to consistent left-handed cell rotation.

According to Riedel-Kruse *et al.* (2007 and references therein), with respect to bull sperm, the dynamics of the base connecting the cilium to the cell body plays a crucial role in determining the waveform of beating. Thus, cells may control their beating by changing the properties of the basal connection. In the case of *Chlamydomonas*, where much is known about the base (Geimer and Melkonian, 2004, 2005), the two cilia (Fig. 6) are connected to each other through proximal fibers (pcf and mpcf) at the base plate and distal fibers (dcf) at about 250 nm from the base (Fig. 6). As a result, the dynamics of the two cilia are connected, and the cell may use this connection to control or influence the beating pattern. In addition to the distal fibers, which link the dynamics of two cilia, there are other components such as the nuclear basal body connectors (NBBCs) in the basal body region that may play a significant role in

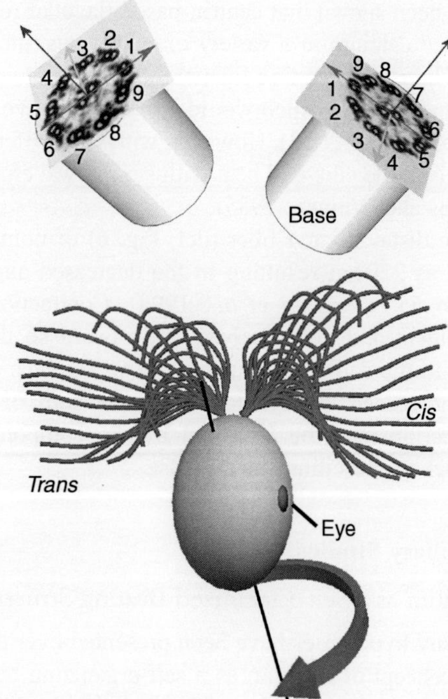

Fig. 5 Attachment of the cilia to the cell body with C2 symmetry (from animation of Jyothish Vidyadharan and the axoneme cross section image of Fig. 2). Note the left-hand rotation of the cell, the ciliary beating patterns and the eye. (See Plate no. 11 in the Color Plate Section.)

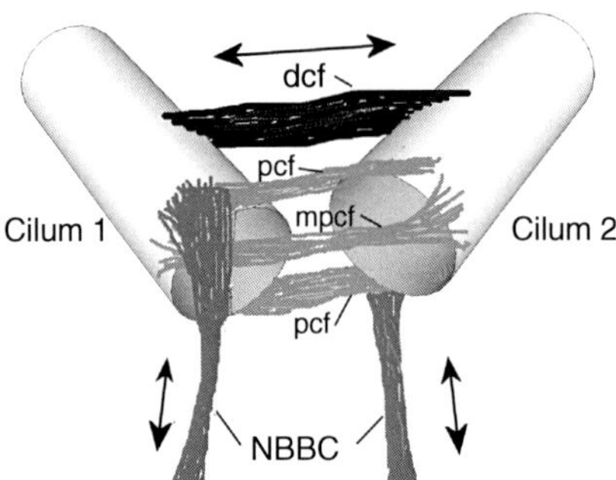

Fig. 6 An angled view through the basal bodies and the base of the cilia of *C. reinhardtii*. dcf, distal connecting striated fiber joining the two basal bodies; pcf, proximal connecting fiber joining the two basal bodies; medial proximal connecting fibers joining the two basal bodies; NBBC, nucleus-basal body connectors (Geimer and Melkonian, 2004, 2005; Silflow and Lefebvre, 2001). The NBBC and pcf are contractile. Note the offset of ends of the two cilia at the base. (See Plate no. 12 in the Color Plate Section.)

cellular control of the beating. The NBBC and dcf contain centrin (caltractin), which shows calcium-sensitive contractile or elastic behavior (Geimer and Melkonian, 2005). It has been shown that centrin-based flagellar roots are contractile under conditions of elevated calcium in a variety of eukaryotes, including *Chlamydomonas* and *Tetraselmis*. There is also a fine filament that runs between the centers of the proximal ends of each basal body, which would be very sensitive to the relative motion of either cilium (O'Toole *et al.*, 2003). However, with very different compliance machinery in a similar organism, the cilia still beat with similar waveforms and the cell still shows phototaxis (Hoops and Witman, 1985).

The distal striated fiber (dcf, Fig. 6) of nominal length 280 nm can contract to as much as 220 nm resulting in the decreased angle between the two cilia from about 65° to 55° (Hayashi *et al.*, 1998). Contraction of the NBBC would also aid this movement and pull the base end inward so that the cilia can more easily exit the holes in the cell wall. This change will induce a force in addition to the force due to sliding caused by dyneins. Now the sum of the forces due to base sliding and connection must be balanced by the component of the total hydrodynamic force parallel to the cilium at the base.

C. Applications of Ciliary Studies

1. Modeling of a Cilium as a Self-Organized Beating Structure

Many hypotheses have been presented over the years to explain the ciliary beating. The concept of a cilium as a self-organizing beating structure (SOBS) appears fairly well accepted (Brokaw, 1985, 2005, 2009; Camalet and Jülicher, 2000; Lindemann, 1994a,b, 2002, 2003, 2007; Lindemann and Mitchell, 2007; Lindemann *et al.*, 2005; Machin, 1963; Riedel-Kruse *et al.*, 2007). According to this concept, cilia beat spontaneously with no biochemical signaling control using only local information. In other words, the axoneme structure with its sliding doublets, together with the motor characteristics and its rates of attachment and detachment, either due to load or the gap between a motor and a doublet, can lead to a positive feedback mechanism that spontaneously generates and sustains beating. The exact origin of the beating of the SOBS still remains a source of controversy. According to one SOBS hypothesis (Camalet and Jülicher, 2000), referred to as the *load-dependent motor detachment model*, a positive feedback is induced in the following manner. As the sliding velocity of a doublet increases, the force generated by a motor decreases according to Fig. 4. However, the rate of detachment of the motors also decreases, because the load on them is less, resulting in the net increase in the number of attached motors as the load decreases. The increase in the number of attached motors is sufficiently high to cause an increased net force per unit length of the axoneme (force per motor times the number of attached motors) as the sliding velocity of the microtubule increases. This in turn causes further increase in the sliding velocity. The restraining forces due to bending rigidity of the doublets and passive elements of the axoneme eventually balance this motor-generated force to produce a regular, sustained beating.

The load-dependent detachment model is not the only one explaining SOBS. The *geometric clutch model* proposes that transverse forces acting on the outer doublet microtubules regulate the activity of dyneins to produce the ciliary beat cycle, by either facilitating the engagement or prying the doublets apart (Lindermann, 2003). Such a model would exploit the dependence of motor characteristics with the distance between the motor and the binding sites as suggested by Fig. 7. The *sliding-velocity control model* hypothesizes that turned-on dynein motors remain on as long as they maintain a high enough sliding velocity. Moving against an elastic resistance, their velocity gradually slows and when too low these active dyneins turn off. The bending

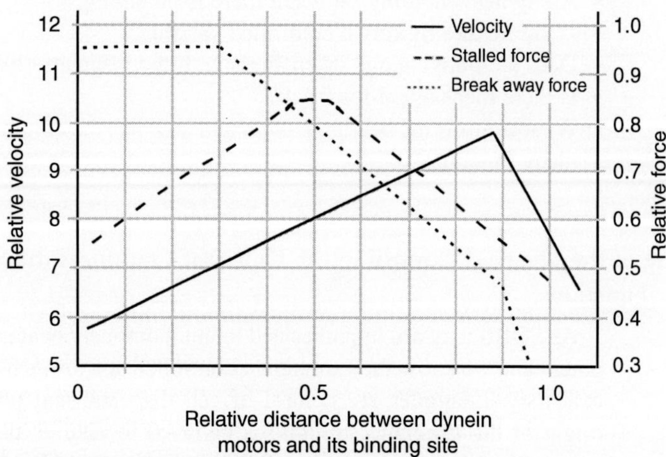

Fig. 7 Inspired by the motor characteristics of the similar piezoelectric linear motors, these are the hypothesized characteristics of the dynein–tubulin linear motors.

direction of the cilium then reverses by initiating the activation of dynein motors on the opposite side of the cilium (Brokaw, 1975, 1991, 2005). The *local curvature model* hypothesizes that the reversal occurs when the curvature reaches a critical magnitude (Brokaw, 1985, 2002).

Recently, Riedel-Kruse *et al.* (2007) compared images of planar flagellar beating of a bull sperm with those predicted by various models of SOBS and concluded that the waveforms were most satisfactorily fit by the load-dependent motor detachment model. They also found that the sliding of the microtubules at the base joining the flagellum to the sperm head plays a significant role in determining the waveform. These investigators concluded that beating patterns in bull sperm are therefore determined by a combination of motor activity and base properties. There are however significant differences between the bull sperm beating patterns examined by Riedel-Kruse *et al.* (2007) and those of *Chlamydomonas.* The "breast stroke" (ciliary) beating (Fig. 1) of the latter differs significantly from the "whip-like" (flagellar) beating of sperm. In *Chlamydomonas* the axoneme is connected to the cell body via the basal body, whereas in bull (mammalian) sperm the axoneme is connected via outer dense fibers to the connecting piece. Nevertheless, it is hoped that the analysis of flagella/ciliary beating in the context of *Chlamydomonas* can be extended to the cilia of other organisms and to sperm flagella including that of humans.

In spite of the extensive information collected on the axoneme's structure and motors and *in vitro* experiments with dyneins pulling doublets, many questions remain unanswered. For example:

- What makes a cilium beat?
- Do dyneins sometimes bind simultaneously on the P and R sides?
- How does the dominant activity switch between the P and R sides?
- Are dyneins holding on when there is no sliding?
- Is the flexure rigidity a controlled variable?
- Does a cilium change its stiffness with the viscosity (load) of the environment?
- Is there mechanical feedback?
- What controls the beating pattern and how has the cilium been adapted to perform its many functions?

2. Modeling of the Overlaid Control in the Cilia that Optimizes the SOBS to Achieve Its Multiple Functions

As SOBS they are hypothesized to autonomously beat at high frequency, but on the other hand we know they are highly controlled at a lower frequency by signals from the cell body. Examples are control of cell dispersal and phototaxis. The evolutionary origin of the axoneme structure is believed to date as far back as the last common eukaryotic ancestor more than 2 billion years ago (Baldauf, 2003; Stechmann and Cavalier-Smith, 2003). The evolutionary process must have evolved this structure in a way that optimized its control to adapt it to the varied functions of contemporary cilia.

A hypothesized scenario is that the axoneme structure came first, dyneins were added which created an SOBS with a helical beating pattern, and then the central pair and spokes (Silflow and Lefebvre, 2001; Smith, 2007) were added later to control it (Satir *et al.*, 2007) so that it would beat more efficiently in a plane rather than helically and be responsive to external signals.

3. Modeling for Signal Processing Within the Cell Body that Multiplexes External and Internal Inputs, Chooses Modes of Behavior and Then Controls the Cilia

Cilia are the output devices for much of the behavior of *Chlamydomonas* and as such are an assay for the sensory information (external signals and metabolic) processing that happens within the cell. Because cilia provide the cell with the ability to steer relative to light (phototaxis), their study provides insight into how the cell regulates phototaxis. Besides, the photoreceptor for light phototaxis is influenced by many environmental conditions so insight can be gained into how a cell processes multiple inputs. It is even thought that there may be too few signaling intermediates to obtain all the behavioral choices the cell has. A key feature of signal transduction is a large variety of environmental and internal stimuli mapping onto relatively few intracellular second messengers, yet maintaining specificity of response (Zaccolo, 2002). Perhaps there are more "second messengers," for example, internal pH, redox potential, Mg^{2+}, and ATP, which are implicated in phototaxis in addition to Ca^{2+}, cAMP, and IP_3. With stimulus–response functions (SRFs) one can clearly demonstrate their role.

Phototaxis is a common behavior in microorganisms and is the progenitor of animal eyes (Foster, 2009) and hence also animal brains, so that the organization and primitive controls developed at this cell level potentially continue into cellular control mechanisms of multicellular organisms. Observation of ciliary beating then is a window into the dynamical system of phototaxis (see Section III.E.13) and multiple intracellular feedback loops of cell signal processing network functions associated with a cell's behavior. Questions such as what are a response and how does a cell decide on which behavior and how multiple inputs are multiplexed can be addressed. Thus the *Chlamydomonas* phototaxis system is a great model for the systems biology of cell behavior.

4. Modeling for Function of a Sensory Cilium

Mechanical or pressure sensing occurs in *Chlamydomonas*, but has not been extensively studied (with a few exceptions, Wakabayashi *et al.*, 2009; Yoshimura, 1996, 1998). In addition, cilia are known to be mechanical flow sensors. One might anticipate that a flow-sensing cilium in the kidney might be using control of its stiffness to extend the dynamic range of its flow range. Whether it does so or how is not known. Similarly, the cilia in trachea need to detect and adapt to different viscosities.

5. As an Assay of Cellular Metabolism

The ciliary beating frequency (CBF) is simply related to the ATP level (Zhang and Mitchell, 2004). Hence, using the CBF, the control of the level of ATP within the cell body may be assayed with tenths of a second temporal resolution.

6. Model for Toxicology or Medical Performance Assessment

As far as we are aware this aspect has not yet been exploited. Ciliary beating may be studied for clinical reasons as defective cilia are associated with genetic diseases and their assay may be diagnostic medically. Assay of the effect of some pharmaceutical agent (so-called safety pharmacology or toxicology) can most readily be studied by observation of how cilia beat in the presence of the agent.

II. Rationale

The purpose here is to review methods of special applicability to the study of ciliary beating or to explain methods that have already been applied in particular to the beating of *Chlamydomonas* cilia. The vast literature of methods available to analyze ciliary beating will not be covered. It is sufficient here to give a list of helpful sources of varying mathematical sophistication to assist in the analysis task.

III. Techniques

A. Recording of Ciliary Beating Data

Because of different techniques used to record ciliary beating, analysis divides into two resolutions. At low spatial resolution information includes the CBF, stroke velocity, and the relative phase. These data are likely to have been recorded for very long times and hence has the potential for submillisecond temporal resolution (Section III.E.2). At high spatial resolution information includes the local curvatures of each cilium and derived forces along each cilium. Because of the initial difficulty with developing automated data collection and analysis, at present, there are much less data available at high resolution. This level of resolution is increasingly accessible and likely to be the norm in the future. It is worthwhile to discuss how this information may be analyzed (Section III.F). Of course, time-series analysis applies to both.

1. Low Spatial Resolution, Very High Temporal Resolution—Quadrant Photodiode

Chlamydomonas steers relative to a light source by differential control of beating of its two cilia. The phototactic response to the light stimulus has been extensively studied by Ruffer and Nultsch (1985, 1987, 1990, 1991, 1995, 1998) using visual comparison of film images and in our laboratory (Foster *et al.*, 2006; Josef, 2005; Josef *et al.*, 2005a,b, 2006) using an electro-optical detector (Fig. 8).

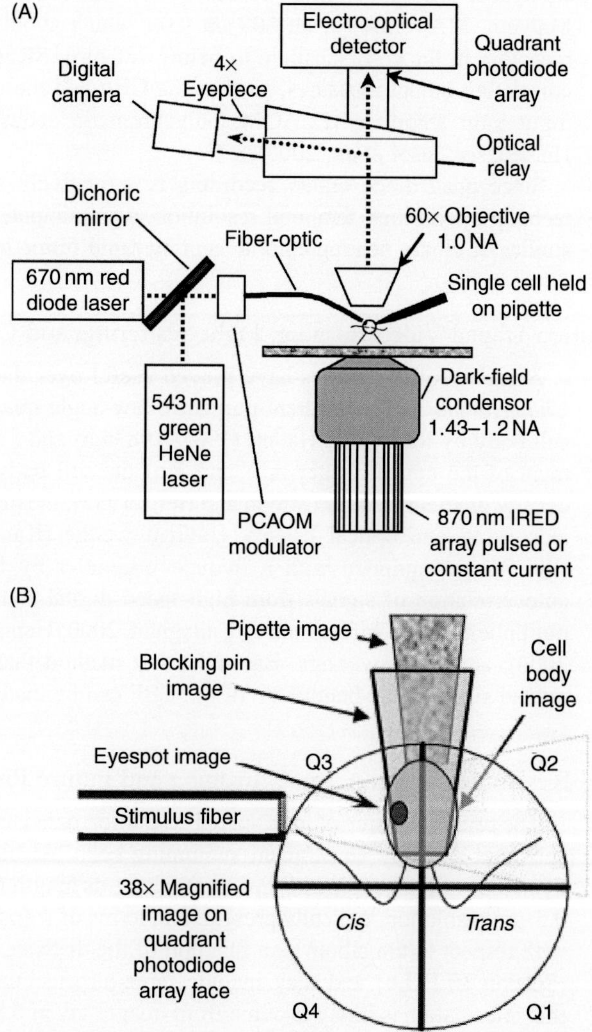

Fig. 8 The ciliary-monitoring apparatus of Josef *et al.* (2006). A. Schematic drawing of the single-cell ciliary monitor with IR-diode illumination and digital camera. A digital camera synchronized with the pulsed IR (870 nm) diode array and ciliary beating frequency provided observation of the cilia so they could be oriented to the imaging plane, as shown in (B). During data collection, the IR diode array operates in constant current mode, supplying uniform dark-field illumination of the cilia. An optical fiber with an acousto-optical modulator (PCAOM) delivers amplitude-modulated green (543 nm) stimulation to the cell.

This ciliary monitor perturbs, records, and analyzes ciliary beating of a held cell for many hours. The software and hardware allow computer control of amplitude and wavelength of light and the processing of ciliary movement. Josef *et al.* (2005a,b, 2006) stimulated *Chlamydomonas* with an amplitude-modulated 543 nm laser (green HeNe) light. To observe beating they used eleven 870 nm CW IRLEDs (continuous wave infra-red light emitting diodes, 30 mW each, TSFF5200; Vishay Intertechnology, Malvern, MA, USA) or an 808 nm laser which avoids stimulation of all the photoreceptors in the cell (sensitive to below 780 nm). SRFs were obtained by temporally correlating output measures, such as the CBF and the maximum stroke velocity, with input light stimulant (PCAOM, polychromatic acousto-optical modulator, 543 nm, HeNe laser, Josef *et al.*, 2006).

Since quad diode ciliary recording is a relatively simple and easy to implement technique with high temporal resolution, we anticipate it will remain useful for most studies of ciliary beating complementing rapid frame imaging of cilia.

2. Low-Resolution Digital Video Imaging, Light Scattering, and Other Techniques

A variety of techniques have proved useful over the years to measure the CBF of *Chlamydomonas* for different purposes: low-angle quasi-elastic light scattering of the cell body by Racey and Hallett (1981, 1983a,b) and Schaller *et al.* (1997); the movement of the optical center in *X* and *Y* of the cell body using two orthogonal optical density gradient wedges (Smyth and Berg, 1982), the movement of cell body along one axis using one optical density gradient wedge (Kamiya, 2000); measured with a stroboscope equipped with a frequency counter by Kamiya and Okamoto (1985); autocorrelation of signals from high-speed digital cameras, photodiodes, and photomultipliers (e.g., Chilvers and O'Callaghan, 2000; Hennessy *et al.*, 1986; Schipor *et al.*, 2006); and laser tweezers. Basically, any method that can detect fluctuations of an optical signal at the bandwidth of the CBF can be made to work.

3. High Spatial Resolution Digital Camera Imaging and Future Prospects

The primary source of high spatial resolution data is images of ciliary beating of held cells and spatially constrained swimming cells. At high spatial resolution, the local dynamics of the bending of the cilium along its length may be calculated to better than 0.5-μm resolution, typically presented in terms of ψ (psi plots—the local tangent angle with respect to the cilium as a function of the distance down the cilium from its base) (Section III.B.3.b) and may include three-dimensional (3D) information. Currently, this information is likely to come from images taken 0.8–5 ms apart. Josef *et al.* (2006) used a mechanical polar-coordinate stage capable of all six degrees of freedom to hold and manipulate the cells. It positioned the captured cell, located the eyespot, and tilted and rotated so that both cilia were in focus in the image plane. This mechanical stage has been replaced with a computer-controlled polar-coordinate stage with piezoelectric motors. The new stage brings the cell into a reference orientation within 0.5 μm of the center of the polar coordinates of the stage using two motors under joystick control.

Experience with the quad photodiode 4000-Hz ciliary monitor (Josef *et al.*, 2005a) has found that recording 20–25 images per beat cycle is about optimum for determination of ciliary motion from images. Since wild-type *Chlamydomonas* beats at ~55 Hz at 20°C, 1200 frames/s would be about optimum. This instrument could be extended for stereo observation with the addition of a second camera, since a 60 × 1.0 numerical aperture (NA) microscope objective image can be split at the rear focal plane into separate left and right optical paths (Srinivasan, 2008) giving two views of the cilia with wide angular separation.

Critical to obtaining useable ciliary images is an illumination light source that can be pulsed up to 1200 pulses/s. The 0.24-μm diameter cilia are illuminated in the Foster laboratory in dark field by light scattering with a pulsed 20 W 808 nm fiber-coupled laser diode without significantly heating or stimulating of cellular photoreceptors. This illumination produces maximally high-contrast images which are easier for automatic analysis than other optical methods but are not as bright as phase contrast. The 808 nm wavelength is a good compromise between water and cell photoreceptor absorption. A spinning small-angle diffuser is used to mitigate the small interference of the laser due to its partial coherence at the cilia. For optimum illumination, ~0.12 ms 808 nm pulses at 200 W is recommended. Such bright exposures of the rapidly moving cilia minimize motion blur and noise of ciliary images and maximize image contrast for image processing.

Since 2004 the best camera seems to be a fast frame back-illuminated EMCCD (electron multiplying charge-coupled device) camera based on the e2V Technologies CCD60 sensor. At 808 nm an EMCCD camera is about 25 times more sensitive than an image intensifier–CCD combination (ICCD) of only a few years ago. At 808 nm the primary difference is the ~3% quantum efficiency for an affordable 1000 Hz ICCD compared to an effective 37% quantum efficiency for the EMCCD. The microscope in Foster's laboratory is supported by a digital signal processor (DSP)-based multiprocessor core with a processing power of ~16 GFLOPS. The system can acquire data on ciliary motion continuously for hours, with high spatial and temporal resolution, synchronization with versatile light stimulation of the cell, and with semiautomatic computer processing of images.

4. Atomic Force Microscopy

In addition to measuring CBF (Teff *et al.*, 2008) accurately, atomic force microscopy (AFM) can measure directly the force produced by each cilium during the beat. The method provides a very nice direct confirmation of the result expected from the hydrodynamic analysis.

B. Analysis of Ciliary Beating Data

1. Introduction to Dynamic Systems Biology and Time Series

Ciliary beating is a temporal phenomena, whether in relatively steady-state conditions, for example, constant light of constant quality, or in dynamic conditions, for example, in response to a pulse of light. Consequently, analysis of ciliary beating falls into the general category of time-series analysis for which there are many textbooks and articles (Aoki, 1987; Jenkins and Watts, 1968). The goal here is provide the needed *Chlamydomonas* references, outline useful concepts needed to understand those references, and illustrate the concepts with examples in a way that will hopefully reveal the value of each technique. While doing so, it is hoped that the discussion will also provide a useful framework for understanding the cilia and its control by the cell body, and raise awareness of the controlling networks in cilia. Ciliary beating is so overlaid with controls that the underlying uncontrolled SOBS, which is fundamental to how cilia work, remains as of 2009 imperfectly understood. When beginning analysis it is important to be open to the unexpected, but it is also helpful to know what to look for and hence some suggestions on that are offered here.

2. Control System Models

In the flagella/cilia literature (Otter, 1989) control has meant the "physiological, biochemical, and hydrodynamic properties" of the cilia that maintain a steady-state pattern of beating due to an unchanging pattern of passive and active sliding between adjacent outer doublet microtubules and constancy of the base compliance (to be explained below). Control has also been used to mean "response state" control referring to a transient or altered physiological states in which the pattern of beating is altered in response to some type of stimulus either transiently or continuously altered as long as the stimulus is present. These views of control may be unified into the dynamical systems model (Dorf and Bishop, 1998; Kantz and Schreiber, 2003; Poincaré, 1892), which pictures all dynamic events (maintenance of steady state and deviations from it) as temporal trajectories in a phase or parameter space. In this model, all the variables that affect response, namely, chemical concentrations, temperature, light, pressure, and hydrodynamics (local forces) are separate dimensions in this parameter or phase space. In this parameter space all possible states of a system, the interconnected network of the cell, are represented with each possible state mapped to a unique point in the multidimensional parameter space.

Here *dynamic* means that the present value of the output depends on the *history* of the input, and depends little, if at all, on the present value of the input. At any instant in time, the system lies at some point in its parameter space. In this multidimensional space, a stimulus moves the system to a new point in that parameter space by changing one or more variables, such as light intensity. Moving along a trajectory in that parameter space means that different variables change dynamically with time. Therefore, a response to a short stimulus consists of relaxing from the stimulated position along a trajectory of changing variables that may, for example, steer the cell. The response is transient if dynamically the system finds an attractor or fixed stable point in the phase space on an observable timescale. An attractor is a set of specific variables to which a dynamical system (i.e., one that is time dependent) evolves after a long enough time (Wuensche, 2004). A basin of attraction is all those positions in the phase space that the system has that will evolve to that attractor. If the system is disturbed within that basin of attraction it will fall back to that attractor. The attractor may be a point or a curve or something more complex. If the disturbance is relatively small, then we have the control corresponding to maintenance of a steady state. If the disturbance is large, then we have response state control. There can be regions of phase space (particular sets of parameters) that are avoided, so-called repellors. There can be adjacent attractors with a saddle in between whose height may vary or the two attractors may merge to different degrees. Consequently, it could be that the system may relax to position near a saddle such that random fluctuations may cause the trajectory to fall toward one or another attractor. Nature could in principle use this situation for behavioral optimization (see Section III.E.8).

An issue is whether entities like attractors and trajectories are useful concepts in the context of a cell signal processing network. Some would argue that a living cell must constantly act upon itself and its environment to achieve survival objectives, which are not found in the most commonly thought of mechanical or electrical dynamical systems, that is, the system is optimized differently. For a living cell similar inputs must result in similar outputs, in other words, there must be considerable constraints on the dynamical system. We believe this is a useful common framework to integrate how a system like phototaxis and cilia work in *Chlamydomonas* and is closely connected to the state-space representation.

Classically, a dynamical system is analyzed in the frequency domain with transfer function models (Dorf and Bishop, 1998; Jenkins and Watts, 1968) and Bode plots. The approach is popular because Bode plots are easy to draw and interpret (see Sections III.E.2 and 13). An alternative is to use the time-domain or the state-space representation approach common in control engineering (Aoki, 1987), usually depicted by a series of differential equations. Although both are equivalent mathematically, each has its own advantages. In the state-space model, the parameters mentioned above become the state variables and the state vector is equivalent to the location of the system in the parameter space discussed above. One should become familiar with both the frequency- and time-domain analysis approaches.

For the case of ciliary beating in *Chlamydomonas*, we propose a modular attractor network architecture to interpret cell-signaling states that we see, for example, those corresponding to the different types of phototaxis. Wuensche (2004), in particular, has introduced subclusters without full connectivity as possibly being the optimum choice for the type of network architecture required to integrate fast biochemical signaling (it also seems the most probable for a biological cell). In Fig. 9 a few of the subclusters are depicted. They consist of groups of proteins that interact, often connected by second messengers. Such a network maximizes the number of possible attractor states. It is important for cell survival to have a variety of stable behaviors to choose among. Obviously, the cell is physically modular with two attached, but significantly different and independent cilia, mitochondria, cytoplasm, eye, endoplasmic reticulum, and chloroplast. Furthermore, there are biochemical as well as electrical subsystems and one must adopt a systems engineer perspective of multidomain modeling in which modules are analyzed individually along with their interconnections. Such ideas exist

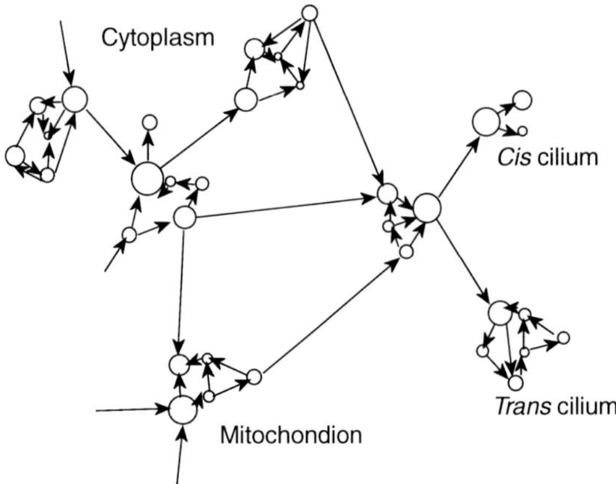

Fig. 9 The figure shows a small piece of a biosignal-processing network inspired by the suggestion of Wuensche (2004) that such networks would work most effectively to produce distinct basins of attraction if they were made of subclusters, just as you would expect naturally for a cell. We add here the fact of many compartments which help to bring about subclusters.

in other contexts such as in the works by Bar-Yam and Epstein (2004) and Zhou and Lipowsky (2005). In the words of Wuensche (2004): "In a basin of attraction field, perturbations to network states will reset the dynamics, which may then jump to another basin, or to a different position in the same basin. Stability requires a high probability of returning to the same basin, whereas adaptability or differentiation requires appropriate jumps to other basins in response to specific signals. Perturbations or external signals are most likely to affect attractor states, because that is where the dynamics spends the most time."

A more explicitly constrained model of dynamical systems comes from control systems theory (Aström and Murray, 2008). A subtype of control notable for biological applications is the self-adaptive control system model. The *Chlamydomonas* control system for phototaxis is not yet adequately described to distinguish which if any of the following features are important, but to research its properties these possibilities need to be considered: (1) a passive adaptive system or module that functions well by being designed to have an inherently low sensitivity to changes in the environment; (2) an active, dynamic input signal adaptive control system or module in which the cell measures input variables and modifies the control system in accordance with these changes where no sensing of the system response, say efficiency of its phototaxis is necessary (i.e., operates open loop); (c) an active dynamic control system which has inherently a model reference, which it compares transfer function parameters and modifies its transfer function accordingly; and (d) an active dynamic control system which converts input signals to a more desirable form to achieve the desired response (Davies, 1970). All these possibilities could have been achieved by the self-organizing process of cellular evolution and should be considered when developing few parameter functional descriptions as discussed in Section III.E.13. A self-adaptive control system is likely to be characterized by many feedback loops. Furthermore, for successful multivariable control it must reach an optimum rapidly since delays can lead to instability. In *Chlamydomonas* one might look for: (1) some reduction of current dynamic characteristics to a useable form such as cAMP concentration that controls many system parameters and (2) some actuator or adaptive processor tailored into a conventional feedback control system which would operate even with its failure, but normally giving multiloop modular feedback control. Unlike brain or learning systems which are designed to recognize familiar features and patterns in a situation and respond on the basis of past experience, which requires considerable logic circuits and is limited by memory storage, single cells are thought to have these self-adaptive systems designed to modify themselves in the face of disturbances. A very complex but important issue is that the adaptivity is likely to introduce additional long-term non-linearity and hence stability can then be an issue. In our experience, forcing instability of response in *Chlamydomonas* requires extreme measures, although not impossible.

Dynamical systems tend to be dominated by multiple feedback loops and *Chlamydomonas* is not an exception. In the feedback, two or more subdynamical systems are connected together under each other's influence so their dynamics are coupled making analysis of the whole dynamical system complex. Hence, analysis almost requires knowledge of formal methods (Aström and Murray, 2008). Feedback occurs on a wide range of scales from the local involving only a few proteins to the whole behavioral system of phototaxis, where there is sensing (an error signal is determined), information processing (a decision is made on what mode of behavior), and actuation in terms of ciliary beating control in order to steer (Fig. 10). Feedback is also responsible for cell homeostasis which controls CBF and enables the multiplexing of sensory inputs. Feedback, as extensively used by biological cells, also can create dynamic instability causing oscillations and even runaway behavior (see Section III.E.12).

In summary, a cellular signal processing network (exclusive of the receptor input and effector output layers to which it is multiplexed) is considered as containing different cell states (called "attractors" in network theory corresponding to positions in a parameter space of all the system variables) to which the system relaxes following stimulation. The state of the cell is dynamically a point moving in a trajectory in this

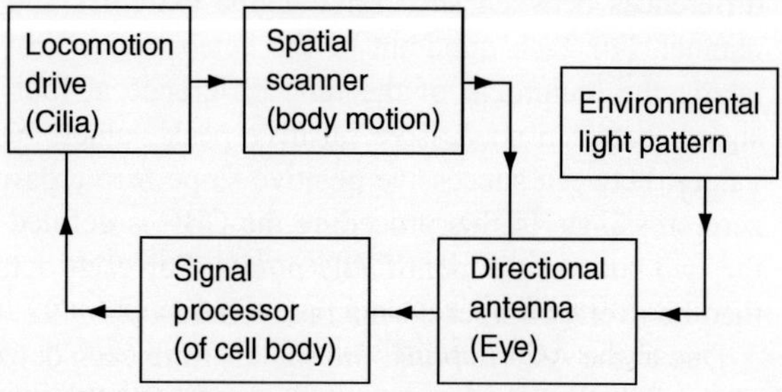

Fig. 10 The figure shows the phototaxis feedback loop as we envisaged 30 years ago (Foster and Smyth, 1980).

parameter space. Pulse stimuli jump locations in parameter space and the relaxation back in a basin of attraction to a favored location in parameter space is the response. Falling into the "negative phototaxis" attractor results in the ciliary response shapes that give one negative phototaxis. The key feature is the temporal response of the dynamics approaching an attractor (the idea that an attractor sends out a programmed response is too cumbersome and unlikely for a single cell). The falling into an attractor within its basin of attraction is similar each time, so responses are similar as well. This network of attractors arises from self-organization of the inputs and their connection to outputs. To this concept "self-adaptation" is added, which may have evolved from the initial self-organization. For example, some cells can apparently measure an input and the response depends not only on the signal but also the noise level (Bezrukov and Vodyanoy, 1997, stochastic resonance). Possibly, the cell has a reference SRF to which the current SRF is compared and corrected. A cell also has a measure of its well-being, which could feedback on the phototaxis–ciliary control system.

3. Extraction of Useful Parameters from Raw Data

a. From Quad Photodiode Signals of Ciliary Beating. The value of the direct raw signals is quite limited, because of the almost total dominance of the signal due to the beating cycle itself. However, extracted parameters such as the CBF, the "stroke velocity" of each cilium, the phase difference between the two cilia, the angle between the cilia, and the time in the cycle that switches to and from ciliary synchrony has proved very informative.

The current practice is to AC couple the raw signal so that the drifting DC level is removed, apply an analog band-pass filter, 10 and 80 Hz, and digitize at 4000 Hz. Finally, a digital finite impulse response filter is applied with a cut off at 120 Hz (Amnuanpol, 2009) to further reduce the noise.

Many parameters have been derived (24 in Josef *et al.*, 2005a,b). For example, the CBF is the number of beating cycles the cilium performs each second. One beat cycle is comprised of an effective stroke and a returning recovery stroke. The effective stroke consists of the motion of the tip of the cilium from a maximally anterior position with respect to the cell body to a maximally posterior position. The recovery stroke is the returning stroke. The beat frequency for each cilium can be determined from the time differences between successive signal extremes and zero crossings in the following manner. For each quadrant of the detector, the beat frequency can be computed by taking the reciprocal of the time difference at four easily identifiable points of the signal: between successive positive peak values, between successive negative peak values, between successive positive slope zero crossings, and between negative slope zero crossings. In this procedure the CBF is defined at the midpoint in time between the two successive identifiable points. For each detector quadrant, CBF values may then be averaged over 24-ms time windows.

Due to the AC coupling, the signals from each detector quadrant are first derivatives of the light levels, the root-mean-square (RMS) amplitude of each signal is proportional to the rate at which a cilium enters or exits a detector quadrant and how completely it enters/exits that quadrant (the magnitude of the stroke). To compute the *trans* cilium overall velocity, the RMS amplitude values for Q1 and Q2 (Fig. 8) were averaged over 24-ms time windows. To compute the *cis* cilium overall velocity, Q3 and Q4 were averaged. Ciliary stroke velocity correlates well with ciliary dominance, which is easy to measure in freely swimming cells, but is much harder to calculate from held cells.

The relative phase between detector output signals quantified the synchrony of the cilia. For each beat cycle, time differences between Q1 and Q4 (the *cis* phase minus the *trans* phase) were found by comparing durations at the four easily identifiable points: positive peak values, negative peak values, positive slope zero crossings, and negative slope zero crossings. If the *cis* ciliary stroke reaches maximally forward a little earlier than the *trans* stroke, then the relative phase will be positive. The relative phase was taken as occurring at the midpoint between each identifiable point and the time differences averaged over a 24-ms time window. Relative phase can be determined by multiplying the time difference for each window by the corresponding CBF and converting to degrees. This process was repeated for Q2 and Q3.

Change in the orientation of the cilium may be assayed by the anterior phase minus the posterior phase for each cilium. This relative phase of the Q1 phase minus the Q2 phase for the *trans* cilium, and the relative phases Q4 minus Q3 for the *cis* cilium can be similarly determined to the phase differences obtained for the synchrony of the *cis* and *trans* cilia (see Section III.E.9).

b. From Digital Images. **Computer recognition and tracing of ciliary curves.** Since the bending shapes of cilia are determined by the internal forces, the structure of the cilium, and its hydrodynamic interaction with the surrounding fluid, knowledge of ciliary bending shapes are crucial for full understanding of the underlying mechanisms of ciliary movement. The first step in studying ciliary curves has been to represent the cilium by its midline. This analysis use to be done by tracing the ciliary/flagellar images by eye. Later on, images were digitized by eye and computer analyzed (Brokaw, 1984, 1990) and then the cilia were traced semiautomatically from digital images; Baba and Mogami (1985) developed a software (BohbohSoft, http://bohboh-soft.dyndns.org/) which is freely available for this purpose. The operator identifies where the cilium is in an image and places a small arc-shaped box over a portion of the ciliary image and the computer completes the tracing. The method can be done with high accuracy provided the images are sufficiently good. They originally used simulated images with 50 nm/pixel, but later the software has been very successfully applied to real images (e.g. Kinukawa *et al.*, 2005).

In order to apply computer automation to indefinite sequences of images (100,000 or more), it is not feasible to manually identify each cilium. However, if the cell is held on a micropipette, it is possible to analyze continuous beating sequences for indefinite durations. Our computer program only requires manually determining the single approximate base origin of the cilia and tracing the first image in the sequence (potentially using BohbohSoft or by eye) to initialize several parameters. Since the individual frames do not have to be viewed, the processing task is fast. If the computer loses track, the operator may reset.

One approach for software image analysis (Foster, unpublished). Each image of a cilium is the two-dimensional (2D) projection of the 3D cilium viewed from some direction. In a single image this viewpoint will be different for the *cis* and *trans* cilium. The first step is image enhancement and segmentation or identification of where the cilia are in the image. To find the portion of the cilium that has moved, adjacent temporal images may be differenced, effectively removing the constant background. To see the relatively nonmoving portion of the cilium, the adjacent images may be added, reducing background noise. The "moving" and "unmoving" cilia may then be identified by range and domain filtering (Tomasi and Manduchi, 1998) from the differenced and summed images. This filtering accentuates the connectiveness of the ciliary image. Each neighborhood in the image may then be averaged and thresholded to identify regions where there are higher than normal intensities. A second copy of each image may be formed from the smaller eigenvalue of the covariance matrix of each neighborhood a procedure that emphasizes lines (McLaughlin, 2000); the smaller the eigenvalue, the narrower the line. This second copy may also be thresholded discarding regions without lines (cilia) and areas below this threshold in the first image may be removed (Fig. 11) (Srinivasan, 2008; Srinivasan *et al.*, 2008). Line and point noise are relatively uncorrelated so considerable cleaning up of the image results even with respect to a relatively poor image.

The second and more difficult step is to find the axes of the now segmented cilia automatically with the computer. The Foster approach depends on a robust fuzzy c-means algorithm (Frigui, 1999; Lam and Yan, 2007; Yan, 2001, 2004). If the enhanced images are of sufficient quality, then the Bohbohsoft program may also be used by automatically finding the next cilium by prediction knowing the point in which the two cilia connect in the cell body. Calculations for finding the cilium are currently speeded up by prediction of successive positions based on knowledge of how a wave propagates down the bending cilium and the four-phase switch-point model (Brokaw *et al.*, 1982; Satir, 1985) and its parameterization. The program, which is under continuous development, currently makes use of the facts that: (1) both cilia begin from a single point internally in the cell, (2) waves are propagated outward from the cell body (hence there is no difficulty in following a ciliary or flagellar beat pattern), and (3) the known sequence of events: principal bend initiation followed by its propagation, recovery bend initiation followed by its propagation, and repeat. More predictive and interpretive details will probably be included in the future, such as the fact that the ciliary shape in 3D seems to consist approximately of adjoining regions of circular arcs and straight lines.

Representation of the traced cilia. The fixed geometry of the cilium relative to the cell body makes it possible to fix local coordinates to the structure. We assign the osculating plane (defined by the **t** and **n** vectors) of the Serret–Frenet formalism (Fig. 12) (Nutbourne and Martin, 1988) to the ciliary bending plane, which is represented by the orthogonal tangent and normal vectors of the space curve. The twisting of the cilia can be assigned to be in the normal plane (Fig. 12) (defined by the **b** and **n**

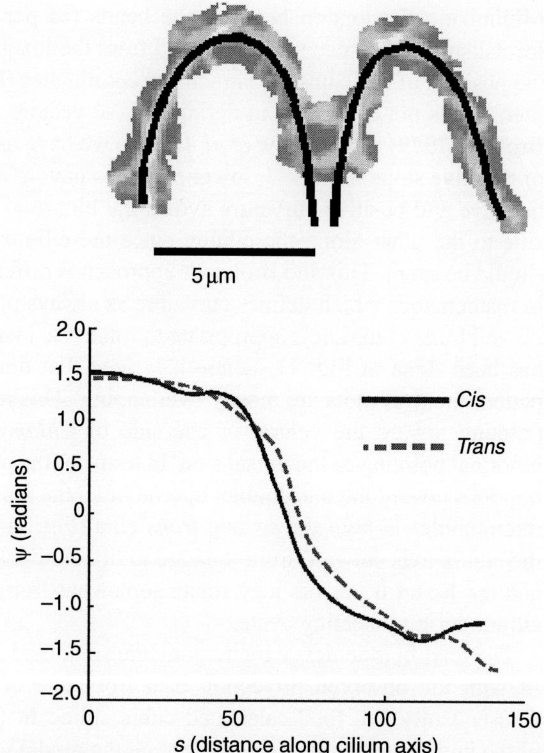

Fig. 11 Normal synchronous beating of *Chlamydomonas* cilia analyzed automatically by computer. Exposure of 0.5 ms (ψ is the tangent angle). The *trans* and *cis* curves are overlapped by rotating the view of one by 180°. (See Plate no. 13 in the Color Plate Section.)

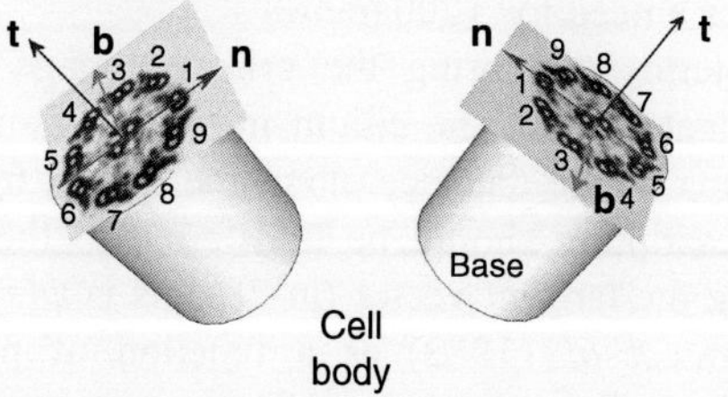

Fig. 12 The figure shows the Frenet–Serret coordinate system on the *cis* and *trans* cilia with the tangent **t** direction pointing outward from the base and maintaining the orientation convention discussed in the text. The images at the ends of each cilium are based on Fig. 2.

vectors) as expected, and the mastigonemes lie in the rectifying plane (defined by the **b** and **t** vectors). However, this formalism is pathological or very noisy if the curvature of the space curve representation goes to near zero, that is, the centerline of the cilium is nearly straight. Of course, the physical structure has a coordinate system built in.

Minimizing the torsion between the bends (as per the Bishop or parallel transport formalism) of the space curve derived from the images does the best job of estimating the position of these unseen intraciliary coordinates (Hanson, 2006). Unfortunately, the literature is not consistent in defining these vectors. Following the convention set by Brokaw (1979) and Brokaw *et al.* (1982), we have assigned the principal bend (region of negative slope in Fig. 11, lower panel) as having negative curvature. Allowing both negative and positive curvature avoids the binormal anomalously switching from one side to the other along the cilium, since the cilium is not twisting much the torsion should be small. This and Brokaw's approach is different from the common convention in mathematics which defines curvature as always positive. Also in comparison of the *cis* and *trans* cilium, it is appropriate to rotate the image of one or the other to match as has been done in Fig. 11, where it is seen that during a fairly symmetrical beating pattern the $\psi(s)$ plots are mostly overlapping. This results in the *cis* cilium's binormal pointing toward the ventral or eye side of *Chlamydomonas* and the *trans* cilium's binormal pointing to the dorsal side. In terms of the ciliary structure, the normal vector, **n**, points toward the one doublet microtubule; the binormal, **b**, toward the three doublet microtubules in both the *cis* and *trans* cilia (Fig. 11); and the tangent, **t**, points along the cilium axis outward from the base to tip. Note that these are the nominal directions and the **b** and **n** vectors may rotate somewhat from these orientations relative to the cilium during a beating cycle.

The final fit is achieved by iterating the smoothed residual errors of the estimated 3D fit with the observed projection data from one or two (stereo) views if available. Consequently, the final calculated cubic spline fit (of $\psi(s)$ and $T(s)$) is independent of the initial guess and not dependent on the model used to predict its position. Twenty images per beat cycle are about optimum for accurate determination of the local ciliary velocities $\mathbf{V}(s)$ from adjacent images for the fastest beating cilia. Note there are three components to the velocities, in the tangential, normal, and binormal directions. The optimum frame rate depends on the anticipated CBF, for example, 60 Hz beating implies a need for 1200 frames/s.

In terms of storing the primary image data, it is best to store the $r(s)$, the coordinates along the cilium in the lab frame (r) as determined from the raw data in terms of the distance along the cilium from the base, s. To work with the data, the description should be transferred to the ciliary coordinate system. We suggest storing the tangent vector (in 2D this is $\psi(s)$, the direction is the shear angle in 2D, Brokaw *et al.*, 1982) as a function of position, s. $\psi(s)$ is the integral of the curvature along the cilium. The curvature is the change in the tangent angle with the position. The tangent angle preserves the orientation at the base and is less noisy than the curvature because it is not differentiated. The curvature (s) with an initial base orientation is sufficient to completely describe the cilium in 2D. ψ is an angular measure of the amount of shear displacement between flagellar tubules as a function of length along the flagellum, $\Delta^c(s,t)$, that is, it is proportional to the cumulative displacement of the one doublet microtubule versus the five to six doublets relative to the base, and is also the integral of the curvature in the local osculating plane defined by the local normal and the tangent. To know the absolute cumulative displacement, one has to include the displacement of the doublets at the base, $\Delta_0(t)$. In 2D, ψ is calculated from a smoothing spline fit (Wahba, 1990) of the tangent angle of the cilium relative to the lateral axis (the square of the second derivative of the angle is minimized). In 3D, one must include $T(s)$, the net twist of the cilium relative to the base, or the integral of the torsion as a function of the position along the cilium, s. We expect that $T(s)$ has a very slow dependence on s and hence it is probably safe to strongly smooth the $T(s)$ function minimizing the second derivative of $T(s)$ using a smoothing spline. In 3D, if the data in the laboratory coordinates are used to do the calculations, it is necessary to fit quartic or quintic splines to the data in order to obtain the higher derivatives needed to calculate the torsion. Finally, for hydrodynamics calculations needed to calculate the external and hence internal forces within the cilia, knowledge of the local motion is also necessary, namely, $\mathbf{V}(s)$ (the local velocity as a function of the position along the ciliary length (see Section III.F). The most useful form to summarize these data is in the form of cubic splines which permit relatively easy calculation of any other needed parameters depending on the goals of the study and are a relatively compact representation. Other workers have different preferences for presentation. Collaborators of Baba, for example, display the local curvatures as a function of the distance along the cilium.

Extraction of the relevant parameters from these curves. It is necessary to partially process the raw data as discussed above, because there is no consensus on what are the relevant parameters of ciliary beating due to the divergent views of how cilia work. Since the relevant parameters depend on the model, a current goal is to identify the most meaningful parameters. Most of the physicists working on cilia support different SOBS hypotheses, in which the temporal local forces, local flexure rigidities, basal sliding, compliances, and orientation are the meaningful parameters. A different descriptive approach comes from Brokaw *et al.* (1982) where the CBF, the interval between initiation of sliding in reverse bend and initiation of sliding in principal bend, the curvatures in the principal and recovery bends, and the rate of sliding in the principal (P) and recovery (R) bends are considered the relevant primary parameters. The nature of proposed beating control is very different for the different models as well. In the SOBS hypothesis a number of control sites are considered. These include the location of dynein motors that are actively sliding, holding on, or passively sliding, which affects the motion and flexure rigidities along the cilium. Furthermore, if there are compliance changes at the base they could be crucial. One should be aware that the varying viewpoints (not reviewed here) strongly influence the analysis.

Nevertheless, it is possible to use principal component analysis (Jolliffe, 2002) to identify the relevant parameters that are controlled under a rich variety of conditions. The procedure is to systematically vary environmental parameters so that one has the components expressed in the data set and then apply the principal component analysis.

The 2D projection analysis of the ciliary image has dominated the field due to its simplicity. Ideally, 3D reconstruction of ciliary traces should be used for analysis of models. One reason is that the shape of 3D objects is simpler in 3D than in projection. A circular arc in 3D becomes the more complex ellipse in 2D projection. Another reason is that the 2D approach loses the information of ciliary torsion. One approach to get 3D information is to obtain stereo images and use principles of stereology to reconstruct the 3D image (Srinivasan, 2008). In this procedure the two image projections are scaled to the same height (they are forced to have the same y-axis). The disparity in the horizontal axis position is used to perceive depth (Teunis and Machemer, 1994). The 3D representations of cilia in the laboratory frame are not suitable for analysis and modeling. They need to be transformed into the local cilium frame as a function of the position along the cilium. Hence, they are uniformly interpolated, and smoothed, using cubic splines as a function of position along their length. The smoothed representations are unit-speed parameterized. Torsion, curvature, cilium base orientation, and position are calculated using the Frenet formulae. These may be parameterized as already discussed. It is important to keep foremost in mind that evolutionarily one would anticipate that the radial symmetry of the axoneme and "helical" beating arose first (Brokaw, 2002) and only later was the axoneme modified for planar beating as in *Chlamydomonas*. Therefore, retention of some circumferential pattern of control should be expected.

An alternative approach to analysis of 2D images is to assume a simpler structure, that is, assume an ellipse is really a circular arc and process the image as if this was strictly true, calculating the $\psi(s)$ and $T(s)$ functions accordingly. For a relatively planar beating pattern such as seen in *Chlamydomonas*, this approach could work moderately well. This assumption of a simpler 3D structure can also be used to reduce ambiguities during 3D reconstructions of stereo images.

c. From AFM. AFM makes it possible to directly measure the force exerted by the cilia during a beat cycle under different conditions. Since from Fig. 4, the velocity of dynein motors is proportional to ATP and the CBF is also proportional to the ATP (Zhang and Mitchell, 2004), it is perhaps not surprising that Teff *et al.* (2008) found that the force produced is proportional to the CBF.

d. From Electron Microscopy. Mitchell (2003) has shown that electron microscopy (EM) snapshots of moving cilia can be obtained and has determined the critical information that the C1 side of the inner pair of microtubules always faces the outside of a bend (Fig. 13). Furthermore, Lindeman and Mitchell (2007) have found that the diameter in the bending plane is larger in the bends than in the straight parts, contrary to the expectation you would have if you just bent a bundle of stands. Both these results have important theoretical ramifications for how cilia may control their bending even though the cilia were not bending at the time of the microscopy.

C. Analysis of Spontaneous Unstimulated Responses

1. Intermittency

It is possible for a cell to show seemingly "random" behavior; however, its analysis may lead to new insight. An interesting example is the stochastic resonance previously

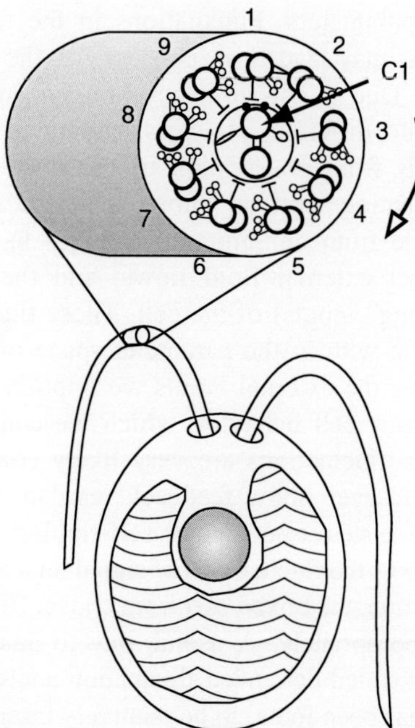

Fig. 13 Drawing of the doublet position and central pair orientation (indicated by C1 being outward in a bend) in a principal bend during the recovery stroke (from Fig. 9 of Mitchell, 2003).

mentioned, which improves the ability to hear. An example for *Chlamydomonas* is the potential for intermittency types of behavior. In nonlinear systems, near a stationary point, a system can exhibit shifts randomly between two phases, called laminar and turbulent, when the underlying system and its parameters remain constant, that is, intermittency (Ott, 1993; Schuster, 1988). A reasonable hypothesis is that the synchronous and nonsynchronous periods of ciliary beating represent the laminar and turbulent phases of intermittency and that the switching between them occurs with minor changes or simply fluctuations in the underlying parameters. A power law characterizes the distribution of the dwell times in the turbulent state in intermittency. Therefore, a test of this hypothesis to explain these switches in behavior is the manifestation of a nonlinear intermittent process. One can: (1) measure the duration of many such nonsynchronous periods to determine if they have a Gaussian, exponential, or power-law distribution and (2) construct the phase-space set (Kantz and Schreiber, 2003). The cell has the potential of tuning the probability of the shifting by changing how much the attractors are merged together.

2. So-Called Spontaneous Response and Consequent Correlated Behavior

A biological cell does not experience a truly steady state because the size of the system is so small that large fluctuations occur naturally in the variables or internal parameters. Fluctuations in the free concentrations of calcium or hydronium ions may lead to responses with the system continually falling into the same attractor. Due to intermittency, the system may continually fall with some probability into different attractors leading to different ciliary beating. One can learn about this from observation of the unstimulated behavior, by monitoring steady ciliary beating with no known external stimuli. The stimuli to cause responses may come from unmonitored changes in the environment, from pressure sensors that detect external fluid flows, and the metabolism (sometimes referred to as "well-being" inputs) of the cell. These fluctuations or signals move the position of the system with in the parameter space of variables already described in the same way as for the external inputs we impose. However, unlike the external signals we use to assay cell behavior, which we contrive to be relatively uncorrelated, these signals or fluctuations are very likely correlated. Further, due to cell homeostasis they could even show feedback regulation.

Such a system often shows self-similarity as shown in Fig. 14 in which the upper graph shows the fluctuations of signal on a long timescale and the lower graph shows a small section, the boxed part from above. Both signals look fairly statistically similar if scaled appropriately. A useful way to analyze spontaneous beating for being self-similar is called detrended fluctuation analysis (DFA) (Goldberger *et al.*, 2002). This method has been found to be relatively insensitive to the nonstationarity of the system. Although Josef *et al.* (2006) have shown the system to be quite stationary, for example,

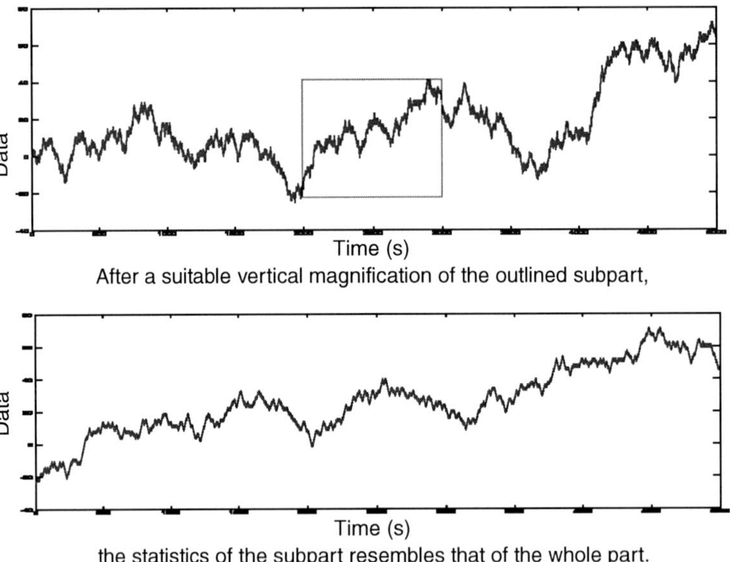

Fig. 14 The figure shows the principle of self-similarity, a small part of the large data set looks statistically like the large part with suitable scaling (Amnuanpol, 2009).

the mean and variance of the process does not change over time, the DFA approach appears to give better results than alternatives.

3. Self-Similarity Analysis

In response to external uncorrelated stimuli, most responses do not correlate with the stimulus for longer than a few hundred milliseconds. On the other hand, internal processes might modify behavior through the same routes as external stimuli, but are undoubtedly correlated for much long times than the deliberately uncorrelated external stimuli. Examples are processes that control ATP and calcium ion concentration. Mathematical methods using time-series analysis may be used to study these processes. We have examined the statistical properties of cilia beating over time in *Chlamydomonas* in the dark and light. We investigated the way these statistical properties preserve across short to long timescales, analogous to the spatial self-similarity in which geometrical structures preserve across short to long length scales. Using DFA to study the temporal self-similarity, Amnuanpol (2009) found that the beat frequency, stroke velocity, and relative phase of the cilia shows persistent positive correlation in the dark for at least 200 s. Using these techniques, one can determine what processes are involved in each correlation and whether they are due to drift or diffusion.

Intuitively, the time-series data **x(t)** are temporally self-similar if the mean and correlation statistics of its subparts, under suitable horizontal–vertical magnification, resemble those of the whole part. More formally, a time series $x(t)$ which is invariant under scaling time by a factor of a, $t \rightarrow t/a$, and scaling x by a factor of $a^{-\alpha}$, $x \rightarrow a^{-\alpha}x$, exhibits temporal self-similarity:

$$x(t) = a^{-\alpha}x\left(\frac{t}{a}\right) \quad (1)$$

Correspondingly, in the frequency domain the scaling relation is $x(\omega) = a^{-\alpha}x(a\omega)$, the scaled time $t' \equiv t/a$, and the scaled variable $x'(t') \equiv x(t/a)$. The ith moment $M_i = \langle x^i(t) \rangle$, which is the time average of the ith powers of $x(t)$, is scaled as $M_i = a^{i\alpha}M_i'$. Any probability distributions can be expanded in terms of moments. The steady-state probability distribution is scaled as $P(X) = a^{\alpha}P'(X')$.

The solution of the functional equation [Eq. (1)] is a power-law function, $x(t) \approx t^{-\alpha}$. In the relaxation process, the probability for a system in an excited state normally decays exponentially with time after switching off the time-dependent external force. On longer timescales it slowly decays with a stretched exponential or a power-law function (Fruenfelder *et al.*, 1988). By this example, the temporal self-similarity may not hold on short timescales. The range of a scaling exponent α characterizes the statistical behavior of time series. With respect to white noise, which is uncorrelated between any two different times, $\alpha = 0.5$. The deviation from 0.5 signifies the correlation present in the data. A value of α between 0 and 0.5 indicates antipersistence, that is, the large values of

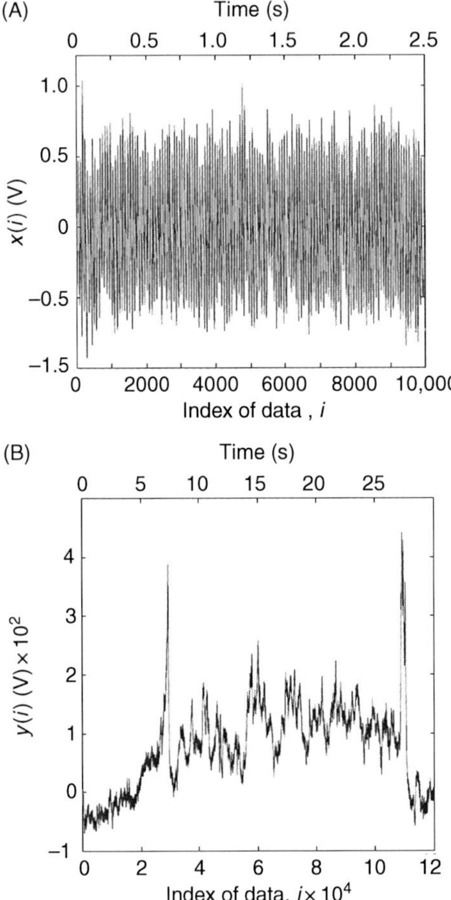

Fig. 15 (A) Time series $x(i)$ of voltage signals from *Chlamydomonas* ciliary beating recording from photodiodes with the number of data points $N = 120{,}000$. (B) Integrated time series $y(i)$ (Amnuanpol, 2009).

x are likely followed by small values and *vice versa*, whereas values of α between 0.5 and 1 indicate persistence, that is, present values of x are likely maintained. This scaling exponent α is extracted from time series by DFA (Peng *et al.*, 1994). The original time series $x(i)$ (Fig. 15, upper panel) is mapped to an integrated time series $y(i)$ (Fig. 15, lower panel), whose fluctuation is more apparent, by:

$$y(i) = \sum_{j=1}^{i} \left(x(j) - \langle x \rangle \right) \tag{2}$$

where $\langle x \rangle$ is the mean. The N data points of integrated time series are divided into subparts each of which contains n data points. Thus there are N/n subparts. For each subpart, the fluctuation around the trend represented by linear interpolated values is computed, yielding the overall fluctuation:

$$F(n)=\left\{\frac{1}{N}\left(\sum_{i=1}^{n}\left(y(i)-y_{\text{lin}}^{(1)}(i)\right)^2+\sum_{i=n+1}^{2n}\left(y(i)-y_{\text{lin}}^{(2)}(i)\right)^2\right.\right.$$
$$\left.\left.+\cdots+\sum_{i=N-n+1}^{N}\left(y(i)-y_{\text{lin}}^{(N/n)}(i)\right)^2\right)\right\}^{1/2} \quad (3)$$

where $y_{\text{lin}}^{(1)}, y_{\text{lin}}^{(2)}, \ldots, y_{\text{lin}}^{(N/n)}$ are the linear interpolated values of the first, second, . . ., (N/n)th subparts, respectively. Typically, the fluctuation $F(n)$ grows with the subpart size n. The slope obtained by linear fitting the logarithmic values of F and n is an estimate of the scaling exponent α. The integrated curve divided, respectively, into two and six subparts is shown in Fig. 16. The summed filled area between the linear fit curves and the fluctuating integral curve is taken as the measure of fluctuation.

The self-similarity analysis approach discussed above in detail is just one example of an analysis technique designed to investigate the signal processing within a biological cell when direct observation of internal variables are not available and one must rely on indirect analytical approaches to the internal processes within the cellular signal processing network. Other methods include memory time analysis (Amnuanpol, 2009), phase-space reconstruction analysis (Amnuanpol, 2009), and nonlinear cascade analysis, for example, linear–nonlinear–linear or nonlinear–linear–nonlinear modeling (Korenberg, 1991). These approaches require knowing only inputs and outputs and sometimes only knowing an output.

4. Analysis of Stable Beating

We found that the CBF is strongly feedback regulated (see below) in at least three ways. Its constancy may be stated in terms of its standard deviation of CBF. In terms of CBF the degree of its correlation depends on the timescale. So this is a means of finding deterministic feedback control and its timescale, since feedback makes it possible for responses to become highly correlated.

D. Analysis of Externally Stimulated Responses and Calculation of Their Stimulus–Response Functions

1. Stimulus–Response Functions and Their Identification

The temporal correlation of output measures or final outputs or any observable intermediates, with inputs yields a SRF (a plot of one of the parameter space variables vs time as in characterization of phase-space trajectories). They encapsulate in condensed form all the deterministic correlated attributes of the dynamical response system with respect to the measured variable. Different measures will have different SRFs. The obtaining of SRF assists in identification of the rules and geometry of the

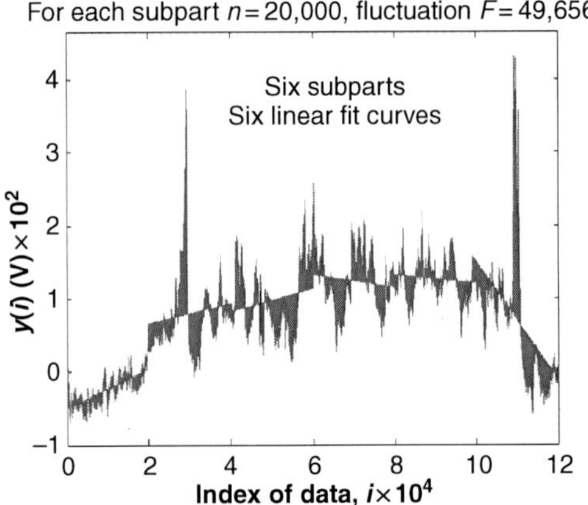

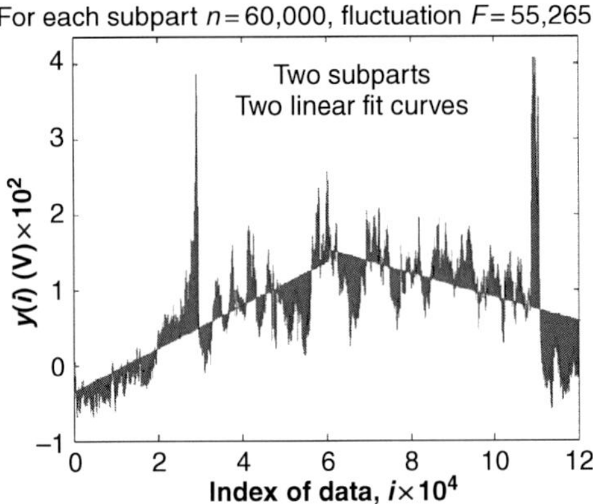

Fig. 16 The figure shows the data of Fig. 15B subdivided into six and then two subsets, part of the procedure in calculation of the fluctuation as discussed in the text (Amnuanpol, 2009). (See Plate no. 14 in the Color Plate Section.)

parameter space. Stimuli may not only be due to green or red light, which can be meaningfully modulated up to 100 Hz (see Section III.E.12), but also to electrical (Yoshimura *et al.*, 1997) or mechanical (Yoshimura, 1996) stimuli. Both electrical and mechanical stimuli may be modulated at as high temporal frequencies as can the light stimuli. Temperature may be modulated up to only about 20 Hz, because the modulation is limited by the thermal diffusion coefficient of water (0.14 mm^2/s, about a third of the heat energy moves beyond 100 μm in 18 ms).

Usefully SRFs predict for any arbitrary input what a deterministic response will be for a variable. The cell response can be predicted for any arbitrary light temporal pattern such as what a cell sees as it rotates along a swimming path. In this way, one can predict what a cell would do under any natural circumstance. For example, one may predict how it would respond if it were rotating at specific angles relative to the light. Another important and generally unappreciated virtue of an SRF is that one can study why the response parameters have the values they have, by observing what they predict would happen if the SRFs were perturbed from their nominal values. SRFs can provide particular insight into how cilia work and how mutants differ from wild type as they can be determined with both high temporal resolution and reproducibility.

One approach to reducing the noise in a measured response is to smooth or average adjacent temporal data points with loss of temporal resolution. A more useful approach is to average the correlations between continuous stimulation and output response maintaining high temporal resolution without smoothing. This averaging of correlations for long times is legitimate because *Chlamydomonas* ciliary beating is stationary for at least 2 h as shown by Josef *et al.* (2006). The correlations are between the response and the stimulus given up to 600 ms earlier for fast responses and to 40 s earlier for slow responses. This averaging improves temporal resolution and increases sensitivity to small changes of the deterministic linear and nonlinear responses caused by the modulated inputs.

The SRF is most frequently presented as an impulse response. For a linear system with only a first-order response, this corresponds in the time domain to the magnitude of the response to an infinitely short unitary pulse (delta function) as a function of time. For a nonlinear system, the SRF may be represented by values on additional time axes. For example, for a second-order nonlinear system, in addition to the first-order impulse response which is now the linear component of the nonlinear system (not the response to a single impulse) determined under the assumption of linear superposition of response to inputs, there is the second-order impulse response corresponding to the interaction of two unitary impulses at the input that alter the response relative to the linear expectation described by the first-order impulse response. While not easy to represent visually, this can be extended to higher order influences. Typically, orders higher than three are not needed to describe a biological dynamical system. Procedurally, one calculates the linear impulse response, subtracts the output it predicts from the observed response, and then calculates the second-order response on this residual output. Any real system is necessarily to a degree nonlinear (they violate the principle of superposition) and dynamic; hence, nonlinear SRFs must be determined. While linear system identification as discussed above is useful, the responses of *Chlamydomonas* are demonstrably nonlinear even over limited operating ranges nonlinear system identification is needed. Since Josef *et al.* (2006) determined the nonlinear component as well as the linear, they were able to show the linear model accounts for about 80% of the observed variance.

Our primary technique is correlation of responses with modulated stimuli: light, pressure, mechanical, chemical, temperature, ions, etc., or internal intermediate variables. The nonlinear (violates principle of superposition) SRF is the relationship between some measured activity (any observable intermediate that is dynamic or behavior) and any stimulus or measurable input for as long as the response is observable. This measurement can be simultaneous multiple inputs and multiple outputs providing explicit evidence of the interaction of these pathways. General practice is to use "external" nonlinear systems identification (Nelles, 2001) to represent output relative to input when the SRF cannot be associated with internal components. The technique is very efficient and wastes no time calculating speculative internal processes. With this correlation specifying the average nonrandom response of the system, we can simulate what the cell would see as it rotates in its environment as determined by the eye's directivity and then predict the cell's swimming. The whole process may be subdivided. For example, the electric field across the plasma membrane probably elicits the immediate (shorter than 300 ms) response of the cilia to light. Comparing the electric field response to light with that of motions of each cilium will determine what each cilium does in response to the electric field signal. We can use this representation to evaluate how much of the behavior is captured by a model such as a simplified linear model or internal state models (reviewed by Poon and Merfeld, 2005).

Once internal modular components have been identified they may be incorporated into "*internal* state" models, keeping "external" representations for the unknown parts. To gain a preliminary appreciation of the dynamics of the network, a model may be calculated with linear and nonlinear modules with a minimum of parameters consistent with the cell compartments, electric field, and diffusive signaling to the two cilia in the case of *Chlamydomonas* (Foster *et al.*, 2006; Josef *et al.*, 2006). Identified nonlinear modules will aid in ordering the signal network.

2. Responses to Square-Wave Stimulation

Classically, a square-wave stimulus is probably the first applied to identify how a system is responding. The temporal derivative of the response serves as a first-order estimate of an impulse response. The responses are typically the first indications one has that the system is not linear, in that the response to a step-up stimulus is nearly always not the negative of the response to a step-down stimulus. This is not surprising because of the primarily biochemical nature of biological systems and the decidedly nonlinear nature of electrical signaling in a cell. Chemicals do not have negative concentrations like a voltage which can be positive and negative. The degree a protein is phosphorylated is determined by kinases and phosphatases, which are separate systems with their own kinetics. There may in fact be many feedback processes selectively increasing or decreasing a specific variable.

3. Responses to Sinusoidal Stimulation

a. Transfer Function Analysis, Impulse Response and Frequency Domain Analysis. One way to measure the linear SRF is to measure the responses to a series of sinusoidal waves of different temporal frequencies, that is, measure its frequency response, in a similar way as an audiologist measures hearing. From the gain and the phase delays as a function of frequency (the Bode plot) one may simply calculate the impulse response. This is a very useful and worthwhile check on the Gaussian white noise (GWN) method below, but is not nearly as efficient or as low noise. One may compare Figs. 3 and 4 using sinusoidal illumination with Figs. 6 and 7 using GWN illumination in Josef *et al.* (2006). Furthermore, sinusoidal stimuli form an important part of "loop analysis" (a method for studying the stability of modules interconnected by feedback) (Aström and Murray, 2008). This approach allows for identification of stability margins of those local loops.

b. Use of Sinusoidal-Wave Stimuli to Accurately Measure Delays Due to Transport or Chemical Cascade Delays. A plot of the delay of sinusoidal response to a sinusoidal stimulus over several orders of magnitude of varying frequency is an accurate way to measure the time delay in a system. Often a delay may be characterized by a single number, for example, the phase delay, ϕ, in radians equals $-2\pi f\tau$, where f is the measurement frequency (cycles/second) and τ is the phase delay in seconds (see Section III.E.11).

c. Response Waveform Analysis. Since the response to a sine-wave stimulus is another sine wave if the dynamical system is linear, analysis of the response waveform is quite revealing. For example, the response might be the square of the input waveform or a much higher power. Cell biochemistry is quite nonlinear in detail although for reasons of needed stability, it may casually appear relatively linear.

d. Sinusoids Approximately Simulate the Natural Signal for Phototaxis. Since the cell rotates ~2 Hz, a stimulus of this frequency will emphasize the most important frequency component of the natural signal. See Section III.E.14 for an example in which the actual signal is only very approximately sinusoidal, but nevertheless maintains the same fundamental frequency.

e. Identification of Negative Feedback by Breaking with Positive Feedback. It could be that some parameters appear to be tightly held, that is, lie at the bottom of a steep-well attractor. Any small disturbance immediately brings the system back to a specific level implying the existence of a feedback network. However, such a condition may be explored more fully using loop analysis by forcing the system via modulating a critical variable over a wide range of frequencies. If the attractor (or the behavior) is sensitive to the variable, then at some frequency the phase margin of the feedback loop will be exceeded and the system will dramatically depart from stability as negative changes to positive feedback. For examples see Section III.E.12.

4. Identification of Stimulus–Response Functions with White Noise Stimulation

a. Use of Gaussian White Noise as a Stimulus and Use of Correlation Analysis. With a computer, one may generate a random signal that has a Gaussian distribution of intensities (GWN). Stimulation with GWN is the most efficient form of stimulation for identification or characterization of a cell signal processing network, that is, the SRFs. The efficiency comes from the fact that all temporal frequencies are tested at once rather than separately. This favored stimulus is extremely rich. The stimulus variable [e.g., light intensity, I, or, for spanning a much greater dynamic range, $(I-I_0)/(I+I_0)$] is modulated by a pseudorandom signal with an amplitude probability density that is Gaussian (normal distribution), and power spectral density that is essentially flat up to some bandwidth limit that extends just up to an octave or so beyond the system bandwidth as excessive stimulus bandwidth increase noise.

b. Achieving Submillisecond Temporal Resolution Even Though Data Are Sampled at 24-ms Intervals or Longer. It is sometimes not appreciated that, provided that times of data sampling are accurately known and the signal is sampled for a long time, temporal resolution of response may be derived that is much higher than the sampling interval. If one naively gave a pulse stimulus and synchronously sampled response every 20 ms following the stimulus, one would have 20-ms resolution. However, if instead one gave pulses at known, but random times, relative to the 20-ms sampling times, with sufficient repetitions, one can readily achieve better than 1 ms resolution in the SRF relationship. For a phenomenon like CBF where one cycle (~18 ms) is much longer than the latency (<1 ms), this approach is necessary to follow the control. If we repeated the pulse stimulus every second for 2 h, we would have 7200 repeats, which would increase our sensitivity to a signal by ~85 times. In actuality, we do something equivalent, which is to computer generate a random signal that has a Gaussian distribution of intensities (GWN). The noise reduction is the same as for pulses, but this GWN stimulus does a better job because the stimulation stays within the dynamic range in which the cell normally encounters without shocking the cell (equivalent to entering sunlight from a darkroom) with a large pulse stimulus.

c. Nonlinear External Nonparametric SRFs: An External System Identification Method Using Parallel Cascade Method. Simple electronic devices of undergraduate education may be characterized by voltage inputs and outputs that can range from positive to negative. However, biological signaling networks involve concentrations of enzymes, second messengers, and photons that are always positive. Nature's solution is to make one biochemical system to increase, say methylation or phosphorylation, and makes a second biochemical system with different kinetics, say demethylation or dephosphorylation, to decrease the level of active biochemical substances (Voit, 2000). Nature also makes use of rectified electrical signals as well as the biochemical systems. Analysis tools should work toward reflecting these component properties to more easily identify the biochemical and electrophysiological components.

There are many "external" or "nonparametric" nonlinear systems identification methods (Kantz and Schreiber, 1997; Marmarelis, 2004; Nelles, 2001; Westwick and Kearney, 2003; Thompson, 1999) to represent output relative to input when the SRF cannot be associated with internal components. Josef (2005) has used the parallel cascade method of Michael Korenberg (1991). The parallel cascade topology is simple. An individual cascade consists of a linear dynamic element, L, following by a nonlinear static one, N (this arrangement is commonly known as a Wiener cascade). Typically, a system will be represented by several LN cascades in parallel, such that each cascade (with L as input element) receives the experimental stimulus and the outputs of all cascades (specifically their

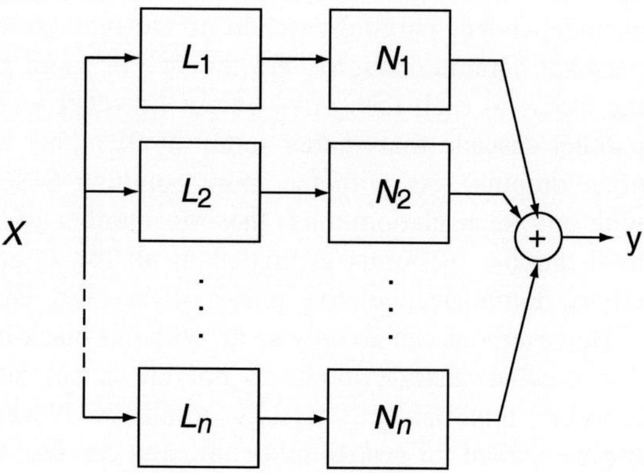

Fig. 17 Parallel cascade architecture as discussed in the text.

N elements) are summed to constitute the estimated or predicted output of the system.

Figure 17 shows the parallel cascade system identification architecture for a single-input single-output (SISO) system. Each path consists of a linear dynamic element followed by a nonlinear static (memoryless or instantaneous) element. L_1 is computed by cross-correlation of the output y (specifically using the experimental response y to what is usually chosen to be a pseudorandom GWN input signal x) with the input x; then N_1, which is usually represented as an ordinary polynomial [alternatively, splines (Wahba, 1990) may be used], is computed by linear least-squares regression between the output y and the output of L_1. The simplicity of determining the nonlinear elements, even when of high-order nonlinearity, is one of the advantages of parallel cascade identification (PCI). A benefit of PCI is that Volterra kernels can be obtained directly from it. This alternative representation of the input–output relation is the most useful form for interpretation. Insofar as the system is approximately linear, one can employ linear system theory in the time and frequency domains using the first-order Volterra kernel as impulse response, and its Fourier transform as transfer function. These can be fit, most easily in the frequency domain—where calculus conveniently reduces to algebra—using analytical models of the system dynamics. Then one can generalize the linear system results to model the nonlinear dynamics as well. Furthermore, one can make analytical models fit by complex-valued nonlinear least squares (nonlinear regression) with full error analysis (using variance–covariance matrices for data and for parameters), and using formal error propagation for any formulas that depend on multiple parameters so derived. Besides allowing standard errors to be quoted with the parameters, this error analysis also enables hypothesis testing.

In the lab the multi-input single-output PCI has been implemented and a straightforward extension is to multi-input multi-output (including, as a special case, single-input multi-output). The code extension for multi-output is much simpler than that for multi-input, in that one simply provides, for each output, an independent parallel cascade of the type shown in Fig. 17 above (Korenberg, personal communication), and the two or more parallel cascades branch out from the inputs to each respective output (in other words, each output has its dedicated parallel cascade that shares some or all inputs with the parallel cascades for the other outputs). As with the more common SISO case of PCI, one must ensure, with any such elaborations, that the number of data points acquired exceeds the total number of points estimated in all the L_i and N_i. Multiple outputs could be ciliary dominance, relative phase of the cilia, and the *cis* CBF.

However, one can go only so far with this black-box system-identification approach. The parallel cascade model is not the actual internal architecture of the system; however, importantly it is fully predictive. Nevertheless, system analysis methods can be applied for system modeling, and can lead to clues concerning the underlying biochemical and cellular dynamics of the system. Of course, with the kernels and transfer functions obtained one is probing the rate-limiting kinetics of the system. Reactions that proceed on a much faster timescale are, in effect, instantaneous and are thus inaccessible to these system identification methods, in view of the experimental bandwidth limits. That is not a significant disadvantage, though, since we are mainly interested in the processes that govern the manifest dynamics because it is these processes that nature mostly uses for control, and these kinetic processes correspond to the rate-limiting reactions. System identification is particularly useful when combined with genetic dissection of behavior, as was done with analysis of single and double mutants of *Phycomyces* (Poe *et al.*, 1986) and has begun to be done with *Chlamydomonas* single mutants as discussed in Section III.G.1, and potentially double mutants in the future.

5. Breakdown of Transfer Functions into Few Parameter Models

The above external nonparametric models are purely descriptive and their usefulness lies in their being very compact and relatively quick to compute, although not unique descriptions. To gain physical and biological insight, it is necessary to follow up on these descriptions with parametric functional models with a minimum of parameters that model most of the data and provide the same predictive power. Biological dynamical systems are typically formed by interconnected interacting modules. The different compartments such as cilia and eye come to mind, but within these compartments are smaller interconnected modules that form the basis of the signal processing. Each of these smaller modules might have three, four, or more proteins that have internal feedbacks, an input, and an output. Usually, they can be characterized by very few parameters along with a simple transfer function inspired by Bode plots. If a module is nonlinear in character it can be located within the network. If modules are linear they cannot be uniquely placed unless intermediate signals can be identified, but their identification remains useful as an intermediate step in the full understanding of the network or control system.

E. Application of Ciliary Analysis Techniques to *Chlamydomonas* Ciliary Beating Recorded With a Quad Photodiode

1. Self-Similarity Analysis

In Fig. 18 the scaling exponent (slope of the line) (see Section III.C.3) of the CBF is close to that of the stroke velocity and the *cis–trans* phase difference below 24 s, but above this time it changes its value. Both the stroke velocity and the *cis–trans* phase difference seem to share the same scaling exponent of about 2/3. The value is greater than 1/2 implying that there is more than average persistence of the current value. A possible explanation is that due to the high concentration of molecules within the cell there is subdiffusion rather than regular diffusion, where it is assumed each molecule is able to move independently. For random diffusion the mean squared distance is proportional to time, t^{α} where $\alpha = 1$. For subdiffusion, $\alpha < 1$, the concentrations are more correlated than if they could move freely. The scaling component can be equivalently thought of as the exponent of the decay of autocorrelation. It is slower where the exponent is $-2/3$ compared to -1 for the random case, and the noise spectrum is $1/f^{1/3}$. For the

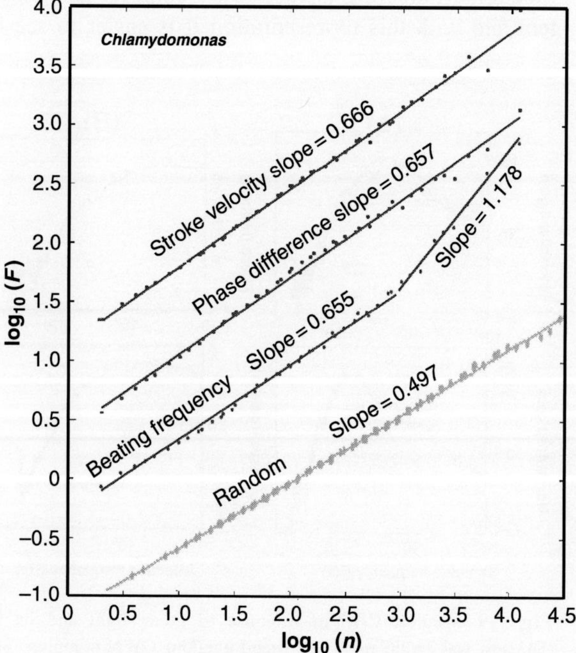

Fig. 18 Results for a 192 s time series of *Chlamydomonas* ciliary beating recorded using the quadrant photodiode technique. The series has been analyzed for the stroke velocity (0.666 slope of fluctuation vs log time), CBF (the slope is 0.655 and 1.178 for times shorter and longer than 24 s, respectively), and *cis—trans* phase difference (0.657) (Amnuanpol, 2009).

CBF the increase in the scaling exponent at long timescales is also observed in the intertrade time series of the stock market (Ivanov *et al.*, 2004) and probably for the same reason, namely, that some monitor recognizes what is happening at a certain timescale (next day in the stock market case) and acts accordingly. In the stock market, investors after 1 day recognize the market is going up or down and push it more in the same direction.

2. Uniqueness of *Trans* Versus *Cis* to Rhodopsin Activation

With our low spatial resolution quad photodiode ciliary monitor, we measured the correlation of the ciliary responses with stimuli. Most classical analysis is actually done in the frequency domain because of the simplicity of its interpretation. This approach is represented by Bode plots (A and B) showing the CBF (Fig. 19) and stroke velocity responses (Fig. 20). The output response was correlated with a 196-s-long input light stimulus (Josef, 2005; Josef *et al.*, 2005a,b). The slopes of the gain plot and the phase changes provide directly the nature of the functional processes. For example, the slope at high frequency gives the number of processes involved in the cascade of the processing and the functional modeling is easy. The temporal response (C) is mathematically equivalent and with this representation it is easier to see how the cell would respond as

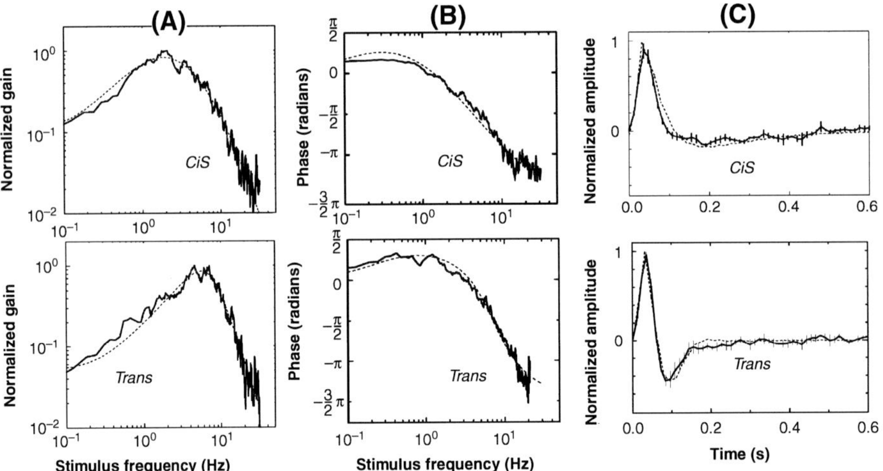

Fig. 19 Typical CBF in response to green light and its linear modeled response. Green light (543 nm) was modulated by a pseudorandom GWN stimulus. The *cis*- and *trans*-CBF for each record were separately cross-correlated with $\log_{10}$ (stimulus intensities) to produce respective impulse responses. The *cis* and *trans* responses were then separately averaged [solid lines in (C) marked *cis* and *trans*]. The Fourier transform of the averaged impulse responses yielded the gains (A, solid lines) and phases (B, solid lines). The dashed lines show a linear model of the responses (from Fig. 6 of Josef *et al.*, 2006).

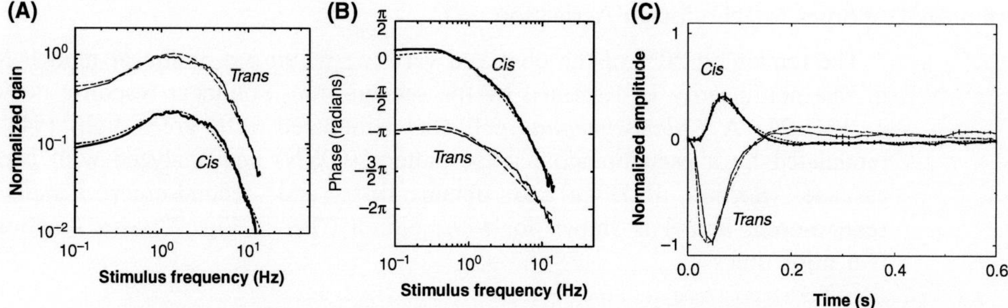

Fig. 20 Typical ciliary stroke velocity response to green light and its linear modeled response of the negatively phototactic strain 806. Green light (543 nm) was modulated by a pseudorandom GWN stimulus. Stroke velocities for the *cis* and *trans* cilia for each record were independently cross-correlated with $\log_{10}$ (stimulus intensities) to produce impulse responses that were then separately averaged in (C) for *cis* cilium (thick solid line) and *trans* cilium (thin solid line). The Fourier transform of the averaged impulse responses produced gain (A, solid lines) and phase (B, solid lines). The dashed lines show a linear model of the responses (from Fig. 7 of Josef *et al.*, 2006).

it rotates, but functional modeling is less transparent. The *trans* and *cis* cilia are different. They have their own unique as well as common responses. For example, the CBF and stroke velocity (SV) SRF are quite different as seen comparing Figs. 19 and 20. On the other hand, response to changes in availability of ATP (not shown here) appear to be similar (Adulrattananuwat, 2009). Note in particular that the *trans* cilium has a much higher frequency response for CBF than the *cis*. Further that the phase of the *trans* cilium is about π different from the *cis* and that the delay is markedly, actually reproducibly, longer than for the *trans* cilium.

Such results provide insight into what the signal processing system does to optimize the cell's phototaxis. By averaging over 196 s, high temporal resolution is obtained in spite of the noisiness of individual ciliary responses. Note that the observed delays to response are short relative to a single beating cycle of ~18 ms. Furthermore, these measured responses show significant *cis–trans* differences which contribute to how the cell steers. Note there is an initial delay of the *cis* cilium stroke velocity in Fig. 20C and further the *cis* cilium has a slower CBF response than the *trans* in Fig. 19A. The peak times for *cis* and *trans* CBF are ~40 and ~30 ms, respectively, whereas the peak times for the stroke velocity are 80 and 50 ms. Why the system has *cis* slower than *trans* for both responses is not known. Figure 20C also shows that steering for phototaxis involves briefly altering the relative effectiveness of the stroke velocity of the *cis* relative to the *trans* cilium. The changes are as if the shape of the *trans* beat exclusively involves the outer dynein arms and the *cis* beat shape is exclusively determined by the inner arms, but with the frequency and power contributed by the outer arms (Josef *et al.*, 2005b). Linear approximations (dotted lines) of the SR functions are shown in Figs. 16 and 17. One should note in this example that 80% of the observed variance is captured by the linear model.

3. Nonlinear Responses to Rhodopsin Activation

The remaining 20% of the observed variance requires a nonlinear model. Most of this nonlinearity is described by the second-order nonlinear response depicted in Fig. 21. A *Chlamydomonas* cell was stimulated with green light (543 nm) modulated by a pseudorandom noise pattern (GWN) and analyzed with parallel cascade (Section III.D.4.c) to obtain first- and second-order kernels. A second-order kernel is shown for *trans* cilium CBF in Fig. 21 as a function of two prior times.

4. Steering Phototactic Responses to Green Light

Chlamydomonas steers relative to a light source by differential control of beating of its two cilia. The seven known behavioral responses are as follows: helical and super-helical negative and positive phototaxis, diaphototaxis, nondirectional, and photoshock (switch from ciliary to flagellar beating mode) (Fig. 22) (Foster and Smyth, 1980; Yoshimura *et al.*, 1997). The eye (to track the light direction) on the cell body signals the two cilia. The different responses of the cilia as shown in Fig. 20 make it possible for the cell to steer.

From the literature and preliminary work, turning response seems to be an increase in the curvature and rate of sliding in the R (reverse bend side) of one cilium versus a decrease in rate of sliding in P bend of the other cilium. One hypothesis is that all these modes of ciliary beating can be simulated or achieved by combinations of increasing or decreasing the rate of sliding on the P or R sides and controlling the timing and location of these changes. A different hypothesis is that the dynamic stiffness is alternatively varied to affect the changed phototactic beating patterns. To complete the picture one should add control of circumferential and length propagation of dynein activity and whether a response is transient or sustained (Josef *et al.*, 2005a).

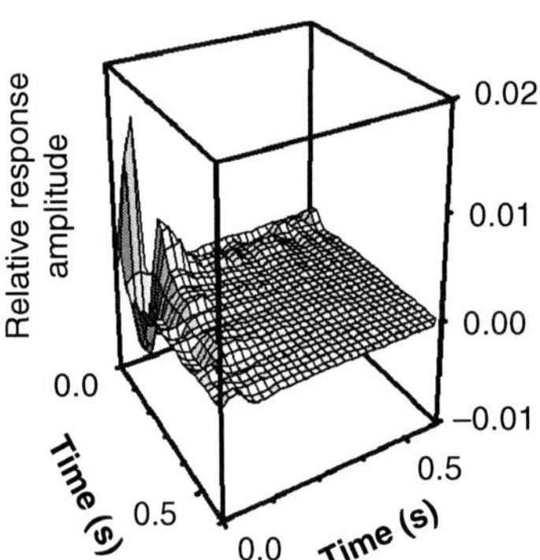

Fig. 21 The figure shows the second-order kernel, that is, a nonlinear response, for the CBF of the *trans* cilium (Josef, 2005).

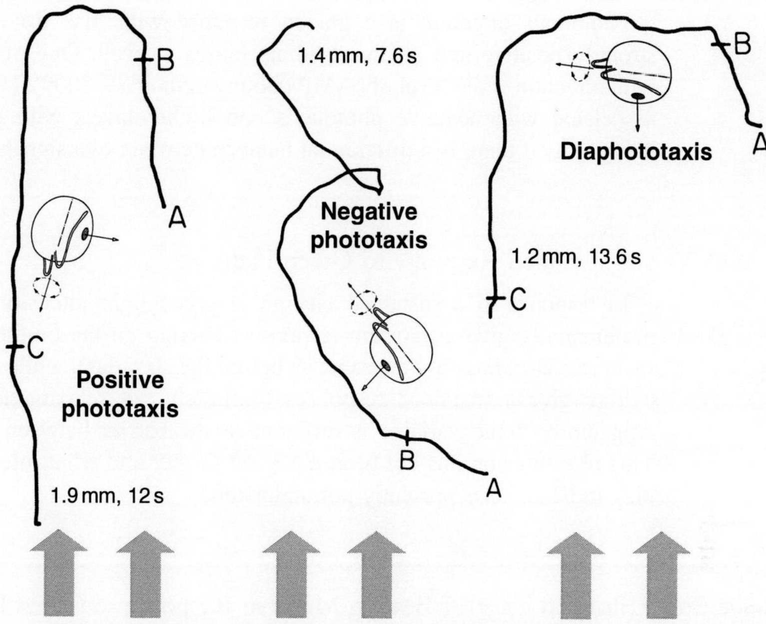

Fig. 22 The figure shows three of the seven types of phototaxis: helical positive, superhelical negative, and diaphototaxis (modified from Fig. 2 of Foster *et al.*, 2006). Light is on only from B to C. The tracing comes from the original data of Robert Smyth using Howard Berg's cell tracking machine (Foster and Smyth, 1980). (See Plate no. 15 in the Color Plate Section.)

5. Phototaxis Direction Control

Chlamydomonas always rotates in a left-handed helix (Foster and Smyth, 1980), because of the twofold rotation symmetry (C2) of the cilia and the tilt of their power stroke. In its helical swimming path the more effective cilium goes to the outside of the path (Isogai *et al.*, 2000). The eye of *Chlamydomonas*, which is near the *cis* cilium (by definition), will be on the outside of the helix if the *cis* cilium exerts a larger torque compared to the *trans* cilium. What determines direction, toward or away from light, is whether the *cis* transiently dominates the *trans* with light on, or with light off. In negative taxis the *cis* transiently dominates with light on, and *trans* dominates with light off (Fig. 20) (Isogai *et al.*, 2000; Josef *et al.*, 2006). In positive taxis the reverse is true. How the cell switches from among its phototaxis options and why a cell sometimes chooses superhelical versus helical paths are not yet clear, although it could be just a matter of light intensity in which at lower light the superhelical track allows longer evaluation of the light orientation than the helical track would. Calcium concentration appears to control this ciliary dominance. For positively phototactic demembraned cells, *cis* is dominant when extracellular [Ca^{2+}] is below 20 nm, *trans* dominant above 20 nm Ca^{2+} (Kamiya and Witman, 1984). Elucidation of the connection between the observed behavioral transient ciliary dominance and the reported calcium dependence of ciliary dominance of demembranated cells has not yet been reported.

Phototaxis direction is a graded response with bias from strongly negative to strongly positive and all intermediate biases as well. One of the factors correlated with direction is the level of cAMP (Boonyareth *et al.*, 2009). Low levels of cAMP are associated with negative phototaxis and higher levels with positive phototaxis. It behaves as if there is a differential balance between two signals

6. Transient Versus Sustained Response to Green Light

In response to a sustained change in green light intensity (Fig. 23) some cells preferentially give a transient response (relaxing on the timescale of the experiment to an attractor, possibly the same as before the stimulus), while other cells of the same culture give a sustained responses (definitely not relaxing to the same attractor). Apparently, some variable is different or the barrier between two attractors is low. This phenomenon has not been analyzed further and what role or advantage it might play in behavior is presently not understood.

7. Change from Ciliary to Flagellar Beating Mode in Response to Green Light Stimulus

When the calcium concentration is raised in the cilium to greater than 0.1 mM (Hyams and Borisy, 1978) typically due to a green light initiated calcium action potential in the cell, then the waveform changes from ciliary (asymmetric breaststroke) to flagellar (symmetric whip-like) beating (Yoshimura *et al.*, 1997). According to Lindemann (2007), switching modes from ciliary to flagella beating only requires a change in the probability of dynein attachment on the P (principal bend side; see Fig. 2). This could be accomplished by the use of the radial spokes (Deiner *et al.*, 1993). From an analysis point of view, since the waves are propagated from the base it is very easy to follow the transition from one mode to the other with computer-processed ciliary images. For the low spatial resolution quad diode detector, accurate identification of the beating mode change is problematic since it can be mistaken for cell stopping.

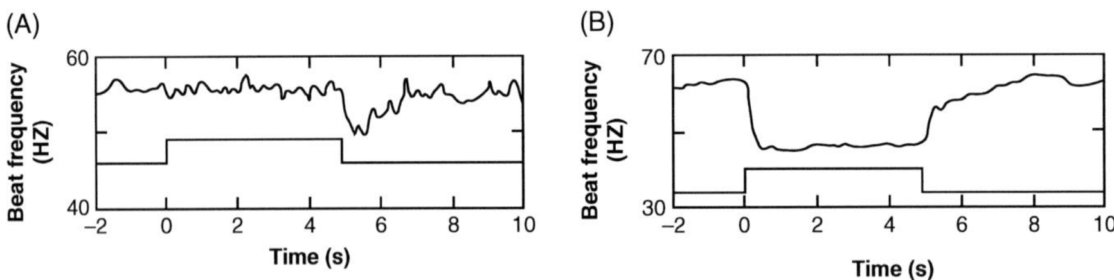

Fig. 23 The figure shows an example of sustained versus transient response for the same step light stimulus (from Fig. 3 of Josef *et al.*, 2005b).

8. Rate of Intermittency, Attractor Merging, Optimization of Dispersal, Modification with Green Light Stimulation

Each compartment is likely to have its own signal processing attractors. The *trans* cilium switches between a normal beat (like the *cis* cilium) and a shallower beat with higher frequency (Fig. 3C, D of Brokaw *et al.*, 1982). The shallow beat is characterized by a much slower rate of sliding in the P bend and a low shear angle change. We have collected statistics on these brief periods that have long been reported (Rüffer and Nultsch, 1985, 1987) during which the CBF of the *trans* cilia is suddenly increased and the synchrony between the *cis* and *trans* cilia is lost for a number of beat cycles (Fig. 24B–D, Josef, 2005). It is particularly prominent in the wild-type strain (1117). Comparatively, the mutant 806 (*agg*1) which has a low cAMP level (Boonyareth *et al.*, 2009) shows less of this behavior. Figure 24A shows two shifts in frequency in strain 806. One hypothesis is that these nonsynchronous periods represent a change in the control mechanism, for example, a consequence of a change in a parameter such as Ca^{2+} or cAMP response within the cilium. The behavioral consequence is that the swimming motion of 806 cells is more ballistic on a timescale of 10 s (Section III.H).

The probability of these transient events changes with green light stimulation as has been reported (Josef *et al.*, 2005a, from Fig. 9). This stimulus-dependent behavior is shown in Fig. 25.

9. Base Orientation Angle of the Two Cilia

As a measure of the base orientation, Josef *et al.* (2005a) used the relative phase between the forward and back quadrant photodiodes looking at the same cilium. It indicates a change in the cilium angle as a function of light intensity (Fig. 26). The response is *not* transient and in the series of responses shown the shift is a linear function of light intensity. The shift probably reflects the fact that the distal fiber (dsf, Fig. 6) connecting the *cis* and *trans* cilia and the NBBCs to the base of each cilium contain centrin (caltractin) contract with higher levels of calcium. Together they cause a linear change with light intensity (Fig. 26).

10. Response to Light Level Changes of Red Light

Figure 27 illustrates that the CBF response to 670 nm red light. The light-on response is markedly different from the light-off response (Josef, 2005; Adulrattananuwat, 2009). There is a first-order step-on response presumably due to the increase in ATP that photosynthesis promotes. One hypothesis is that as a consequence of this additional source, the ATP that is produced from the mitochondria is suppressed and its restoration after the red light is turned off is relatively slow resulting in an ~10 s delay (Fig. 27). As long as the red light is on the CBF is sustained at a higher level in this step experiment, providing another example of response that is not transient.

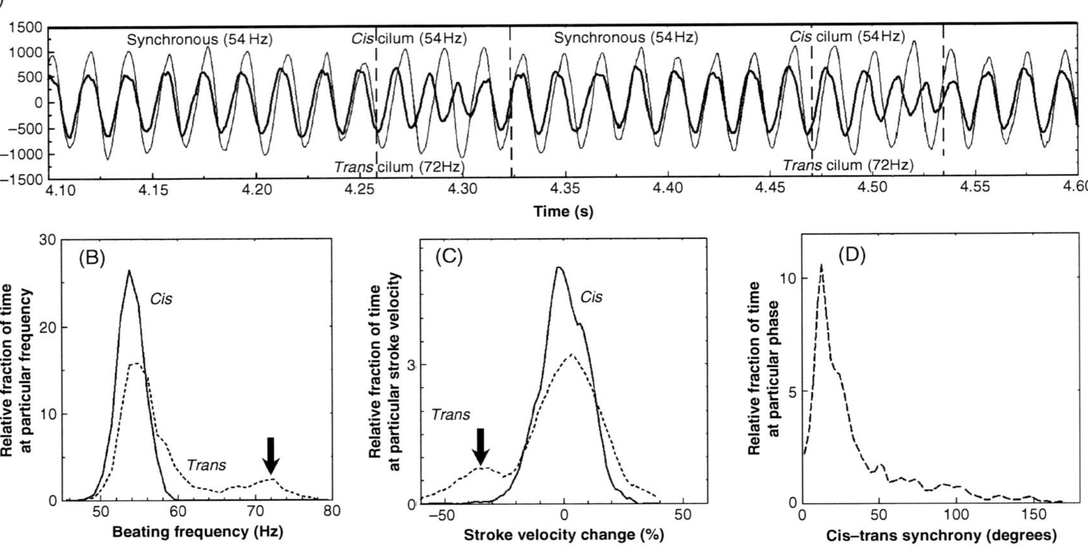

Fig. 24 The figure shows an example of intermittent shifts of the *trans* cilium to a higher CBF. In (A) the raw beating from the quadrant photodiode is shown, thin line for *cis*, thick line for *trans*. The higher CBF makes it possible to sneak in an extra beat before the cilia again become synchronized. This is in strain 806 (*agg*1) which has restrained frequency shifting relative to wild type. In (B) is the fraction of time the cilium spends at a particular CBF, in (C) is the fraction of time at a particular stroke velocity, and in D is the fraction of time the *cis* and *trans* cilia have a particular phase relationship (Josef, 2005).

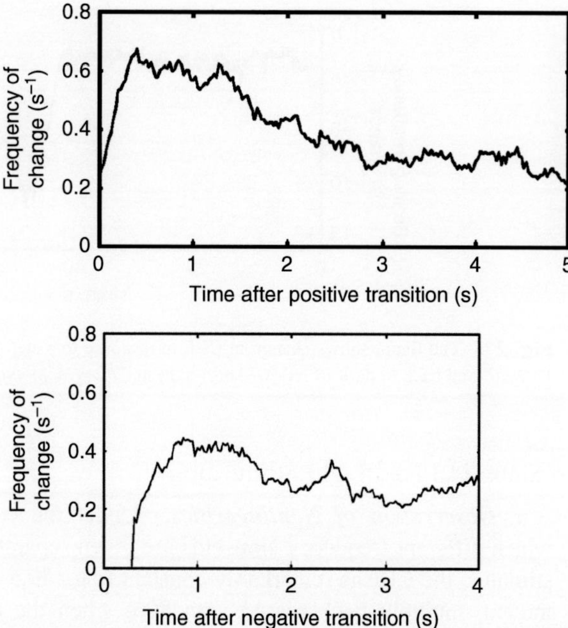

Fig. 25 Shows some of the dynamical changes in the intermittent shifts of the *trans* cilium to higher CBF in response to changes in light level, either increasing in the upper panel or decrease in the lower panel. The green square-wave stimulus was off to 1.5 W/m^2, *Chlamydomonas* strain was 806. The frequency of occurrence for these momentary pattern changes was determined by repeating the green square-wave stimulus one hundred times (Josef, 2005).

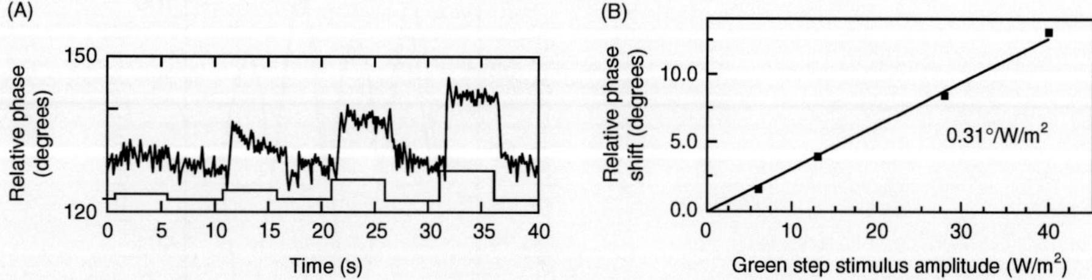

Fig. 26 Shows the change in the angle of *Chlamydomonas* cilia with respect to the cell body measured by the change in phase of the anterior quadrant minus the posterior quadrant signal as a function of green light intensity. The cell was subjected to 20 consecutive graded steps up from 0 to 6, 13, 28, and 40 W/m^2 as shown overlying the response in (A). The records were averaged to obtain phase difference between the anterior portion and posterior portion of the *trans* cilium stroke (Q1 and Q2). Part (B) displays the relative phase change from the dark (no green stimulation) baseline as a function of green-step stimulus amplitude (from Fig. 8 of Josef *et al.*, 2005a).

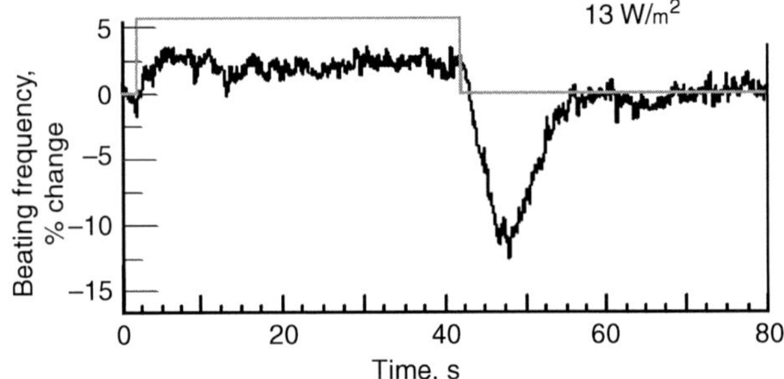

Fig. 27 The figure shows change in CBF in response to a step-up and step-down changes from dark to 13 W/m^2 and back to dark of red (670 nm) light in *Chlamydomonas* 806 strain (Adulrattananuwat, 2009).

11. Response to Sinusoidal Red Light Stimulation

a. Observation of Nonlinearities. Under the conditions of the experiment, in which different frequency sinusoids are given separated by a moderate constant light stimulus, the CBF is remarkably constant (e.g., 49.5 ± 0.6 Hz observed with the *cpc*1 mutant, multiply backcrossed with 806) when the red light is constant, while the response to the red sinusoidal light comes out appearing as if the deviation from the steady-state response is squared (derived from waveform analysis) (Fig. 28). If the SRF were linear the observed response to a sinusoidal input signal would be sinusoidal, so if it is not, it is valuable to carefully analyze the waveform to identify the nonlinearites. In this particular case two nonlinearities are observed, namely, rectification and squaring, but other nonlinear transformations are frequently seen. These nonlinearities are valuable in ordering the processes that takes place in the cell as their placement in a model

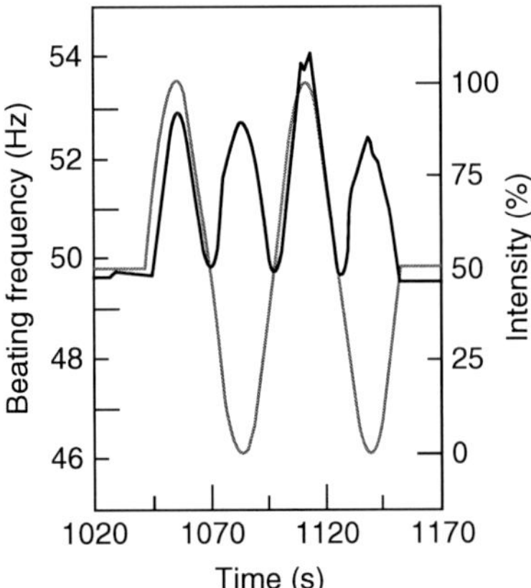

Fig. 28 The figure shows the CBF in response to red light that is modulated at 0.018 Hz. The sinusoid is the modulated red light. The half-wave rectified and squared curve is the actual CBF (Adulrattananuwat, 2009).

has very different effects depending on the order of the individual component processes. How the response in Fig. 28 comes to be this way it is actually complicated (Adulrattananuwat, 2009), but it illustrates the very strong response to modulation of the ATP level in the cilium.

b. Measurement of Delay by Determine Phases of Response to Sinusoids. In the case discussed above, one can learn the delay between the stimulus and the response by measuring the phase delay of response to the succession of sine waves. As can be seen in Fig. 29 the phases are fit very well with a single parameter, namely, by $\phi(f) = -2\pi f\tau$, where τ is the constant delay time of 1.18 ± 0.02 s relative to the modulated red 670 nm light stimulus. This delay represents the total time for light to activate the chloroplast to produce energy-carrying molecules, a chemical cascade and transport mostly by diffusion to the cilium, and finally the conversion of 3-phosphoglycerate into ATP in the cilium to be used by the dynein motors.

12. Breaking Feedback Control

An implication of the remarkably constant CBF reported above is that it must be feedback controlled. When feedback is suspected, more can be learned about the system by attempting to break the control, in particular, one can immediately learn something about the kinetics of that feedback. In Fig. 30 there are very strong responses of *Chlamydomonas* cilia to rapid modulation of red light in the range of 7–100 Hz, which presumably modulate ATP in the cell at these frequencies. This is the range that feedback regulation of ATP could break down in the mitochondria. At lower

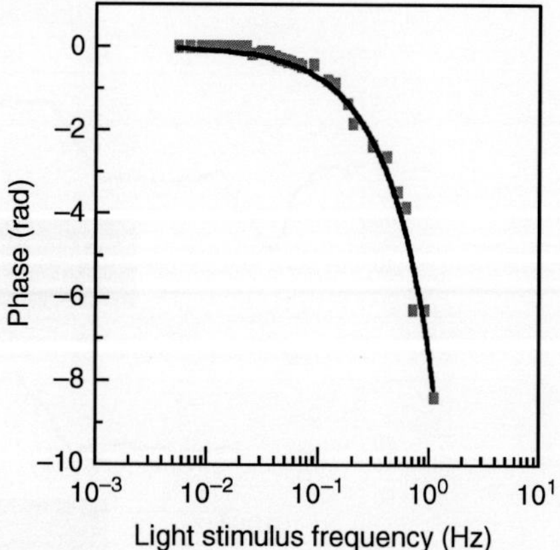

Fig. 29 The figure shows the phase delay plot for modulated red light from 0.006 to 1.1 Hz. The curve is a single parameter fit for the delay giving a net delay of 1.18 ± 0.02 s. This delay is from the red light being absorbed by chlorophyll in the chloroplast to the change in CBF at the cilium. It includes the time for synthesis of metabolites in the chloroplast, transport across that membrane, diffusion through the cytoplasm, transport up the cilium, and synthesis of ATP in the cilium (Adulrattananuwat, 2009).

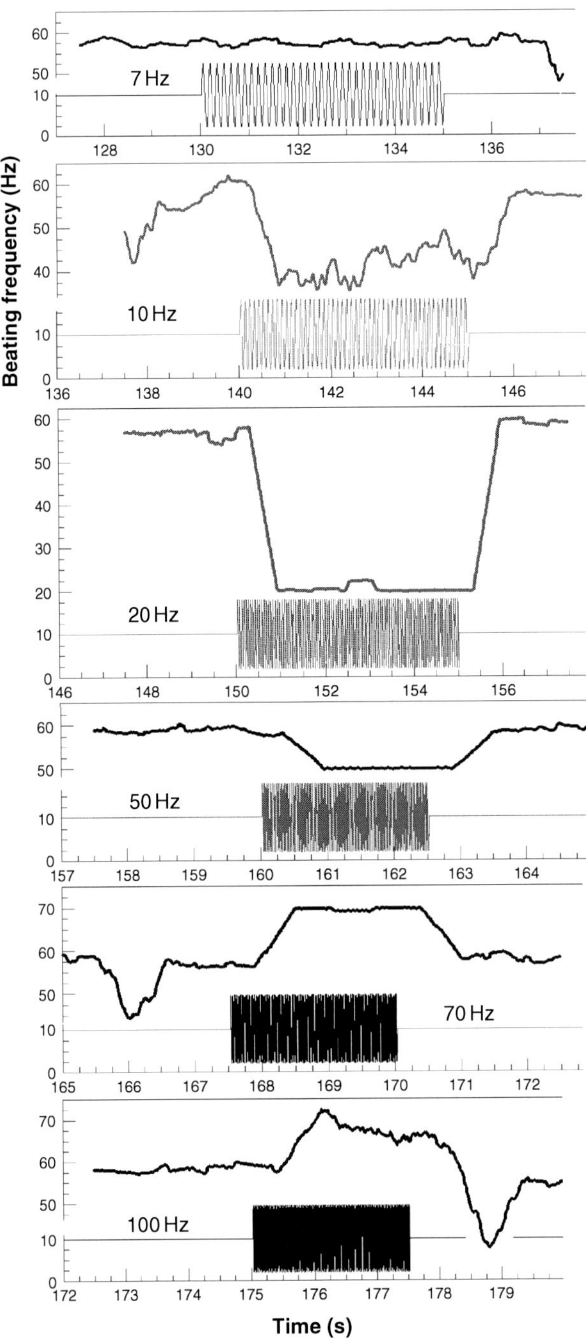

Fig. 30 The figure shows the change in CBF in response to modulation of red light in the range from 7 to 100 Hz (Adulrattananuwat, 2009).

constant light levels the CBF and stroke velocity are relatively constant, in fact, very constant as already noted. However, the CBF (Fig. 30) drops precipitously to the very low-frequency range of ~15 Hz CBF at the 10 and 20 Hz stimulus frequency. In this range, the stroke pattern (not shown) becomes much exaggerated as well giving a very large signal as the cilium crosses from one quadrant to another. On the other hand, up at 70-Hz stimulation frequency, the CBF raises well above normal to what is probably maximum to 68 Hz at this temperature. The amplitude of the stroke pattern at this high frequency becomes reduced as expected (Adulrattananuwat, 2009).

13. Intracellular Signal Processing Associated with Phototaxis, a Functional Representation

One of the purposes of analyzing ciliary beating data is to understand how the cilia are controlled. The processing of signals makes it possible for a cell to swim relative to a source of light and to control in a graded way its bias toward or away from light. From a functional viewpoint how the system does both attributes is presented in the following figures. The proposed functions are obtained directly from the Bode plots like those shown in Figs. 19 and 20 and responses to sinusoidal inputs in the frequency range of cell rotation. The Bode plot approach is popular because of the relative ease of their interpretation.

To illustrate the ideas it is useful to consider a simpler cell than *Chlamydomonas*. If in a gain–frequency plot there is a +1 slope as shown in Fig. 31 (for *Escherichia coli*), it implies directly that the signal processor involves differentiation of all frequencies (*f*), typical for sensory responses. Figure 31 shows a pure differentiation gain function, $G_1(f) \propto 2\pi f$ with phase $\phi(f) = \pi/2$ (not shown) seen in isolation from 0.001 to 0.1 Hz (<log frequency −0.8). Superimposed on this differentiation is the effect of additional filters effective at higher frequencies. These multiply the differentiated signal which is likely to be the first component in the signal processing pathway as it deals with the wide range of level of input signal in the most effective manner. Due to the noise and the similarity of different interpretations, the higher

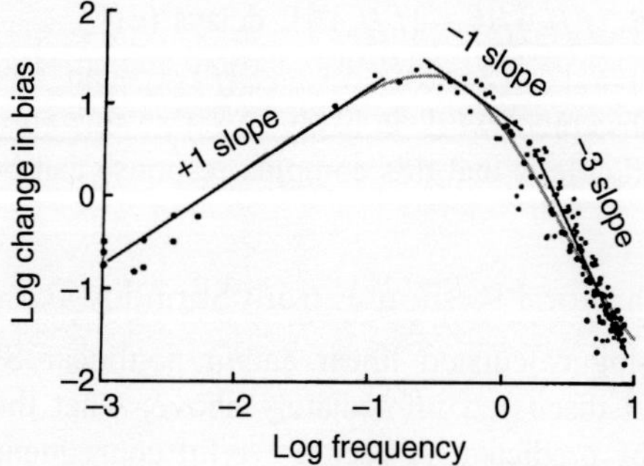

Fig. 31 The change in the clockwise-counterclockwise bias in flagellar rotation of *E. coli* resulting from modulation of attractant, plotted as log (change in bias) versus log (frequency) of temporal modulation of the concentration of α-methylaspartate. The low-frequency data set comes from exponentiated sine-wave stimuli, while the high-frequency data comes from Fourier transforming the impulse response (from Berg, 2004).

frequency interpretation is less precise. The simplest interpretation might be two simple second-order low-pass filters of the form $G_2(f) = [1 + (2\pi f T_1)^2]^{-1}$, $\phi(f) = -2\tan^{-1} 2\pi f T$, where T_1 is ~0.8 s for the lower frequency filter taking the +1 to −1 slope and the higher frequency filter taking the −1 to −3 slope where T_2 is ~0.11 s. The total gain, G, is the product of the gains, the total phase shift is the sum of the individual phase shifts. Corresponding to the differentiator and the two filters, there are undoubtedly biochemical modules that perform these three functions. In this particular case the differentiator is well known (Tu *et al.*, 2008).

As one might anticipate a eukaryotic cell like *Chlamydomonas* is an order of magnitude more complex than a prokaryote. The 93 sensory transduction genes probably involved in signal processing just in the cilia imply a moderate level of complexity (Pazour *et al.*, 2005). By comparison, only nine proteins have been identified in the chemotaxis signal processing of the bacterium *E. coli* (Berg, 2004). The number of photoresponse options identified so far is seven rather than the two of *E. coli*. Both diffusive biochemical and fast electrical networks are integrated to carry signals and the system involves the interaction of at least five compartments (cytoplasm, two cilia, mitochondria, and chloroplast). Nevertheless, one can do a similar analysis in terms of our modular attractor network architecture model (Fig. 32, for the response to green light). First the photons are integrated and then there is a nonlinear log or modulation function compression of the green light signal followed by a differentiator, followed by a splitting into two pathways, one leading to a rapid pathway (seen in WT and *agg*1 mutant) and one slow delayed pathway (primarily in WT only). The slow signal insures maximum response at the time of the decrease in light intensity for *Chlamydomonas* rotating about 2 Hz, leading to positive phototaxis. It is the balance between these two response pathways that results in the graded phototaxis response depending on conditions. While the bacterial signal processing makes prominent use of a differentiator and two second-order low-pass filters, *Chlamydomonas* adds several new types of filters, namely, linear lead high-pass filters ($G(f) = [1 + (2\pi f T)^2]^{1/2}$, $\phi(f) = \tan^{-1} 2\pi f T$), quadratic lag second-order low-pass filters ($G(f) = \{[1 - (f/f_n)^2]^2 + [2(\zeta f/f_n)]^2\}^{-1/2}$, $\phi(f) = -\tan^{-1} 2\zeta(f/f_n)/[1 - (f/f_n)^2]$), delays ($\phi(f) = -2\pi f \tau$), and a first-order low-pass filter ($G(f) = [1 + (2\pi f T)^2]^{-1/2}$, $\phi(f) = -\tan^{-1} 2\pi f T$).

Subsequent work is likely to refine this current working model, but what is fairly remarkable is that this complex response can be reduced to so few parameters.

14. Prediction of Behavioral Responses from Stimulus–Response Functions

Using calculated linear and/or nonlinear SRF (Section III.D) or the parameters model discussed immediately above, what the cell would do to known conditions can be predicted. A very powerful consequence of this result, which is often overlooked, is that perturbation of these SRF is likely to lead to understanding why the SRF are the way they are. In Fig. 33 the relative sensitivity of the eye to light coming from different directions is plotted. From this curve, one may calculate the stimulus that a cell would receive if it rotated with its longitudinal axis normal to the light direction.

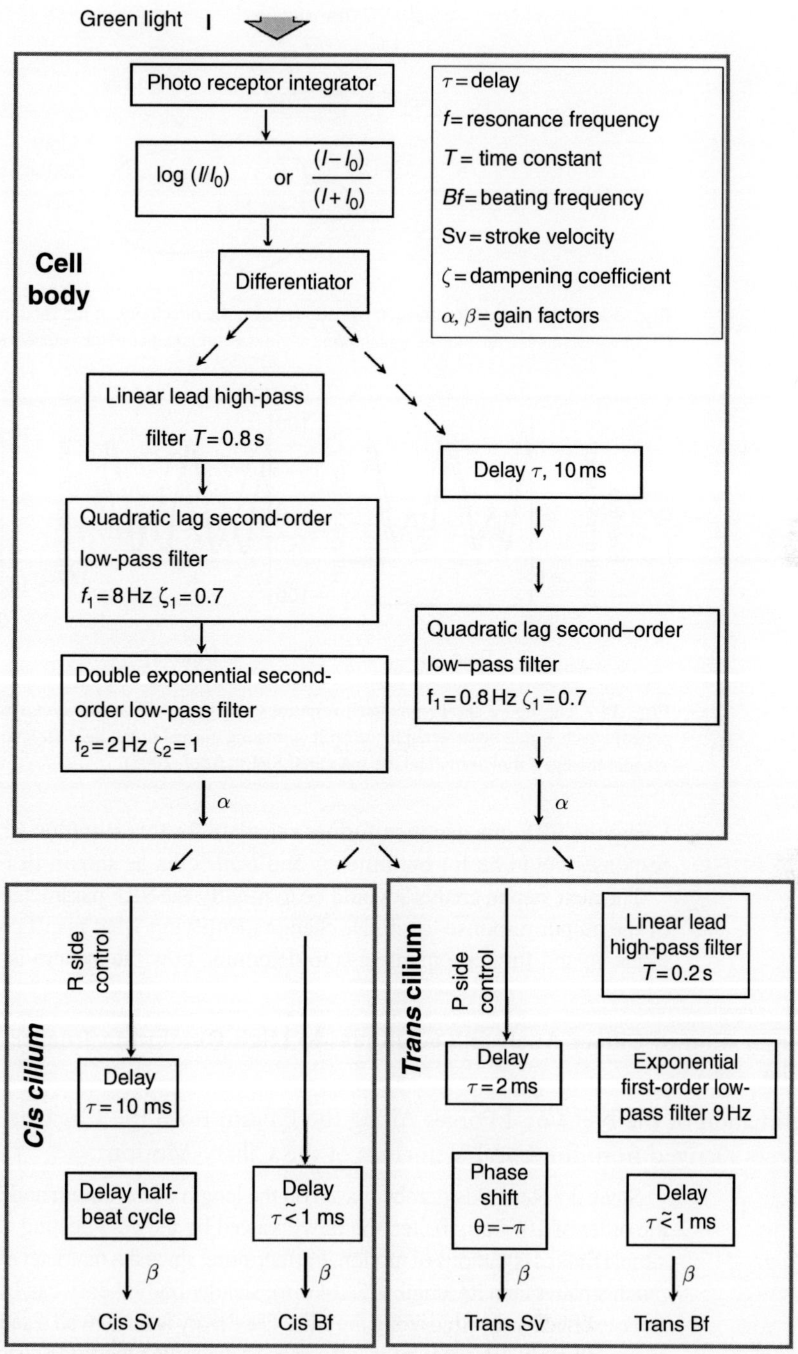

Fig. 32 Model of the systems biology of signal transduction in *Chlamydomonas* from green light stimulation to ciliary responses, see text.

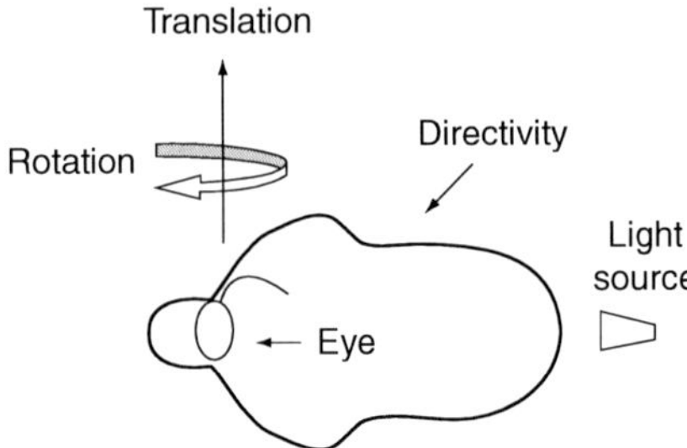

Fig. 33 Polar plot of the fraction of useable light, the directivity, at the rhodopsin light antenna or eye of *Chlamydomonas* as a function of light incident angle as a consequence of the antenna or eye structure (Josef, 2005).

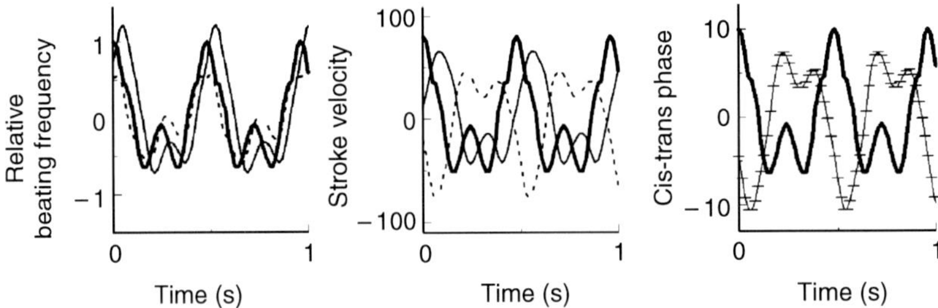

Fig. 34 The figure shows predicted responses to light as *Chlamydomonas* rotates. The predicted light pattern which would be detected by cell if it is rotating normal to the light direction at about 2 Hz taking into account the eye's directivity and the measured SRFs (Josef, 2005).

Using the SRF one can then further calculate for this stimulus what the specific ciliary response would be for both the *cis* and *trans* cilia as shown in Fig. 34.

The next step in analysis would be to modify the SRF parameters, judge the sensitivity of the output response to similar change identifying what is critical in the SRF and with mutants, and further experiments to determine how the system is working.

F. Application of Ciliary Analysis Techniques to High Resolution Images of *Chlamydomonas* Cilia

1. Estimation of the Net Local Forces Along the Cilium from the Calculated Hydrodynamic Forces Derived from the Local Velocities of the Ciliary Motion

Since the Reynolds number based on the length of the cilium and its stroke velocity is on the order of 10^{-3} or smaller, the flow induced by the cilia beating can be described by the simple Stokes equations of motion. Furthermore, since the diameter of a cilium (~240 nm) is much smaller than its length (12–14 μm), slender body theory can be used to determine the force exerted by the fluid along the cilia. The theory is fairly well established and a number of studies show in detail how it may be used to determine the force distribution (see Batchelor, 1970; Higdon, 1979; Hines and Blum, 1983; Mackaplow and Shaqfeh, 1998). The velocity at a point in the fluid is expressed as a line integral (along the cilium) of the force distribution times the velocity induced by the force (Green's function for Stokes equations, also knows as the Stokeslet). The Green's function derived by Higdon (1979) is chosen so that the no-slip boundary condition on the surface of the cell body (approximated as a sphere of diameter about 8 μm) is automatically satisfied. In addition to the Stokeslet distribution along the centerline of the cilium, a source dipole distribution may be added to improve the accuracy without significantly increasing the computational effort. Care should be taken that boundaries are kept at least several millimeters away from the cell for otherwise they will have to be taken into account.

As noted by Batchelor (1970), approximating the velocity by means of a line distribution of Stokeslet is valid in the outer region, away from the surface of the cilium. In the inner region, close to the surface of the cilium, one must construct another approximation that satisfies the no-slip boundary condition at the cilium surface taking into account its cross section shape. The requirement of the matching of the inner and outer region approximations in the overlap region (at distances from the surface large compared with the cilium diameter, but small compared with its length) gives rise to an integral equation for determining the force distribution along the cilium length. Batchelor (1970) used this matching procedure for slender bodies with circular and elliptic cross sections. The cilium cross section of *Chlamydomonas*, however, is complex.

Chlamydomonas has mastigonemes (fairly stiff ciliary hairs, 16 nm diameter, ~950 nm long; Bergman *et al.*, 1975; Brennen, 1976; Witman *et al.*, 1972) on its cilia that increase the swimming speed 25–40% and decrease the CBF ~10% compared to bare cilia (Nakamura *et al.*, 1996). These ciliary hairs protrude in two regularly placed rows 180° apart in the rectifying (defined by **b** and **t**) plane (Fig. 12) with the distance between their bases lying between 96 and 120 nm (Bergman *et al.*, 1975). To account for the mastigonemes, an elliptical cross section approximation might be used provided one had measured values of the tangential and normal drags of actual cilia, in and normal to their beating plane to estimate the appropriate ellipticity.

A simpler model, which assumes that the hydrodynamic force at a cilium cross section is simply proportional to the velocity of the cilium there (Lighthill, 1976), is often used because it gives good qualitative insights and avoids solving the integral equation derived via slender body theory. However, if subtle differences in the motor activities along the cilium (relevant to studies of control of the cilia) are to be studied, it is necessary to solve a more accurate model based on the slender body theory that includes the effect of the cell body.

2. Motor Forces and Sliding

The hydrodynamic force $F_j(s,t)$, together with the cilium position $r_j(s,t)$, s being the distance along the cilium measured from its base and t the time, may be used to determine the motor forces along the cilium. Since the inertial effects are negligible, the hydrodynamic force is balanced by the elastic forces arising from the bending of the doublets and the internal forces consisting of active forces due to motor activity and passive restraining forces due to nexins. Nexins connecting adjacent doublets provide an elastic resistance to longitudinal sliding with a complex nonlinear dependence on stretch (estimated to be 16–100 pN/μm, Lindeman *et al.*, 2005). It is assumed that each doublet is inextensible, but flexible since a single microtubule has been measured to be anisotropic with a shear modulus of ~1.4 MPa and a Young's modulus of ~100 MPa (Kis *et al.*, 2002). The sum of the internal forces at any cross section of an axoneme must vanish even though some doublets are under tension and others under compression.

For planar beats it has been shown that the 9 + 2 structure of axoneme can be equivalently replaced by two elastic filaments (groups of doublets) separated by distance a and bending rigidity κ, both of which can be estimated from the known geometry and mechanical properties of the axoneme ($\kappa \approx 800\,\text{pN}\,\mu\text{m}^2$, Deiner *et al.*, 1993, Riedel-Kruse *et al.*, 2007). Let f be the internal shear force acting along the filament on the inner side of the bend and $-f$ the shear force on the outer one. A general equation for force balances on a pair of inextensible elastic elements is derived by Camalet and Jülicher (2000). Their analysis can be extended to show that:

$$af(s,t) = \kappa(s,t)\partial_s C(s,t) + n_j(s,t)\int_s^L F_j(s',t)\,ds' \tag{4}$$

where C is the curvature, n_j is the unit normal to the filament in the plane of the beating, ∂_s represents the derivative with arc length s, L is the cilium length, and F_j is the hydrodynamic force per unit length. Since all the quantities on the right-hand side of the above equation can be determined from imaging and hydrodynamic analysis, f can be determined along the cilium length.

It is possible that the local bending rigidity or flexure is under control and alterable by internal viscosity, mechanical feedback, or cell command. Antibodies that bind microtubules together increase the flexure rigidity of a cilium (Okuno *et al.*, 1981). Consequently, a signal that increased the fraction of dyneins bound would stiffen the cilium and conversely a signal that decreased the fraction of dynein bound would make the cilium more flexible, markedly changing the shape of the ciliary beat.

All SOBS models assume that the motor force is a function of doublet sliding. The relative sliding of the filaments is given by:

$$\Delta(s,t) = \Delta_0(t) + \Delta^c(s,t), \text{ where } \Delta^c(s,t) = \int_0^s aC(s',t)ds' = a\big(\psi(s) - \psi(0)\big) \tag{5}$$

and $\Delta_0(t)$ is the relative sliding of the two filaments at the base ($s = 0$). Note that since $\Delta^c(s, t)$ can be determined from the ciliary images, the total sliding along the cilium can be determined to within a single parameter $\Delta_0(t)$. This parameter has been assumed to be zero by a number of investigators in the past, an assumption that may not be valid according to recent analysis (see the next section).

3. Modeling of Basal Connection

According to Riedel-Kruse *et al.* (2007) and as previously suggested (Vernon and Woolley, 2002, 2004) $\Delta_0(t)$ and the dynamics of the base connecting the cilium to the cell body likely plays a crucial role in determining the waveform of beating in bull sperm. Thus, cells may control their beating by changing the properties of the basal connection. In the case of *Chlamydomonas*, the two cilia (Fig. 6) are connected to each other through proximal and distal fibers. As a result, the dynamics of the two cilia are connected, and the cell may use this connection to control the beating pattern.

In addition to the distal fibers, which link the dynamics of two cilia, there are other components in the basal body region that may play significant role in cellular control of the beating. Some of these components contain centrin which shows calcium-sensitive contractile or elastic behavior (Geimer and Melkonian, 2005; Salisbury, 1983; Salisbury *et al.*, 1984). It has been shown that centrin-based flagellar roots are contractile under conditions of elevated calcium in a variety of algae, including *Chlamydomonas* and *Tetraselmis* (Salisbury and Floyd, 1978; Salisbury *et al.*, 1984, 1986, 1987). Hayashi *et al.* (1998) showed that the contraction of centrin is responsible for changing the base angle in a ciliary to flagellar beating transition in *Chlamydomonas*. Josef *et al.* (2005a) have reported that the average angle between the cilia decreases with light intensity, the expected result for a calcium-dependent contraction. Riedel-Kruse *et al.* (2007) modeled the component of the base force parallel to the cilium at base as given by:

$$F_\mathrm{B} = \gamma_s \left(\frac{d\Delta_0}{dt} \right) + k_s \Delta_0$$

where γ_s and k_s are, respectively, basal friction and stiffness. In the case of bull sperm with a single cilium, the force F_B can be determined by equating it to the component of the total hydrodynamic force parallel to the cilium at the base, thus providing sufficient information for determining γ_s and k_s.

The base structure in *Chlamydomonas* is different than in bull sperm. The axoneme of each cilium extends to the attached basal bodies connected by a distal striated fiber, which can contract. Any change in the angle θ between the two cilia will therefore induce a force in addition to the force F_B due to sliding. We may assume that this additional connection force is given by $F_\mathrm{C} = b[\cos(\theta/2) - \cos(\theta_0/2)] \approx -b'(\theta - \theta_0)$, where the (complex) stiffness b is related to the strength of distal fiber and $b' = b\ \sin(\theta_0/2)/2$. Now the sum of the forces due to base sliding and connection must be balanced by the component of the total hydrodynamic force parallel to the cilium at the base.

4. Comments on the Relative Timescales of Ciliary Phenomena

Some of the relevant timescales are as follows: time for the momentum to diffuse over the length of the cilium is ~0.1 ms, the time for a small molecule to diffuse across the diameter of the axoneme is ~0.4 ms, time $\tau$for attachment/detachment of ATPase motors is ~1 ms, and the time for a wave produced by viscous drag acting on an elastic axoneme to travel the axoneme length ($\mu L^4/\kappa$, μ being the viscosity of water) is about 25 ms. Apart from observation of delays, which can be submillisecond, in providing a local stimulus, changes in the beating pattern occurring on a timescale faster than 25 ms are therefore masked. A consequence of this relation for the time to propagate a wave down the axoneme is that lowering the stiffness, for example by not having outer dynein arms which could contribute to stiffness would lower the CBF. This could be the explanation for the lower CBF observed for such mutants. The response to the electrical signal from the cell body is much slower. It has 1% of peak gain at 55 Hz, the CBF, and peaks at ~4 Hz (Capano *et al.*, 2008). This is sufficiently fast for phototaxis because the natural light modulation frequency due to cell rotation is ~2 Hz (Foster and Smyth, 1980; Isogai *et al.*, 2000).

G. Methods to Perturb Ciliary Responses and Therefore Learn More About Their Function

1. Use of Mutants to Aid Ciliary Analysis

To obtain results that are as unambiguous as possible, it is recommended to use strains that are as isogenic as possible. Table I shows some potential mutants that one could use to study how cilia beat.

2. Use of Pharmacological and Environmental Manipulation

Since the impulse responses of the CBF and stroke velocity have a short delay, the signal processing for these responses in the cell body must be due to ion channel activity. Consequently, with the availability of channel blockers, this is an excellent opportunity to sort out some of the details of this part of the signaling network. Furthermore, ciliary beating can be influenced by many environmental variables such as light and external ion concentrations of sodium, potassium, and calcium as well as internal variables that may be influenced externally such the internal pH, redox potential, Mg^{2+}, and ATP concentrations, along with internal standard second messengers such as Ca^{2+}, cAMP, NO, IP_3, and the membrane potential. All these variables must be evaluated to understand how the cell does its "computations."

Table I
Some Potential *Chlamydomonas* Mutants Available for the Study of Ciliary Beating

Mutant	Defect	Experimental value
*oda*1	Lacks outer arm dyneins	Alters CBF
pf 28	Also lacks outer arm dyneins but doublet microtubule sliding responds differently to high calcium than *oda*1	Blocks *cis–trans* difference CBF and alters CBF
oda11	Lacks outer dynein α-heavy chain	Blocks *cis–trans* difference CBF
pf 9-2	Lacks inner arm dyneins	Alters waveform
*ida*1	lacks inner arm I1	Alters waveform
*cpc*1-2	Lacks central pair-associated complex blocking ATP synthesis from 3-phosophoglycerate in the cilium	Removes ATP control of the beating from the cilium
*sup*1	small deletions within the doublet microtubule-binding stalk	Removes calcium control of outer arm dyneins
*pf*30	Lacks the actively "dragging" dynein f heavy chain	Hypothesized to control curvature or cut off
*lsp*1	Lacks sensor for calcium in cilium	Alters response to calcium levels
ptx1	Phototaxis mutant, orients to either direction	
vfl2	Lacks centrin	Alters basal body compliance

Source: Asai and Wilkes (2004): *oda*1, *pf*28; Brokaw and Kamiya (1987): *oda*1, *pf*28, *oda11, pf*9-2; Habermacher and Sale (1996): *pf*28; Harris (1989): *oda*1, *pf*28, *sup*1, *pf*30, *vfl2*; Hayashi *et al.* (1998): *vfl2*; Kagami and Kamiya (1992): *ida*1; Kamiya (1988): *oda*1, *pf*28; Kamiya (2002): *oda*1, *pf*28, *oda*11, *pf*9-2, *ida*1, *cpc*1-2; King (2000): *oda*1, *pf*28, *oda*11, *pf*9-2; King and Dutcher (1997): *ida*1; Kotani *et al.* (2007): *pf*30; LeDizet and Piperno (1995): *pf*28; Mitchell and Rosenbaum (1985): *oda*1, *pf*28; Mitchell and Sale (1999): *cpc*1-2: Myster *et al.* (1997): *pf*9-2, *ida*1; Porter *et al.* (1994): *oda*1, *pf*28; Okita *et al.* (2005): *lsp*1, *ptx*1; Porter and Sale (2000): *oda*1, *pf*28, *oda11, pf* 9-2; Sakakibara and Nakayama (1998): *pf*28; Satir (2003): *oda*1, *pf*28, *oda11, pf*9-2.

3. Viscosity

Studies of the effect of fluid viscosity on ciliary beating have demonstrated the need to separately specify the distinct roles of inner and outer dyneins (Brokaw, 2001; Yagi *et al.*, 2005). *Chlamydomonas* adapts to higher viscosities by making much greater use of the outer dynein motors. Further detailed study at various viscosities is probably justified.

H. Indirect Assessment of Ciliary Function, Comparison to Other Assays of Cell Behavior that Do Not Look at Cilia Directly

There are other assays of cell behavior that are valuable besides studying the beating of cilia. The most common are analysis of electric current which is the signal that controls calcium influx into the cilia and hence ciliary beating and the derived phototaxis. Electrical signals to the cilium may be analyzed in the same manner as CBF, etc. A second common approach is swimming path analysis (Fig. 22), which yields ciliary dominance rather trivially in that the dominant cilium is the one on the outside of a helical path (Fig. 35).

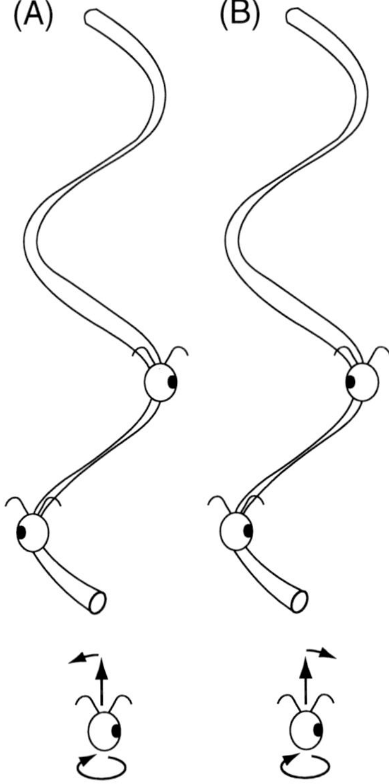

Fig. 35 Relationship of the direction of the eye relative to the cell's helical swimming track and the differential in torque generated by the pair of cilia. (A) When the *cis* cilium (nearest to the eye) generates greater torque than the *trans*, the eye points toward the outside of the helix. (B) On the other hand, when the *trans* cilium generates greater torque than the *cis*, the eye points toward the inside of the helix. This is true for a cell that rotates left handedly (counterclockwise as viewed from the posterior) like *Chlamydomonas* (Foster and Smyth, 1980). Figure from Isogai *et al.* (2000).

1. Analysis of Curved Swimming Paths in 3D

The analysis is mathematically essentially the same as for any curve in space, such as tracing a line down the center of a cilium (Section III.B.4.b). The helical, super-helical, and turning swimming paths can be treated as quartic or quintic smoothing splines along three orthogonal projections in laboratory space. This high order of spline is required to obtain the derivatives of motion necessary for calculation of the path torsion. Other methods (Crenshaw *et al.*, 2000; Walker, 1998) also can be applied. The direction of helical axis of path in 3D (Fig. 22) and its parameters of curvature and torsion, and steering parameters like the change in the curvature/torsion ratio, which causes the cell to turn and the phasing angle [the angle between the planes defined by the tangent to the path (Fig. 5) and the normal (**n**, the direction to the helical axis) with the plane defined by the tangent and the eye direction]. Superposed on the smoothed swimming paths in the case of *Chlamydomonas* will be the perturbing back and forth motion of the beating cycle. This beating motion can be treated as a small periodic motion superposed on the mainly helical path described by the smoothing spline connecting a particular point in each beating cycle. The advantage over the held cell technique is that the ciliary are loaded as for a freely swimming cell and ciliary dominance is now easy to measure from whether the eye is facing inward (*cis* dominant) or outward (*trans* dominant) (Isogai *et al.*, 2000).

Another aspect that may be determined directly from observation of freely swimming cells is how they disperse. The control of the probability of the intermittency described previously determines the dispersal pattern of swimming cells. Two useful measures of this behavior are the McCutcheon index (ratio of the total displacement from the starting point divided by path distance traveled, which can range from 0 to 1) and the Levy slope or exponent which is the power of time proportional to the mean square displacement. If the exponent is one, it corresponds to random diffusion, if two it means the organism is ballistic or swimming straight, the distance rather than the mean squared distance is proportional to time. At intermediate times, like 10 s the exponent is more or less half way in between the extremes.

IV. Future Directions

Chlamydomonas is already a premier model for the study of the mechanism of ciliary beating and their control. Various experimental approaches and analysis techniques described here and elsewhere can facilitate understanding of how cilia work and are being controlled. We can anticipate determining in more detail: (1) the architecture of signal processing related to receptors and the multiplexing of the external sensory inputs as well as their integration with the internal "well-being" metabolic inputs, (2) the intracellular processing including multiple feedback loops, and (c) the output steering and navigation provided by the cilia. *Chlamydomonas* is a superb model for understanding the signal processing of a eukaryotic cell, having several signal outputs, namely its beating cilia, which can be monitored on the millisecond to second scale of the cell's decisions and many mutants known to be altered in signal processing. Its cilia are also excellent models for understanding how cilia beat. Its reverse engineering, beginning with an initial estimate of parameters and following with iterative successive stages of experiments and modeling, is relatively easy. It is close to the ideal concept described by van Riel (2006), "the system inputs can be easily and freely perturbed by the experimenter to reveal a rich ensemble of dynamic responses and that the responses of many of the system variables can be measured accurately. Furthermore … this cycle can be iterated several times without significant extra 'costs'." With existing strains that are significantly isogenic and computer-controlled measurements, modeling and testing with experiment can be carried out iteratively to solve the architecture of signal processing and how the cilia work and are regulated for phototaxis. Finally, the design principles can be discovered by use of developed mathematical models based on measured SRF that can be perturbed and then investigated to ask why the observed parameters have their values.

Acknowledgments

I would like to thank the assistance of my present and former recent students: Suphatra Adulrattananuwat, Sitichoke Amnuanpol, Keith Josef, Ganesh Srinivasan, Ned Tuck, and Jyothish Vidyadharan, and present collaborators: Professors Howard Blair, Hiroshi Higuchi, Michael Korenberg, David Mitchell, Jureepan Saranak, Ashok Sangani, and Pinfen Yang in the preparation of this review, since many of the ideas expressed I learnt from or arose in my interaction with them.

References

Adulrattananuwat, S. (2009). The bioenergetics and dynamics of ciliary responses and the systems biology of phototaxis in *Chlamydomonas*. Ph.D. thesis, Syracuse University.

Amnuanpol, S. (2009). Dislocations in a vortex lattice and complexity of *Chlamydomonas* ciliary beating. Ph.D. thesis, Syracuse University.

Aoki, M. (1987). "State Space Modeling of Time Series." Springer-Verlag, New York.

Asai, D.J., and Wilkes, D.E. (2004). The dynein heavy chain family. *J. Eukaryot. Microbiol.* **51**, 23–29.

Aström, K.J., and Murray, R.M. (2008). "Feedback Systems: An Introduction for Scientists and Engineers." Princeton University Press, Princeton, NJ.

Baba, S.A., and Mogami, Y. (1985). An approach to digital image analysis of bending shapes of eukaryotic flagella and cilia. *Cell Motil. Cytoskeleton* **5**, 475–489.

Baldauf, S.L. (2003). The deep roots of eukaryotes. *Science* **300**, 1703–1706.

Bar-Yam, Y., and Epstein, E.R. (2004). Response of complex networks to stimuli. *Proc. Natl. Acad. Sci. USA* **101**, 4341–4345.

Batchelor, G.K. (1970). Slender-body theory for particles of arbitrary cross-section in Stokes flow. *J. Fluid Mech.* **44**, 419–440.

Berg, H.C. (2004). "*E. coli* in Motion." p. 134. Springer-Verlag, New York.

Bergman, K., Goodenough, U.W, Goodenough, D.A., Jawitz, J., and Martin, H. (1975). Gametic differentiation in *Chlamydomonas reinhardtii*. II. Flagellar membranes and the agglutination reaction. *J. Cell Biol.* **67**, 606–622.

Bezrukov, S.M., and Vodyanoy, I. (1997). Signal transduction across alamethicin ion channels in the presence of noise. *Biophys. J.* **73**, 2456–2464.

Blake, J.R. (2001). Microbiological fluid mechanics: A tribute to Sir James Lighthill. *Math. Methods Appl. Sci.* **24**, 1469–1483.

Boonyareth, M., Saranak, J., Pinthong, D., Sanvarinda, Y., and Foster, K.W. (2009). Roles of cyclic AMP in regulation of phototaxis in *Chlamydomonas reinhardtii. Biologia* **64** (in press).

Brennen, C. (1976). Locomotion of flagellates with mastigonemes. *J. Mechanochem. Cell Motil.* **3**, 207–217.

Brennen, C., and Winet, H. (1977). Fluid mechanics of propulsion by cilia and flagella. *Annu. Rev. Fluid Mech.* **9**, 339–398.

Brokaw, C.J. (1975). Molecular mechanism for oscillation in flagella and muscle (cross-bridges/computer simulation). *Proc. Natl. Acad. Sci. USA* **72**, 3102–3106.

Brokaw, C.J. (1979). Calcium-induced asymmetrical beating of Triton-demembranated sea urchin sperm flagella. *J. Cell Biol.* **82**, 401–411.

Brokaw, C.J. (1984). Automated methods for estimation of sperm flagellar bending parameters. *Cell Motil.* **4**, 417–430.

Brokaw, C.J. (1985). Computer simulation of flagellar movement VI. Simple curvature controlled models are incompletely specified. *Biophys. J.* **48**, 633–642.

Brokaw, C.J. (1990). Computerized analysis of flagellar motility by digitization and fitting of film images with straight segments of equal length. *Cell Motil. Cytoskeleton* **17**, 309–316.

Brokaw, C.J. (1991). Microtubule sliding in swimming sperm flagella: Direct and indirect measurements on sea urchin and tunicate spermatozoa. *J. Cell Biol.* **114**, 1201–1215.

Brokaw, C.J. (2001). Simulating the effects of fluid viscosity on the behaviour of sperm flagella. *Math. Methods Appl. Sci.* **24**, 1351–1365.

Brokaw, C.J. (2002). Computer simulation of flagellar movement VIII.Coordination of dynein by local curvature control can generate helical bending waves. *Cell Motil. Cytoskeleton* **53**, 103–124.

Brokaw, C.J. (2005). Computer simulation of flagellar movement IX. Oscillation and symmetry breaking in a model for short flagella and nodal cilia. *Cell Motil. Cytoskeleton* **60**, 35–47.

Brokaw, C.J. (2009). Thinking about flagellar oscillation. *Cell Motil. Cytoskeleton* **66**, 425–436.

Brokaw, C.J., and Kamiya, R. (1987). Bending patterns of *Chlamydomonas* flagella: IV. Mutants with defects in inner and outer dynein arms indicate differences in dynein arm function. *Cell Motil. Cytoskeleton* **8**, 68–75.

Brokaw, C.J., Luck, D.J.L., and Huang, B. (1982). Analysis of the movement of *Chlamydomonas* flagella: The function of the radial-spoke system is revealed by comparison of wild-type and mutant flagella. *J. Cell Biol.* **92**, 722–732.

Camalet, S., and Jülicher, F. (2000). Physical aspects of axonemal beating and swimming. *New J. Phys.* **2**, 1–14.

Capano, C., Saranak, J., and Foster, K.W. (2008). The systems biology of *Chlamydomonas* phototaxis as determined by electric current and ciliary beating. EMBO Workshop on the Cell and Molecular Biology of *Chlamydomonas*, May 27–June 1, 2008, Hyères-les-Palmiers, Var, France.

Chilvers, M.A., and O'Callaghan, C. (2000). Analysis of ciliary beat pattern and beat frequency using digital high speed imaging: Comparison with the photomultiplier and photodiode methods. *Thorax* **55**, 314–317.

Cortez, R., Cowen, N., Dillon, R., and Fauci, L. (2004). Simulation of swimming organisms: Coupling internal mechanics with external fluid dynamics. *Comput. Sci. Eng.* **6**, 38–45.

Cosson, J. (1996). A moving image of flagella: News and views on the mechanisms involved in axonemal beating. *Cell Biol. Int.* **20**, 83–94.

Crenshaw, H.C., Ciampaglio, D.N., and McHenry, M. (2000). Analysis of the three-dimensional trajectories of organism, estimates of velocity, curvature and torsion from positional information. *J. Exp. Biol.* **203**, 961–982.

Davies, W.D.T. (1970). "System Identification for Self-Adaptive Control." John Wiley and Sons Ltd., London.

Deiner, M., Tamm, S.L., and Tamm, S. (1993). Mechanical properties of ciliary axonemes and membranes as shown by paddle cilia. *J. Cell Sci.* **104**, 1251–1262.

Dillon, R.H., and Fauci, L.J. (2000). An integrative model of internal axoneme mechanics and external fluid dynamics in ciliary beating. *J. Theor. Biol.* **207**, 415–430.

Dillon, R.H., Fauci, L.J., and Omoto, C. (2003). Mathematical modeling of axoneme mechanics and fluid dynamics in ciliary and sperm motility. *Dyn. Contin. Discrete Impulse Syst. Ser. A* **10**, 745–757.

Dillon, R.H., Fauci, L.J., Omoto, C., and Yang, X. (2007). Fluid dynamic models of flagellar and ciliary beating. *Ann. NY Acad. Sci.* **1101**, 494–505.

Dorf, R.C., and Bishop, R.H. (1998). "Modern Control Systems." 8th edn., Addison-Wesley, Longman, Menlo Park, CA.

Foster, K.W., Josef, K., Saranak, J., and Tuck, N. (2006). Dynamics of a sensory signaling network in a unicellular eukaryote. *In* "IEEE Engineering in Medicine and Biology, EMBS '06. 28th Annual International Conference of the IEEE," New York, N.Y., pp. 252–255.

Foster, K.W., Saranak, J., Patel, N., Zarrilli, G., Okabe, M., Kline, T., and Nakanishi, K. (1984). A rhodopsin is the functioning photoreceptor for phototaxis in the unicellular eukaryote *Chlamydomonas*. *Nature* **311**, 756–759.

Foster, K.W., and Smyth, R.D. (1980). Light antennas in phototactic algae. *Microbiol. Rev.* **44**, 572–630.

Foster, K.W. (2009). Eye evolution: Two eyes can be better than one. *Curr. Biol.* **19**, R208–R210.

Frauenfelder, H., Parak, F., and Young, R.D. (1988). Conformational substates in proteins. *Annu. Rev. Biophys. Chem.* **17**, 451–480.

Frigui, H. (1999). A robust competitive clustering algorithm with applications in computer vision. *IEEE Trans. Pattern Anal. Mach. Intell.* **21**, 450–465.

Geimer, S., and Melkonian, M. (2004). The ultrastructure of the *Chlamydomonas reinhardtii* basal apparatus: Identification of an early marker of radial asymmetry inherent in the basal body. *J. Cell Sci.* **117**, 2663–2674.

Geimer, S., and Melkonian, M. (2005). Centrin scaffold in *Chlamydomonas reinhardtii* revealed by immunoelectron microscopy. *Eukaryot. Cell* **4**, 1253–1263.

Goldberger, A.L., Amaral, L.A.N., Hausdorff, J.M., Ivanov, P.C., Peng, C.-K., and Stanley, H.E. (2002). Fractal dynamics in physiology: Alternations with disease and aging. *Proc. Natl. Acad. Sci. USA* **99**, 2466–2472.

Gray, J., and Hancock, G.J. (1955). The propulsion of sea-urchin spermatozoa. *J. Exp. Biol.* **32**, 802–814.

Gueron, S., and Levit-Gurevich, K. (1998). Computation of the internal forces in cilia: Application to ciliary motion, the effects of viscosity, and cilia interactions. *Biophys. J.* **74**, 1658–1676.

Gueron, S., and Levit-Gurevich, K. (1999). Energetic considerations of ciliary beating and the advantage of metachronal coordination. *Proc. Natl. Acad. Sci. USA* **96**, 12240–12245.

Gueron, S., and Levit-Gurevich, K. (2001a). A three-dimensional model of ciliary motion based on the internal 9 + 2 structure. *Proc. R. Soc. Lond. B* **268**, 599–607.

Gueron, S., and Levit-Gurevich, K. (2001b). The three-dimensional motion of slender filaments. *Math. Methods Appl. Sci.* **24**, 1577–1603.

Gueron, S., and Liron, N. (1992). Ciliary motion modeling, and dynamic multicilia interactions. *Biophys. J.* **63**, 1045–1058.

Gueron, S., and Liron, N. (1993). Simulations of three-dimensional ciliary beats and cilia interactions. *Biophys. J.* **65**, 499–507.

Habermacher, G., and Sale, W.S. (1996). Regulation of flagellar dynein by phosphorylation of a 138-kD inner arm dynein intermediate chain. *J. Cell Biol.* **136**, 167–176.

Hanson, A.J. (2006). "Visualizing Quaternions." Elsevier, San Francisco.

Harris, E.H. (1989). "The *Chlamydomonas* Sourcebook." Academic Press, San Diego.

Hayashi, M., Yagi, T., Yoshimura, K., and Kamiya, R. (1998). Real-time observation of Ca^{2+} induced basal body reorientation in *Chlamydomonas. Cell Motil. Cytoskeleton* **41**, 49–56.

Hennessy, S.J., Wong, L.B., Yeates, D.B., and Miller, I.F. (1986). Automated measurement of ciliary beat frequency. *J. Appl. Physiol.* **60**, 2109–2113.

Higdon, J.J.L. (1979). The generation of feeding currents by flagellar motions. *J. Fluid Mech.* **94**, 305–330.

Hines, M., and Blum, J.J. (1983). Three-dimensional mechanics of eukaryotic flagella. *Biophys. J.* **41**, 67–79.

Hoops, H.J., and Witman, G.B. (1985). Basal bodies and associated structures are not required for normal flagellar motion or phototaxis in the green alga *Chlorogonium elongatum. J. Cell Biol.* **100**, 297–309.

Hyams, J.S., and Borisy, G.G. (1978). Isolated flagellar apparatus of *Chlamydomonas*: Characterization of forward swimming and alteration of waveform and reversal of motion by calcium ions *in vitro*. *J. Cell Sci.* **33**, 235–253.

Isogai, N., Kamiya, R., and Yoshimura, K. (2000). Dominance between the two flagella during phototaactic turning in *Chlamydomonas*. *Zool. Sci.* **17**, 1251–1266.

Ivanov, P.C., Yuen, A., Podobnik, B., and Lee, Y. (2004).Common scaling patterns in intertrade times of U. S. stocks. *Phys. Rev. E*, **69**, 56107.

Jenkins, G.M., and Watts, D.G. (1968). "Spectral Analysis and Its Applications." Holden-Day, San Francisco.

Johnson, R.E., and Brokaw, C.J. (1979). Flagellar hydrodynamics. A comparison between resistive-force theory and slender-body theory. *Biophys. J.* **25**, 113–127.

Jolliffe, I.T. (2002). "Principal Component Analysis, Series: Springer Series in Statistics." 2nd edn. Springer, New York.

Josef, K. (2005). The nonlinear phototaxis signaling network of *Chlamydomonas* investigated by observing ciliary responses of individual cells to green and red light. Ph.D. dissertation, Syracuse University, Syracuse, NY.

Josef, K., Saranak, J., and Foster, K.W. (2005a). An electro-optic monitor of the behavior of *Chlamydomonas reinhardtii* cilia. *Cell Motil. Cytoskeleton* **61**, 83–96.

Josef, K., Saranak, J., and Foster, K.W. (2005b). Ciliary behavior of a negatively phototactic *Chlamydomonas reinhardtii. Cell Motil. Cytoskeleton* **61**, 97–111.

Josef, K., Saranak, J., and Foster, K.W. (2006). Linear systems analysis of the ciliary steering behavior associated with negative-phototaxis in *Chlamydomonas reinhardtii. Cell Motil. Cytoskeleton* **63**, 758–777.

Kagami, O., and Kamiya, R. (1992). Translocation and rotation of microtubules caused by multiple species of *Chlamydomonas* inner-arm dynein. *J. Cell Sci.* **103**, 653–664.

Kamiya, R. (1988). Mutations at twelve independent loci result in absence of outer dynein arms in *Chylamydomonas reinhardtii*. *J. Cell Biol.* **107**, 2253–2258.

Kamiya, R. (2000). Analysis of cell vibration for assessing axonemal motility in *Chlamydomonas*. *Methods* **22**, 383–387.

Kamiya, R. (2002). Functional diversity of axonemal dyneins as studied in *Chlamydomonas* mutants. *Int. Rev. Cytol.* **219**, 115–155.

Kamiya, R., and Okamoto, M. (1985). A mutant of *Chlamydomonas reinhardtii* that lacks the flagellar outer dynein arm but can swim. *J. Cell Sci.* **74**, 181–191.

Kamiya, R., and Witman, G. (1984). Submicromolar levels of calcium control the balance of beating between the two flagella in demembranated models of *Chlamydomonas*. *J. Cell Biol.* **98**, 97–107.

Kantz, H., and Schreiber, T. (1997). "Nonlinear Time Series Analysis." Cambridge University Press, Cambridge, UK.

King, S.M. (2000). The dynein microtubule motor. *Biochim. Biophys. Acta* **1496**, 60–75.

King, S.J., and Dutcher, S.K. (1997). Phosphoregulation of an inner dynein arm complex in *Chlamydomonas reinhardtii* is altered in phototactic mutant strains. *J. Cell Biol.* **136**, 177–191.

Kinukawa, M., Ohmuro, J., Baba, S.A., Murashige, S., Okuno, M., Nagata, M., and Aoki, F. (2005). Analysis of flagellar bending in hamster spermatozoa: Characterization of an effective stroke. *Biol. Reprod.* **73**, 1269–1274.

Kis, A., Kasas, S., Babic, B., Kulik, A.J., Benoît, W., Briggs, G.A.D., Schönenberger, C., Catsicas, S., and Forró, L. (2002). Nanomechanics of microtubules. *Phys. Rev. Lett.* **89**, 248101–248114.

Kojima, H., Kikumoto, M., Sakakibara, H., and Oiwab, K. (2002). Mechanical properties of a single-headed processive motor, inner-arm dynein subspecies-c of C*hlamydomonas* studied at the single molecule level. *J. Biolog. Phys.* **28**, 335–345.

Korenberg, M.J. (1991). Parallel cascade identification and Kernel estimation for nonlinear systems. *Ann. Biomed. Eng.* **19**, 429–455.

Kotani, N., Sakakibara, H., Burgess, S.A., Kojima, H., and Oiwa, K. (2007). Mechanical properties of inner-arm dynein-f (dynein I1) studied with *in vitro* motility assays. *Biophys. J.* **93**, 886–894.

Lam, B.S.Y., and Yan, H. (2007). A curve tracing algorihm using level set based affine transform *Pattern Recognit. Lett.* **28**, 181–196.

LeDizet, M., and Piperno, G. (1995). The light chain p28 associates with a subset of inner dynein arm heavy chains in *Chlamydomonas* axonemes. *Mol. Biol. Cell* **6**, 697–711.

Lighthill, J. (1976). Flagellar hydrodynamics. *SIAM Rev.* **18**, 161–230.

Lindemann, C.B. (1994a). A "geometric clutch" hypothesis to explain oscillations of the axoneme of cilia and flagella. *J. Theor. Biol.* **168**, 175–189.

Lindemann, C.B. (1994b). A model of flagellar and ciliary functioning which uses the forces transverse to the axoneme as the regulator of dynein activation. *Cell Motil. Cytoskeleton* **29**, 141–154.

Lindemann, C.B. (2002). Geometric clutch model version 3: The role of the inner and outer arm dyneins in the ciliary beat. *Cell Motil. Cytoskeleton* **52**, 242–254.

Lindemann, C.B. (2003). Structural-functional relationships of the dynein, spokes, and central-pair projections predicted from an analysis of the forces acting within the axoneme. *Biophys. J.* **84**, 4115–4126.

Lindemann, C.B. (2007). The geometric clutch as a working hypothesis for future research on cilia and flagella. *Ann. NY Acad. Sci.* **1101**, 477–493.

Lindemann, C.B., Macauley, L.J., and Lesich, K.A. (2005). The counterbend phenomenon in dynein-disabled rat sperm flagellaand what it reveals about the interdoublet elasticity. *Biophys. J.* **89**, 1165–1174.

Lindemann, C., and Mitchell, D. (2007). Evidence for axonemal distortion during the flagellar beat of *Chlamydomonas*. *Cell Motil. Cytoskeleton* **64**, 580–589.

Machin, K.E. (1963). The control and synchronization of *flagellar* movement. *Proc. Roy. Soc. Lond. B.* **158**, 88–104.

Mackaplow, M.B., and Shaqfeh, E.S.G. (1998). A numerical study of fiber suspensions. *J. Fluid Mech.* **376**, 149–182.

Marmarelis, V.Z. (2004). "Nonlinear Dynamics Modeling of Physiological Systems." Wile-Interscience, Hoboken, NJ.

McLaughlin, R.A. (2000). Intelligent algorithms for finding curves and surfaces in real world data. Ph.D. thesis, The University of Western Australia.

Mitchell, D.R. (2000). *Chlamydomonas* flagella. *J. Phycol.* **36**, 261–273.

Mitchell, D.R. (2003). Orientation of central pair complex during flagella bend formation in *Chlamydomonas*. *Cell Motil. Cytoskeleton* **56**, 120–129.

Mitchell, D.R., and Nakatsugawa, M. (2004). Bend propagation drives central pair rotation in *Chlamydomonas reinhardtii* flagella. *J. Cell Biol.* **166**, 709–715.

Mitchell, D.R., and Rosenbaum, J.L. (1985). A motile *Chlamydomonas* flagellar mutant that lacks outer dynein arms. *J. Cell Biol.* **100**, 1228–1234.

Mitchell, D.R., and Sale, W.S. (1999). *Chlamydomonas* insertional mutant that disrupts flagellar central pair microtubule-associated structures. *J. Cell Biol.* **144**, 293–304.

Myster, S.H., Knott, J.A., O'Toole, E., and Porter, M.E. (1997). The *Chlamydomonas Dhc*1 gene encodes a dynein heavy chain subunit required for assembly of the I1 inner arm complex. *Mol. Biol. Cell* **8**, 607–620.

Nakamura, S., Tanaka, G., Maeda, T., Kamiya, R., Matsunaga, T., and Nikaido, O. (1996). Assembly and function of *Chlamydomonas* flagellar mastigonemes as probed with a monoclonal antibody. *J. Cell Sci.* **109**, 57–62.

Nelles, O. (2001). "Nonlinear System Identification: From Classical Approaches to Neural Networks and Fuzzy Models." Springer, New York.

Nicastro, D., Schwartz, C., Pierson, J., Gaudette, R., Porter, M.E., and McIntosh, J.R. (2006). The molecular architecture of axonemes revealed by cryoelectron tomography. *Science* **313**, 944–948.

Nutbourne, A.W., and Martin, R.R. (1988). "Differential Geometry Applied to Curve and Surface Design." Ellis Horwood Limited, Chichester, England.

O'Toole, E.T., Giddings, T.H., McIntosh, J.R., and Dutcher, S.K. (2003). Three-dimensional organization of basal bodies from wild-type and -tubulin deletion strains of *Chlamydomonas reinhardtii*. *Mol. Biol. Cell* **14**, 2999–3012.

Oda, T., Hirokawa, N., and Kikkawa, M. (2007). Three-dimensional structures of the flagellar dynein–microtubule complex by cryoelectron microscopy. *J. Cell Biol.* **177**, 243–252.

Okita, N., Isogai, N., Hirono, M., Kamiya, R., and Yoshimura, K. (2005). Phototactic activity in *Chlamydomonas* "nonphototactic" mutants deficient in Ca2+-dependent control of flagellar dominance or in inner-arm dynein *J. Cell Sci.* **118**, 529–537.

Okuno, M., Asai, D.J., Ogawa, K., and Brokaw, C.J. (1981). Effects of antibodies against dynein and tubulin on the stiffness of flagellar axonemes. *J. Cell Biol.* **91**, 689–694.

Okuno, M., and Hiramoto, Y. (1979). Direct measurements of the stiffness of echinoderm sperm flagella. *J. Exp. Biol.* **79**, 235–243.

Ott, E. (1993). "Chaos in Dynamical Systems." Cambridge University Press, New York.

Otter, T. (1989). Calmodulin and the control of flagellar movement. *In* "Cell Movement." Vol. 1, pp. 281–298. Alan R. Liss, New York.

Pazour, G.J., Agrin, N., Leszyk J., and Witman, G.B. (2005). Proteomic analysis of a eukaryotic cilium. *J. Cell Biol.* **170**, 103–113.

Peng, C.K., Buldyrev, S.V., Havlin, S., Simons, M., Stanley, H.E., and Goldberger, A.L. (1994). Mosaic Organization of DNA Sequences. *Phys. Rev. E* **49**, 1685–1689.

Poe, R.C., Pratap, P., and Lipson, E.D. (1986). System analysis of *Phycomyces* light-growth response: Double mutants. *Biol Cybern.* **55**, 105–113.

Poincaré, H. (1892). "Les Methodes Nouvelles de la Mechanique Celeste." Gauthier-Villars, Paris.

Poon, C.S., and Merfeld, D. (2005). Internal models: The state of the art (Editorial for special issue: Sensory integration, state estimation, and motor control in the brain: Role of internal models). *J. Neural Eng.* **2**.

Porter, M.E., Knott, J.A., Gardner, L.C., Mitchell, D.R., and Dutcher, S.K. (1994). Mutations in the *SUP-PF-1* locus of *Chlamydomonas reinhardtii* identify a regulatory domain in the β-dynein heavy chain. *J. Cell Biol.* **126**, 1495–1507.

Porter, M.E., and Sale, W.S. (2000). The 9 + 2 axoneme anchors multiple inner arm dyneins and a network of kinases and phosphatases that control motility. *J. Cell Biol.* **151**, F37–F42.

Racey, T.J., and Hallett, F.R. (1981). The effect of temperature, Ca2þ Mg2þ, and Ni2þ ions on the swimming speed of *Chlamydomonas reinhardtii* determined by quasi-elastic light scattering. *Exp. Cell Res.* **136**, 371–378.

Racey, T.J., and Hallett, F.R. (1983a). A low angle quasi-elastic light scattering investigation of *Chlamydomonas reinhardtii. J. Muscle. Res. Cell Motil.* **4**, 321–331.

Racey, T.J., and Hallett, F.R. (1983b). A quasi-elastic light scattering and cinematographical comparison of three strains of motile *Chlamydomonas reinhardtii*: A wild type strain, a colchicines resistant mutant and a backward swimming mutant. *J. Muscle. Res. Cell Motil.* **4**, 333–351.

Riedel-Kruse, I.H., Hilfinger, A., Howard, J., and Julicher, F. (2007). How molecular motors shape the flagellar beat. *HFSP J.* **1**, 192–208.

Rüffer, U., and Nultsch, W. (1985). High-speed cinematographic analysis of the movement of *Chlamydomonas. Cell Motil.* **5**, 251–263.

Rüffer, U., and Nultsch, W. (1987). Comparison of the beating of *cis-* and *trans*-flagella of *Chlamydomonas* held on micropipettes. *Cell Motil. Cytoskeleton* **7**, 87–93.

Rüffer, U., and Nultsch, W. (1990). Flagellar photoresponses of *Chlamydomonas* cells held on micropipettes: I. Change in flagellar beat frequency. *Cell Motil. Cytoskeleton* **15**, 162–167.

Rüffer, U., and Nultsch, W. (1991). Flagellar photoresponses of *Chlamydomonas* cells held on micropipettes: II. Change in flagellar beat pattern. *Cell Motil. Cytoskeleton* **18**, 269–278.

Rüffer, U., and Nultsch, W. (1995). Flagellar photoresponses of *Chlamydomonas* cells held on micropipettes: III. Shock response. *Bot. Acta* **108**, 255–265.

Rüffer, U., and Nultsch, W. (1998). Flagellar coordination in *Chlamydomonas* cells held on micropipettes. *Cell Motil. Cytoskeleton* **41**, 297–307.

Sakakibara, H., and Nakayama, H. (1998). Translocation of microtubules caused by the $\alpha\beta$, β, and γ outer arm dynein subparticles of *Chlamydomonas. J. Cell Sci.* **111**, 1155–1164.

Salisbury, J.L. (1983). Contractile flagellar roots: The role of calcium. *J. Submicrosc. Cytol.* **15**, 105–110.

Salisbury, J.L., Aebig, K.W., and Coling, D. (1986). Isolation of the calcium modulated contractile protein of striated flagellar roots. *Meth. Enzymol.* **134**, 408–414.

Salisbury, J.L., Baron, A., Surek, B., and Melkonian, M. (1984). Striated flagellar roots: Isolation and partial characterization of a calcium-modulated contractile organelle. *J. Cell Biol.* **99**, 962–970.

Salisbury, J.L., and Floyd, G.L. (1978). Calcium-induced contraction of the rhizoplast of a quadriflagellate green alga. *Science* **202**, 975–978.

Salisbury, J.L., Sanders, M., and Harpst, L. (1987). Flagellar root contraction and nuclear movement during flagellar regeneration in *Chlamydomonas reinhardtii. J. Cell Biol.* **105**, 1799–1805.

Satir, P. (1985). Switching mechanisms in the control of ciliary motility. *Mod. Cell Biol.* **4**, 1–46.

Satir, P. (2003). Control molecules in protozoan ciliary motility. *Jpn. J. Protozool.* **36**, 87–96.

Satir, P., Guerra, C., and Bell, A.J. (2007). Evolution and persistence of the cilium. *Cell Motil. Cytoskeleton* **64**, 906–913.

Schaller, K., David, R., and Uhl, R. (1997). How *Chlamydomonas* keeps track of the light once it has reached the right phototactic orientation. *Biophys. J.* **73**, 1562–1562.

Schipor, I., Palmer, J.N., Cohen, A.S., and Cohen, N.A. (2006). Quantification of ciliary beat frequency in sinonasal epithelial cells using differential interference contrast microscopy and high-speed digital video imaging. *Am. J. Rhinol.* **20**, 124–127.

Schuster, H. (1988). "Deterministic Chaos, an Introduction." 2nd edn. Physik-Verlag, Weinheim.

Silflow, C.D., and Lefebvre, P.A. (2001). Assembly and motility of eukaryotic cilia and flagella. Lessons from *Chlamydomonas reinhardtii. Plant Physiol.* **127**, 1500–1507.

Smith, E.F. (2007). Hydin seek: Finding a function in ciliary motility. *J. Cell Biol.* **176**, 12403–12404.

Smyth, R.D., and Berg, H. (1982). Change in flagellar beat frequency of *Chlamydomonas* in response to light. *Cell Motil. Suppl.* **1**, 211–215.

Srinivasan, G. (2008). Instruments and methods for high-speed ciliary imaging. Ph.D. thesis, Syracuse University, Syracuse, New York.

Srinivasan, G., Adulrattananuwat, S., Foster, N.S., and Foster, K.W. (2008). Automated tracing of the axes of beating *Chlamydomonas* cilia. EMBO Workshop on the Cell and Molecular Biology of *Chlamydomonas*, May 27–June 1, 2008, Hyères-les-Palmiers, Var, France.

Stechmann, A., and Cavalier-Smith, T. (2003). The root of the eukaryote tree pinpointed. *Curr. Biol.* **13**, R665–R666.

Sui, H., and Downing, K.H. (2006). Molecular architecture of axonemal microtubule doublets revealed by cryo-electron tomography. *Nature* **442**, 475–478.

Teff, Z., Priel, Z., and Gheber, L.A. (2008). The forces applied by cilia depend linearly on their frequency due to constant geometry of the effective stroke. *Biophys. J.* **94**, 298–305.

Teunis, P.F.M., and Machemer, H. (1994). Analysis of three-dimenstional ciliary beating by means of high-speed stereomicroscopy. *Biophys. J.* **67**, 381–394.

Thompson, E.E. (1999). "Design Analysis: Mathematical Modeling of Nolinear Systems." Cambridge University Press, New York.

Tomasi, C., and Manduchi, R. (1998). Bilateral filtering for gray and color images. *In* "Proceedings of the Sixth IEEE International Conference of Computer Vision," Bombay, India pp. 839–846.

van Riel, N.A.W., (2006). Dynamic modeling and analysis of biochemical networks: Mechanism-based models and model-based experiments. *Brief. Bioinform.* **7**, 364–374.

Vernon, G.G., and Woolley, D.M. (2002). Microtubule displacements at the tips of living flagella. *Cell Motil. Cytoskeleton* **52**, 151–160.

Vernon, G.G., and Woolley, D.M. (2004). Basal sliding and the mechanics of oscillation in a mammalian sperm flagellum. *Biophys. J.* **87**, 3934–3944.

Voit, E.O. (2000). "Computational Analysis of Biochemical Systems: A Practical Guide for Biochemists and Molecular Biologists." Cambridge University Press, New York.

Wahba, G. (1990). "Spline Models for Observational Data." SIAM, Philadelphia.

Wakabayashi, K., Ide, T., and Kamiya, R. (2009). Calcium-dependent flagellar motility activation in *Chlamydomonas reinhardtii* in response to mechanical agitation. *Cell Motil. Cytoskeleton* **66**, 736–742.

Walker, J.A. (1998). Estimating velocities and accelerations of animal locomotion: A simulation experiment comparing numerical differentiation algorithms. *J. Exp. Biol.* **201**, 981–995.

Westwick. D.T., and Kearney, R.E. (2003). "Identification of Nonlinear Physiological Systems." IEEE Press, Piscataway, NJ.

Witman, G.B., Carlson, K., Berliner, J., and Rosenbaum, L. (1972). *Chlamydomonas* flagella I. isolation and electrophoretic analysis of microtubules, matrix, membranes, and mastigonemes. *J. Cell Biol.* **54**, 507–589.

Wuensche, A. (2004). Basins of attraction in network dynamics: A conceptual framework for biomolecular networks. *In* "Modularity in Development and Evolution" (G. Schlosser and G.P. Wagner, eds.), pp. 288–311. University of Chicago Press, Chicago.

Yagi, T., Minoura, I., Fujiwara, A., Saito, R., Yasunaga, T., Hirono, M., Yasunaga, T., and Kamiya, R. (2005). An axonemal dynein particularly important for flagellar movement at high viscosity: Implications from a new *Chlamydomonas* mutant deficient in the dynein heavy chain gene DHC9. *J. Biol. Chem.* **280**, 41412–41420.

Yan, H. (2001). Fuzzy curve-tracing algorithm. *IEEE Trans. Sys. Man. Cybern. B Cybern.* **31**, 768–780.

Yan, H. (2004). Convergence condition and efficient implementation of the fuzzy curve-tracing (FCT) algorithm. *IEEE Trans. Syst. Man Cybern. B Cybern.* **34**, 210–221.

Yoshimura, K. (1996). A novel type of mechanoreception by the flagella of *Chlamydomonas*. *J. Exp. Biol.* **199**, 295–302.

Yoshimura, K. (1998). Mechanosensitive channels in the cell body of *Chlamydomonas*. *J. Membr. Biol.* **166**, 149–155.

Yoshimura, K., Shingyoji, C., and Takahashi, K. (1997). Conversion of beating mode in *Chlamydomonas* flagella induced by electric stimulation. *Cell Motil. Cytoskeleton* **36**, 236–245.

Tu, Y., Shimizu, T.S., and Berg, H.C. (2008). Modeling the chemotactic response of *Escherichia coli* to time-varying stimuli. *Proc. Natl. Acad. Sci. USA* **105**, 14855–14860.

Zaccolo, M., Magalhães, P., and Pozzan, T. (2002). Compartmentalisation of cAMP and Ca^{2+} signals. *Curr. Opin. Cell Biol.* **14**, 160–166.

Zhang, H., and Mitchell, D.R. (2004). *Cpc*1, a *Chlamydomonas* central pair protein with an adenylate kinase domain. *J. Cell Sci.* **117**, 4179–4188.

Zhou, H., and Lipowsky, R. (2005). Dynamic pattern evolution on scale-free networks. *Proc. Natl. Acad. Sci. USA* **102**, 10052–10057.

CHAPTER 12

Assays of Cell and Axonemal Motility in *Chlamydomonas reinhardtii*

Ritsu Kamiya

Department of Biological Sciences, Graduate School of Science, University of Tokyo, Bunkyo-ku, Tokyo 113-0033, Japan

Abstract

Chlamydomonas, an organism that offers a variety of flagella-deficient mutants, has been very important for studies of cilia and flagella. Motility assessment of mutant flagella at various levels helps us understand the function of specific axonemal proteins and structures for flagellar function. Measurements of gross cell movements are useful to assess the overall flagellar activity, analyses of demembranated and reactivated cells ("cell models") enable us to study the regulatory mechanism, and measurements of microtubule sliding velocity *in vitro* provide important information about dynein–microtubule interactions. This chapter describes fundamental techniques for these measurements.

978-0-12-374973-4
DOI: 10.1016/S0091-679X(08)91012-8

I. Introduction

Chlamydomonas reinhardtii has thus far provided the most detailed information about axonemal structure and function. This is largely because this organism offers a variety of motility-deficient mutants, and we can analyze the defects with biochemical and molecular biological techniques. Also important is the fact that we can relatively easily observe their flagellar motility in live cells, in demembranated and reactivated cells ("cell models"), and in isolated axonemes. By quantifying the motility in wild type and mutants, we can infer the function of particular axonemal structure or proteins. For example, waveform analysis of live mutant cells shows that outer arm dynein is important for flagella to beat at high frequency, whereas the inner arm dyneins are important to produce a proper waveform (Brokaw and Kamiya, 1987). Motility assessments of demembranated and reactivated flagella reveal that flagellar movements are controlled by Ca^{2+} and cAMP (Bessen *et al.*, 1980; Hasegawa *et al.*, 1987; Hyams and Borisy, 1978; Kamiya and Witman, 1984). Analyses of microtubule sliding in disintegrating axonemes show that the outer arm dynein and inner arm dynein cause microtubule sliding at different speeds (Kurimoto and Kamiya, 1991; Okagaki and Kamiya, 1986), and the radial spoke/central pair system regulates dynein activity through phosphorylation of an inner arm dynein subunit (Howard *et al.*, 1994; Smith, 2002; Smith and Sale, 1992). Finally, *in vitro* motility assays using isolated dyneins show that various kinds of axonemal dyneins strikingly differ in their motile properties (Kagami and Kamiya, 1992; Kikushima and Kamiya, 2008; Smith and Sale, 1991). As is evident from these examples, flagellar motility studies inevitably require analysis on various levels.

The motility assessment techniques described below are based on our previous studies that solely used *Chlamydomonas*, but most of the techniques can be used for other organisms with some modifications.

II. Motility Assessments in Live *Chlamydomonas* Cells

A. Observation and Video Recording

The first step in motility analysis is observation. For simple purposes such as to determine whether or not the cells are swimming smoothly, most microscopes with a total magnification of ~100× can be used. A conventional charge-coupled device (CCD) camera and a video recorder can be used for recording. When the recorded video image is to be used for quantitative analyses, a microscope scale and a video timer should be also recorded.

For observation of flagella, a dark-field microscope is recommended. Typically, a microscope with 10–40× objectives, a dry condenser, NA 0.8–0.92, and a light source of a 100-W halogen lamp can be used. When clear still images of beating flagella are necessary, either the light source must be replaced by a strobe, or the CCD camera must be replaced by a high-speed camera with an electronic shutter.

An oil-immersion dark-field condenser, NA 1.2–1.4, is recommended for recording single outer doublet microtubules in disintegrating axonemes. The objective, typically 40×, is desirably equipped with an iris although it is not prerequisite as long as the NA is lower than that of the condenser. A sensitive TV camera, such as a silicone-intensified target camera, or an image intensifier and a CCD camera are necessary for recording. An oil-immersion condenser is useful also for observation of the eyespot simultaneously with beating flagella; the eyespot clearly shows up under this illumination.

B. Swimming Velocity

Swimming velocity is an important parameter that reflects the overall activity of flagella. The average propulsive force produced by the beating flagella of a cell is proportional to the product of the swimming velocity, the medium viscosity, and the cell size (diameter). When the waveform can be assumed to be constant, the velocity varies linearly with the beat frequency. In other words, the ratio of the swimming velocity to the beat frequency can be used as a parameter that reflects the waveform.

Swimming velocity of cells is simply measured by tracing cells' swimming tracks on a transparent sheet overlaid on a video monitor. As an alternative method, software is commercially available that automatically tracks multiple moving objects. Such software enables one to measure many parameters of swimming tracks, in addition to the average and standard deviation of swimming velocity, including the direction of the swimming paths and the change in direction or speed.

As in all other motility assays, it is important to pay attention to the temperature as both *in vivo* and *in vitro* movements are sharply dependent on this parameter.

C. Flagellar Beat Frequency

Flagellar beat frequency is another important parameter. It can be measured by adjusting the frequency of the strobe illumination so that the flagella wave propagation looks stopped. Because propagation appears stopped at a common divisor frequency (e.g., 1/2 or 1/3 of the real frequency), care must be taken to choose the highest matching frequency. A problem with using *Chlamydomonas* is that the cell body rotates around the body axis while swimming, and does not permit observation from a fixed angle. To avoid this difficulty, previous studies often used uniflagellated cells, which tend to rotate in a small area with the flagellum beating in the plane of observation (Fig. 1). Uniflagellated cells are available as mutants that bear only the flagellum farthest from the eyespot, the *trans*-flagellum. Several *uni* mutations are available; however, most analyses have been performed with one of the original *uni* alleles (*uni1-1*) (Huang *et al.*, 1982). Any given mutant can be analyzed with a background of a *uni* mutation. A drawback of this method is that we can obtain only the information about the *trans*-flagellum (the flagellum farthest from the eyespot), although the *cis*- and *trans*-flagella are known to beat at slightly different frequencies

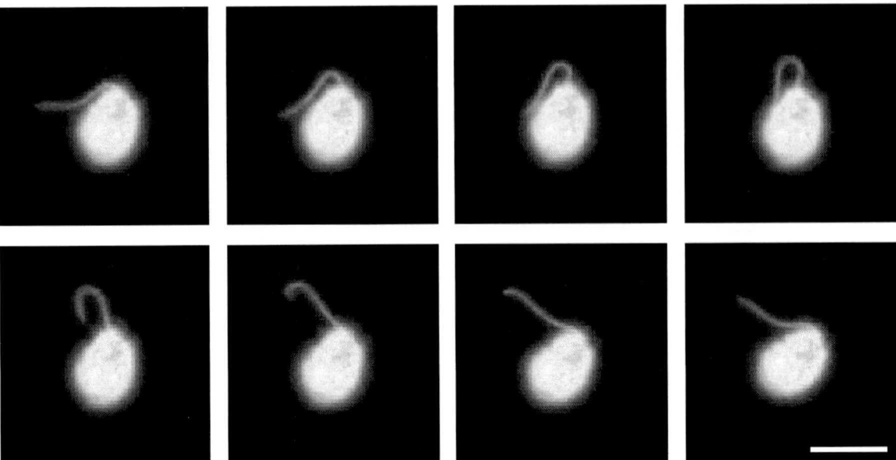

Fig. 1 High-speed video recording of a *uni1* cell. Images are arranged from upper left corner to the lower right corner with an interval of 1/300s. Scale bar = 10 μm.

(Kamiya and Hasegawa, 1987; Ruffer and Nultch, 1987). An alternative method for producing uniflagellated cells is to shear a cell suspension. By choosing cells that appear to have only a single flagellum, and assigning the flagellum to *cis* or *trans* from its position relative to the eyespot, one can obtain information for both flagella (Kamiya and Hasegawa, 1987). In this case, care must be taken that the remaining flagellum is not damaged by the shearing process.

A convenient method for measuring the average flagellar beat frequency in a population of cells is to use a fast Fourier transform (FFT) analyzer (Kamiya, 2000). Images of 10–100 swimming cells in a dark-field microscope are projected on a graded filter, and the transmitted light is detected with a photomultiplier or a photodiode. Because a swimming *Chlamydomonas* cell vibrates back-and-forth once per flagellar beat, we can determine the flagellar beat frequency by analyzing the light noise caused by the vibration of bright cell images. The graded filter converts the back-and-forth movements of cell images into a light intensity fluctuation. We are using a photodetector made with a photodiode, low-noise operational amplifiers (such as OPA111, Burr Brown, Tucson, AZ, USA) and 1-GΩ feedback resistors, attached to a dark-field microscope with a 100-W halogen lamp. FFT is performed using the SIGVIEW software package (http://www.sigview.com/). This method allows us to measure the average beat frequency within a minute. Examples for wild type and *oda1*, a mutant lacking outer arm dynein, are shown in Fig. 2.

D. Flagellar Waveform

For further analysis of flagellar motility, the beat pattern must be recorded and analyzed. In this case also, uniflagellated cells are convenient (Fig. 1). Recording

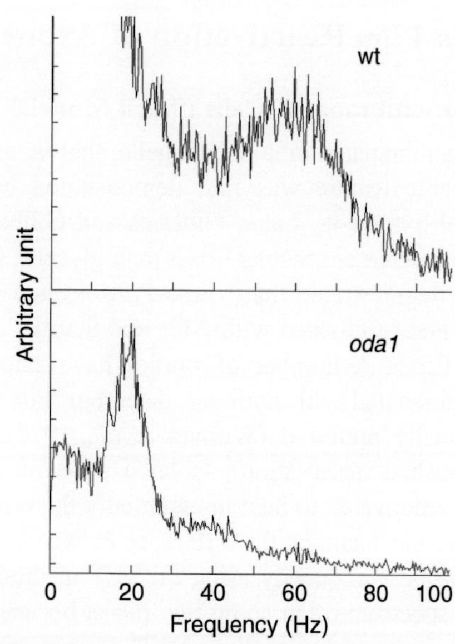

Fig. 2 Power spectra from wild-type (wt) and *oda1* cell suspensions. Swimming cell images were projected onto a graded filter and the intensity fluctuation of the light passing through it was analyzed by FFT. Data were averaged for 1 min. wt cells show a peak at about 57 Hz, which corresponds to the median value of flagellar beat frequencies displayed by many cells. The mutant *oda1*, which lacks outer arm dynein, has a peak at 20 Hz. The noise in the data can be reduced by extending the averaging time. Temperature: 23°C.

can be performed using a high-speed video camera that enables waveform recording for a single beat cycle. Alternatively, when the beating is stable, a strobe or a lower-speed video camera with an electronic shutter can be used to sample successive images from different cycles of beating. For most purposes, a series of beating flagellum images, without quantitative analysis, will give enough information. When quantitative data are required, each flagellum image must be traced on a sheet or on a monitor screen, and digitized manually or by computer. A useful method is to measure the angle (shear angle, θ) between the tangent to a small segment at the position s (measured from the base along the flagellum) and the tangent to the basal segment, and plot $\theta(s)$ for the entire length (Brokaw, 1979; Brokaw *et al.*, 1982). Because $\theta(s)$ is proportional to the sliding displacement between adjacent outer doublet microtubules at point s, the plots produce a single curve that represents the relative microtubule sliding along the flagellar length. Hence the curve is called a shear curve. From several sets of plots for different beating phases over a complete beat cycle, we can determine the amplitude and velocity of back-and-forth microtubule sliding at each position along the flagellum. The curves also give parameters such as the bend angle and curvature in the principal and reverse bends.

III. *In Vitro* Reactivation of Axonemal Beating

A. Reactivation of Demembranated Cells ("Cell Models") and Axonemes

Demembranated cilia and flagella, that is, axonemes, beat when ATP is added. This remarkable feature was first demonstrated by Hoffmann-Berling (1955), who used glycerol extraction. Later, Gibbons and Gibbons (1972) showed that Triton X-100, a nonionic detergent, works better than glycerol for sea urchin sperm. Naitoh and Kaneko (1972) demonstrated that whole *Paramecium* cells can be demembranated with Triton X-100 and reactivated with ATP, and that the ciliary movement is controlled by micromolar Ca^{2+}. A number of studies have since been performed on cilia and flagella demembranated with nonionic detergent. For *Chlamydomonas*, Nonidet P40 has been traditionally preferred (Witman *et al.*, 1978), although Triton X-100 works as well (unpublished observation). Isolated axonemes and axonemes attached to cell bodies can be reactivated to beat in essentially the same manner. With whole *Chlamydomonas* cells, we can examine the difference between the two flagella. In addition, we can easily measure beat frequency using the FFT method described above; in wild-type cells, the power spectrum often show two peaks because the *trans*-axoneme tends to beat faster than the *cis*-axoneme (Fig. 3) The axonemal beat frequency varies with the ATP concentration apparently in a manner consistent with Michaelis–Menten kinetics, as shown for sperm axonemes (Brokaw, 1967). Therefore, double reciprocal plots of frequency versus ATP concentration tend to give linear lines.

For experiments examining the effect of particular factors on axonemal motility, for example that of Ca^{2+} concentration or protein kinase inhibitors, isolated axonemes are better suited than cell models. This is because significant amounts of nonflagellar proteins are contained in cell model samples and they tend to modulate or mask axoneme-specific effects. For example, protein kinase inhibitor activates reactivated

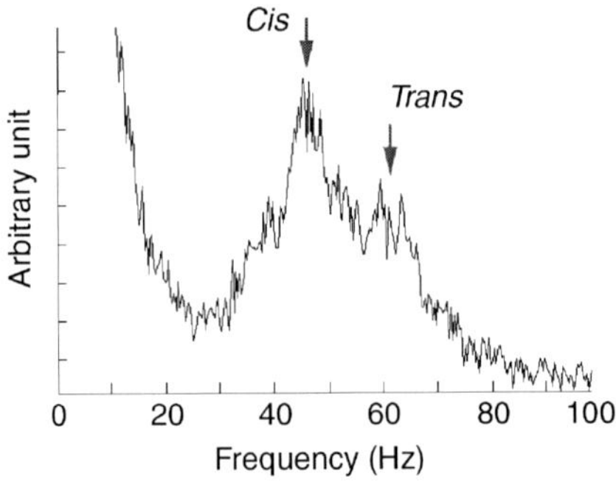

Fig. 3 Power spectrum from reactivated cell models of wild type. Two peaks appear around 47 and 60 Hz, corresponding to the intrinsic beat frequency of the *cis*- and *trans*-flagella, respectively (Kamiya and Hasegawa, 1987). This indicates that individual cell bodies vibrate at two distinct frequencies. In live cells, the two flagella tend to synchronize and beat in an frequency that is intermediate between *cis*- and *trans*-frequencies.

motility of isolated axonemes but not that of cell models (Hasegawa *et al.*, 1987). Also, the concentration of Ca^{2+} effective for converting the axonemal waveform in cell models is about one order of magnitude lower than in isolated axonemes (our unpublished observations; Bessen *et al.*, 1982).

1. Solutions

1. HES: 10 mM HEPES, 1 mM EGTA, 4% sucrose (pH 7.4).
2. HMDEKP: 30 mM HEPES, 5 mM $MgSO_4$, 1 mM DTT, 1 mM EGTA, 50 mM K-acetate, 1% (w/v) polyethylene glycol (MW: 10,000), adjusted with KOH to pH 7.4.
3. NP40: 0.1% Nonidet P40 in HMDEK (HMDEKP minus polyethylene glycol).

2. Methods

Demembranation

1. Culture cells in liquid medium to late logarithmic phase. Check that the cells are well flagellated.
2. Centrifuge 10 ml of the culture in a conical plastic tube at $500\times g$ for 3 min.
3. Resuspend the precipitated cells in 5 ml of HES solution and centrifuge at the same speed. Place the precipitated cells in an ice bath. This washing process is important to prevent flagella detachment; axonemes become detached if $>10^{-4}$ M Ca^{2+} is present.
4. Add 0.3–0.5 ml of the NP40 solution to the cell pellet and gently resuspend the cells with a Pasteur pipette. Check under the microscope that all cells are demembranated; the flagella should now look fainter than in live cells. This "cell model" sample can be stored on ice for up to 3 h and used for reactivation experiments.

Reactivation

5. Add 10 μl of the cell model suspension and 10 μl of an appropriate concentration (10× the final concentration) of ATP to 80 μl of HMDEKP in ice.
6. Take 20 μl of the mixture for observation under the microscope.

3. Tips

1. Use only a small volume of NP40 for demembranation. Denser samples tend to result in better reactivation.
2. For stable reactivation at low ATP concentrations for a long time, use an ATP-regeneration system in addition to ATP. We use 4 mM phosphocreatine and 70 units of creatine phosphokinase.
3. For observations of axonemes attached to cell bodies at $>10^{-4}$ M Ca^{2+}, use the *fa-1* mutant defective in flagellar autotomy (Finst *et al.*, 2000).

4. Observations can be performed as with live cells. However, pay attention to the glass slides as some cause poor reactivation. Coating with a silicone reagent, such as Sigmacoat (Sigma-Aldrich Chemical, St. Louis, MO, USA), is recommended. Silicone coating is also useful to prevent detergent-containing samples from dispersing on the glass surface.

B. Reactivation of Isolated Axonemes

4. Methods

Essentially, the same as for cell models, except that isolated flagella are demembranated instead of cells. Isolation of flagella is as described in Chapter 3 by King (MCB Vol. 92). Washing with HES is not necessary. The effect of Ca^{2+} is more readily examined with isolated axonemes than with cell models; in this case, replace the 1 mM EGTA in HMDEKP reactivation solution with an appropriate Ca^{2+} buffer (e.g., see Bessen *et al.*, 1980; Wakabayashi *et al.*, 1997). At $>10^{-5}$ M Ca^{2+}, axonemes beat with a symmetrical waveform. This waveform corresponds to the flagella-type beat pattern displayed by cells that undergo a photophobic response, a transient backward swimming upon exposure to strong light (Hyams and Borisy, 1978). Movements of reactivated axonemes can be seen in the supplemental movies (http://www.elsevierdirect.com/companions/9780123749734, Supplemental material 1).

IV. Microtubule Sliding by Axonemal Dynein

A. Sliding Disintegration

Axonemal dyneins cause sliding movements between adjacent outer doublets. This was first demonstrated by Summers and Gibbons (1971) using sea urchin sperm flagella. They showed that ATP addition to fragmented axonemes that had been mildly treated with trypsin caused interdoublet sliding. Since then, outer doublet sliding has been observed in many kinds of cilia and flagella. The velocity of sliding was analyzed in the axonemes of sea urchin sperm (Takahashi *et al.*, 1982; Yano and Miki-Noumura, 1980) and various *Chlamydomonas* mutants (Kurimoto and Kamiya, 1991; Okagaki and Kamiya, 1986; Smith 2002; Smith and Sale, 1992). Usually the velocity depends on the ATP concentration in a manner consistent with Michaelis–Menten kinetics (Fig. 4).

1. Materials

1. Axonemes
2. Perfusion chamber

A simple chamber is used that allows solution perfusion and observation. Typically, a chamber can be made from a 18-mm × 18-mm coverslip, a silicone-coated glass slide, and a pair of 5-mm × 18-mm spacers cut from double-sided adhesive tape.

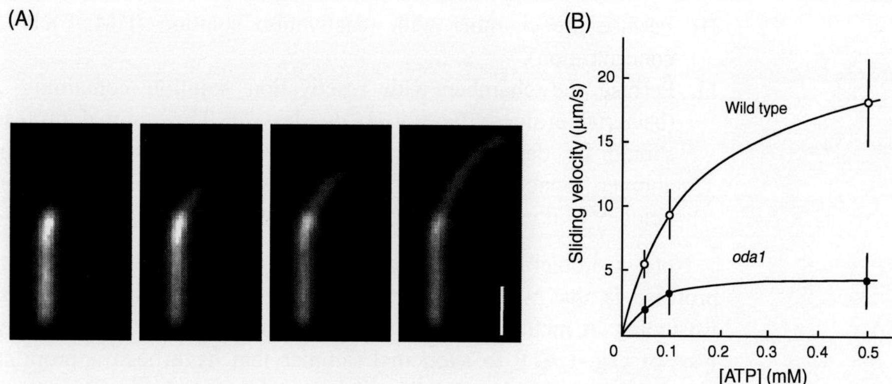

Fig. 4 Microtubule sliding in axonemal fragments in the presence of ATP and protease. (A) An example of a fragment undergoing sliding at 1 mM ATP. Time interval between frames is 0.1 s. The scale bar = 3 µm. (B) ATP concentration dependence of sliding velocity in wild-type (filled circle) and oda1 axonemes lacking outer arm dynein (filled circle).

2. Solutions

1. HMDEKP – polyethylene glycol
2. HMDEKP
3. Reactivation solution (HMDEKP + ATP)
4. Protease (nagarse or elastase) in HMDEK + ATP

3. Methods

Steps 4–8 can be omitted for most experiments.

1. Sonicate isolated flagella in a 1.5-ml microfuge tube so as to produce fragments of appropriate length. Sonication helps observation of sliding because a structure at the proximal end of the axoneme resists sliding; intact axonemes tend to fray apart with all the doublet microtubules connected at the proximal end. A treatment that leaves 20–40% of the total axonemes maintaining their original length is optimal.
2. Collect the sonicated flagella by centrifugation at 18,000 × *g* for 10 min.
3. Demembranate them by suspending the pellet in 0.1% NP40 in HMDEK.
4. Centrifuge at 18,000 × *g* for 15 min.
5. Suspend the pellet in 200 µl HMDEK.
6. Centrifuge at 18,000 × *g* for 15 min.
7. Suspend the pellet in HMDEKP.
8. Adjust the protein concentration to ~0.2 mg/ml by adding HMDEKP. This sample can be stored on ice for several hours.
9. Introduce 10-fold diluted sample into the observation chamber. 10–20 µl should be enough.

10. Perfuse the chamber with reactivation solution (HMDEKP + ATP of desired concentration).
11. Perfuse the chamber with reactivation solution containing 2 μg/ml nagarse (bacterial protease type VIII) or elastase. The optimal protease concentration should be determined experimentally because it varies from one batch to another. Generally, higher concentration of protease results in higher sliding velocity (Kikushima, 2009; Yano and Miki-Noumura, 1980).

Note: A problem with the sliding induction by perfusion with a mixture of ATP and protease is that we cannot predict when sliding will occur. It is inefficient and frustrating. A method for controlling the timing of sliding is to apply ATP by photolysis of caged ATP to axonemal samples that have been appropriately treated with protease. An example of sliding disintegration induced by this method is shown in a supplemental movie (http://www.elsevierdirect.com/companions/9780123749734, Supplemental material 2).

B. *In Vitro* Motility Assays Using Isolated Dynein

Many kinds of motor proteins adsorbed on a glass surface can translocate actin filaments or microtubules in the presence of ATP. This kind of "*in vitro* motility assay," first developed for myosin, was subsequently applied for kinesin, cytoplasmic dynein, and axonemal dyneins. Axonemal dyneins thus far used for such assays include sea urchin outer arm dynein (Paschal *et al.*, 1987), *Tetrahymena* 23S and 14S dyneins (Vale and Toyoshima, 1988, 1989), and various species of *Chlamydomonas* dyneins (Furuta *et al.*, 2009; Kagami and Kamiya, 1992; Kagami *et al.*, 1990; Kotani *et al.*, 2007; Smith and Sale, 1991; Sakakibara and Nakayama, 1998). This kind of assay is expected to reveal functional properties intrinsic to each kind of dynein. As a more advanced assay, force production by a single motor protein can be analyzed using optical tweezers. Most of the previous studies used myosin, kinesin or cytoplasmic dynein, and only a few studies have employed axonemal dyneins (Sakakibara *et al.*, 1999). The method outlined below is a conventional microtubule gliding assay for isolated *Chlamydomonas* dyneins.

1. Materials

1. *Dynein sample*. Prepared by high-performance liquid chromatography or sucrose density centrifugation (see Chapter 3 by King, volume 92) and diluted in HMDE [30 mM HEPES (pH 7.4), 5 mM $MgSO_4$, 1 mM DTT, 1 mM EGTA], or HMDEK (HMDE + 50 mM K-acetate) to give a concentration of 0.02–0.1 mg/ml. For inner arm dyneins, HMDE results in better adsorption to the glass surface. For outer arm dynein, both HMDE and HMDEK work well, but the gliding speed is slightly higher with HMDEK.

2. *Microtubules*. Polymerized from porcine brain tubulin, purified over a phosphocellulose column (Shelanski *et al.*, 1973), and resuspended in HMDE(K) containing 20 μM taxol.
3. *Perfusion chamber, microscope, etc*. This is the same arrangement as described above for the sliding disintegration assay. The internal volume is 10–20 μl.

2. Methods

1. Perfuse the dynein solution into the chamber and incubate for 2–3 min.
2. Perfuse the chamber with HMDE (K) containing 0.5–2 mg/ml bovine serum albumin (BSA).
3. Introduce the microtubule solution (tubulin concentration: ~5 μg/ml), and confirm that the microtubules become attached to the glass surface. If no microtubules are attached, the dynein sample may be too dilute. In that case, repeat step 1.
4. Perfuse the chamber with HMDE(K) containing an appropriate concentration of ATP. Microtubule should start gliding. At >0.5mM ATP, addition of 1–2 mM ADP activates motility in most cases (Kikushima *et al.*, 2004; Yagi, 2000). At very low ATP concentrations, use of an ATP-regenerating system (5 mM creatine phosphate and 70 units/ml of creatine phosphokinase) is recommended.

3. Tips

1. For outer arm dynein, addition of casein instead of BSA results in stable (but slightly slower) gliding.
2. For inner arm dyneins, it is important to lower the ionic strength of the sample solution. Samples obtained by ion-exchange chromatography need to be diluted with a low-salt solution.
3. Some dyneins rotate, as well as translocate, microtubules. To visualize the rotation, use microtubules grown from fragments of outer doublet microtubules, which are slightly curved. Such "tagged" microtubules are also useful for determining the polarity of movement since the outer doublet fragment is mostly positioned at the minus end of the microtubule; to prevent tubulin polymerization from the minus end completely, use a mixture of tubulin and N-ethyl maleimide (NEM)-treated tubulin, instead of tubulin alone (Hyman, 1991).

Acknowledgments

I thank Ken-ichi Wakabayashi (University of Tokyo) and Tomohiro Kubo (University of Tokyo) for providing figures and Susumu Aoyama (University of Tokyo) for movies. I also thank them and Akane Furuta (University of Tokyo), and Toshiki Yagi (Kyoto University) for sharing their protocols.

References

Bessen, M., Fay, R.B., and Witman, G.B. (1980). Calcium control of waveform in isolated flagellar axonemes of *Chlamydomonas*. *J. Cell Biol.* **86**, 446–455.

Brokaw, C.J. (1967). Adenosine triphosphate usage by flagella. *Science* **156**, 76–78.

Brokaw, C.J., and Kamiya, R. (1987). Bending patterns of *Chlamydomonas* flagella: IV. Mutants with defects in inner and outer dynein arms indicate differences in dynein arm function. *Cell Motil. Cytoskeleton* **8**, 68–75.

Brokaw, C.J. (1979). Calcium-induced asymmetrical beating of triton-demembranated sea urchin sperm flagella. *J. Cell Biol.* **82**, 401–411.

Brokaw, C.J., Luck, D.J., and Huang, B. (1982). Analysis of the movement of *Chlamydomonas* flagella: The function of the radial-spoke system is revealed by comparison of wild-type and mutant flagella. *J. Cell Biol.* **92**, 722–732.

Finst, R.J., Kim, P.J., Griffis, E.R., and Quarmby, L.M. (2000). Fa1p is a 171 kDa protein essential for axonemal microtubule severing in *Chlamydomonas*. *J Cell Sci.* **113**, 1963–1971.

Furuta, A., Yagi, T., Yanagisawa, H., Higuchi, H., and Kamiya, R. (2009). Systematic comparison of in vitro motile properties between *Chlamydomonas* wild-type and mutant outer arm dyneins each lacking one of the three heavy chains. *J. Biol. Chem.* **284**, 5927–5935.

Gibbons, B.H., and Gibbons, I.R. (1971). Flagellar movement and adenosine triphosphatase activity in sea urchin sperm extracted with triton X-100. *J. Cell Biol.* **54**, 75–97.

Hasegawa, E., Hayashi, H., Asakura, S., and Kamiya, R. (1987). Stimulation of in vitro motility of *Chlamydomonas* axonemes by inhibition of cAMP-dependent phosphorylation. *Cell Motil. Cytoskeleton* **8**, 302–311.

Hoffmann-Berling, H. (1955). Geisselmodelle und adenosintriphosphat (ATP). *Biochim. Biophys. Acta* **16**, 146–154.

Howard, D.R., Habermacher, G., Glass, D.B., Smith, E.F., and Sale, W.S. (1994). Regulation of Chlamydomonas flagellar dynein by an axonemal protein kinase. *J. Cell Biol.* **127**, 1683–1692.

Huang, B., Ramanis, Z., Dutcher, S.K., and Luck D.J. (1982). Uniflagellar mutants of *Chlamydomonas*: Evidence for the role of basal bodies in transmission of positional information. *Cell* **29**, 745–753.

Hyams, J., and Borisy, G. (1978). Isolated flagellar apparatus of *Chlamydomonas*: Characterization of forward swimming and alteration of waveform and reversal of motion by calcium ions *in vitro*. *J. Cell Sci.* **33**, 235–253.

Hyman, A.A. (1991). Preparation of marked microtubules for the assay of the polarity of microtubule-based motors by fluorescence. *J. Cell Sci. Suppl.* **14**, 125–127.

Kagami, O., and Kamiya, R. (1992). Translocation and rotation of microtubules caused by multiple species of *Chlamydomonas* inner-arm dynein. *J. Cell Sci.* **103**, 653–664.

Kagami, O., Takada, S., and Kamiya, R. (1990). Microtubule translocation caused by three subspecies of inner-arm dynein from *Chlamydomonas* flagella. *FEBS Lett.* **264**, 179–182.

Kamiya, R. (2000). Analysis of cell vibration for assessing axonemal motility in *Chlamydomonas*. *Methods* **22**, 383–387.

Kamiya, R., and Hasegawa, E. (1987). Intrinsic difference in beat frequency between the two flagella of *Chlamydomonas*. *Exp. Cell Res.* **173**, 299–304.

Kamiya, R., and Witman, G.B. (1984). Submicromolar levels of calcium control the balance of beating between the two flagella in demembranated models of *Chlamydomonas*. *J. Cell Biol.* **98**, 97–107.

Kikushima, K. (2009). Central pair apparatus enhances outer-arm dynein activities through regulation of inner-arm dyneins. *Cell Motil. Cytoskeleton* **66**, 272–280.

Kikushima, K., and Kamiya, R. (2008). Clockwise translocation of microtubules by flagellar inner-arm dyneins in vitro. *Biophys. J.* **94**, 4014–4019.

Kikushima, K., Yagi, T., and Kamiya, R. (2004). Slow ADP-dependent acceleration of microtubule translocation produced by an axonemal dynein. *FEBS Lett.* **563**, 119–122.

Kotani N., Sakakibara H., Burgess S.A., Kojima H., and Oiwa K. (2007). Mechanical properties of inner-arm dynein-f (dynein I1) studied with in vitro motility assays. *Biophys. J.* **93**, 886–894.

Kurimoto, E., and Kamiya, R. (1991). Microtubule sliding in flagellar axonemes of *Chlamydomonas* mutants missing inner- or outer-arm dynein: Velocity measurements on new types of mutants by an improved method. *Cell Motil. Cytoskeleton* **19**, 275–281.

Naitoh, Y., and Kaneko, H. (1972). Reactivated triton-extracted models of paramecium: Modification of ciliary movement by calcium ions. *Science* **176**, 523–524.

Okagaki, T., and Kamiya, R. (1986). Microtubule sliding in mutant *Chlamydomonas* axonemes devoid of outer or inner dynein arms. *J. Cell Biol.* **103**, 1895–1902.

Paschal, B.M., King, S.M., Moss, A.G., Collins, C.A., Vallee, R.B., and Witman, G.B. (1987). Isolated flagellar outer arm dynein translocates brain microtubules in vitro. *Nature* **330**, 672–674.

Ruffer, U., and Nultsch, W. (1987). Comparison of the beating of *cis*- and *trans*-flagella of *Chlamydomonas* cells held on micropipettes. *Cell Motil. Cytoskeleton* **7**, 87–93.

Sakakibara, H., Kojima, H., Sakai, Y., Katayama, E., and Oiwa, K. (1999). Inner-arm dynein c of *Chlamydomonas* flagella is a single-headed processive motor. *Nature* **400**, 586–590.

Sakakibara, H., and Nakayama, H. (1998). Translocation of microtubules caused by the alphabeta, beta and gamma outer arm dynein subparticles of *Chlamydomonas*. *J. Cell Sci.* **111**, 1155–1164.

Shelanski, M.L., Gaskin, F., and Cantor, C.R. (1973). Microtubule assembly in the absence of added nucleotides. *Proc. Natl. Acad. Sci. USA* **70**, 765–768.

Smith, E.F. (2002). Regulation of flagellar dynein by the axonemal central apparatus. *Cell Motil. Cytoskeleton* **52**, 33–42.

Smith, E.F., and Sale, W.S. (1991). Microtubule binding and translocation by inner dynein arm subtype-I1. *Cell Motil. Cytoskeeton* **18**, 258–268.

Smith, E.F., and Sale, W.S. (1992). Regulation of dynein-driven microtubule sliding by the radial spokes in flagella. *Science* **257**, 1557–1559.

Summers, K., and Gibbons, I.R. (1971). Adenosine triphosphate-induced sliding of tubules in trypsin-treated flagella of sea urchin sperm. *Proc. Natl. Acad. Sci. USA* **68**, 3092–3096.

Takahashi, K., Shingyoji, C., and Kamimura, S. (1982). Microtubule sliding in reactivated flagella. *Symp. Soc. Exp. Biol.* **35**, 159–177.

Vale, R.D., and Toyoshima, Y.Y. (1988). Rotation and translocation of microtubules in vitro induced by dyneins from *Tetrahymena* cilia. *Cell* **52**, 459–469.

Vale, R.D., and Toyoshima, Y.Y. (1989). Microtubule translocation properties of intact and proteolytically digested dyneins from *Tetrahymena* cilia. *J. Cell Biol.* **108**, 2327–2334.

Wakabayashi, K., Yagi, T., and Kamiya, R. (1997). Ca^{2+}-dependent waveform conversion in the flagellar axoneme of *Chlamydomonas* mutants lacking the central-pair/radial spoke system. *Cell Motil. Cytoskeleton* **38**, 22–28.

Witman, G.B., Plummer, J., and Sander, G. (1978). *Chlamydomonas* flagellar mutants lacking radial spokes and central tubules. Structure, composition, and function of specific axonemal components. *J. Cell Biol.* **76**, 729–747.

Yagi, T. (2000). ADP-dependent microtubule translocation by flagellar inner-arm dyneins. *Cell Struct. Funct.* **25**, 263–267.

Yano, Y., and Miki-Noumura, T. (1980). Sliding velocity between outer doublet microtubules of sea-urchin sperm axonemes. *J. Cell Sci.* **44**, 169–186.

CHAPTER 13

High-Speed Digital Imaging of Ependymal Cilia in the Murine Brain

Karl-Ferdinand Lechtreck[*], **Michael J. Sanderson**[†], *and* **George B. Witman**[*]

[*]Department of Cell Biology, University of Massachusetts Medical School, Worcester, Massachusetts 01655

[†]Department of Physiology, University of Massachusetts Medical School, Worcester, Massachusetts 01655

Abstract

The development and health of mammals requires proper ciliary motility. Ciliated epithelia are found in the airways, the uterus and Fallopian tubes, the efferent ducts of the testes, and the ventricular system of the brain. A technique is described for the motion analysis of ependymal cilia in the murine brain. Vibratome sections of the brain are imaged by differential interference contrast microscopy and recorded by high-speed digital imaging. Side views of individual cilia are traced to establish their

978-0-12-374973-4
DOI: 10.1016/S0091-679X(08)91013-X

bending pattern. Tracking of individual cilia recorded in top view allows determination of bend planarity and beat direction. Ciliary beat frequency is determined from line scans of image sequences. The capacity of the epithelium to move fluid and objects is revealed by analyzing the velocity of polystyrene beads added to brain sections. The technique is useful for detailed assessment of how various conditions or mutations affect the fidelity of ciliary motility at the ependyma. The methods are also applicable to other ciliated epithelia, for example, in airways.

I. Introduction

Ciliated epithelial cells line the surface of the ventricular system of the brain. Aqueducts and foramina connect the paired lateral ventricles in the cerebrum and the midline third and fourth ventricles in the midbrain and cerebellum, respectively. The ventricular system is filled with cerebrospinal fluid (CSF), a watery fluid (0.8 mPa s viscosity at 37°C; Bloomfield *et al.*, 1998) produced by the choroid plexuses, specialized regions of the ventricles. The CSF drains into the subarachnial space and into the spinal cord. Overproduction of CSF, failure to absorb it, or the blockage of its flow through the ventricular system cause hydrocephalus, an accumulation of fluid in the brain. The ependymal cilia move the CSF, but their contribution to the bulk flow of this fluid is limited. Nevertheless, impaired ciliary motility causes hydrocephalus in mice and other small mammals (Banizs *et al.*, 2005; Ibanez-Tallon *et al.*, 2004; Lechtreck *et al.*, 2008; Sapiro *et al.*, 2002; Zhang *et al.*, 2007) and significantly increases the chance of hydrocephalus and ventriculomegaly in humans (Afzelius, 2004; Ibanez-Tallon *et al.*, 2004). A plausible explanation is that ciliary motility is required in mice to keep the interventricular channels open, and contributes to keeping them open in humans, especially during the rapid postnatal growth of the brain (Ibanez-Tallon *et al.*, 2004). Ciliary beating also has been implicated in neuronal guidance (Clarke, 2006; Sawamoto *et al.*, 2006). Juvenile myoclonic epilepsy has been linked to altered ciliary motility, suggesting that defects in ciliary beating can result in neurological diseases (Ikeda *et al.*, 2005; King, 2006; Suzuki *et al.*, 2009).

The efficiency of cilia-based transport depends on the viscosity of the surrounding medium and on ciliary length, beat frequency, bending pattern, and coordination. Most cilia and flagella have a high beat frequency of up to 90 Hz (15–40 Hz for airway and ependymal cilia of mice, 40–60 Hz for sea urchin spermatozoa or *Chlamydomonas*). Therefore, high-speed imaging is required to reveal ciliary bending patterns and aberrations of these patterns. This is now generally achieved by high-speed digital imaging, in which a sequence of digital images is captured by a camera and recorded directly to a computer. The images can then be analyzed one by one or combined to create a digital video as desired.

Rates of up to 500 images/s have been used to analyze ciliary and flagellar movements of single cells. These include sea urchin and mammalian sperm (Ishijima, 1995a, b; Ishijima and Witman, 1987), *Leishmania major* (Gadelha *et al.*, 2007), *Tetrahymena thermophila* (Wood *et al.*, 2007), and *Chlamydomonas reinhardtii* (Ruffer and Nultsch,

1998), free swimming or captured on micropipettes. Beat patterns also have been analyzed for cilia of airway epithelial cells using tissue samples such as brushings (Chilvers and O'Callaghan, 2000; Chilvers *et al.*, 2003) or lung slices (Delmotte and Sanderson, 2006), or using cultured ciliated epithelial cells (Sutto *et al.*, 2004). The techniques used have been described in several methods-oriented publications (Ishijima, 1995a,b; Sanderson and Dirksen, 1985, 1995). In contrast, only a few studies have analyzed ependymal cilia *in vivo* using tissue preparations such as ventricular brushings (Ibanez-Tallon *et al.*, 2004) and primary cell cultures (Weibel *et al.*, 1986). As a result, the motility and bending pattern of ependymal cilia are less well analyzed. In this chapter we describe techniques for high-speed digital imaging and analysis of ciliary motility of the ependyma in brain slices.

II. Materials and Equipment

A. Materials

1. Animals: mice, mutant and wild-type litter mates, preferably between p5 and p8 (animals should be analyzed before hydrocephalus develops to avoid distortion of data by secondary effects).
2. Euthanasia: sodium pentobarbital (50 mg/ml Nembutal sodium solution), syringe, needle.
3. Tissue preparation: scissors, forceps, spatula, razor blades, superglue (Quick Bond Aron Alpha CE-471, Electron Microscopy Sciences, Hatfield, PA 19440, U.S.A.), Petri dishes.
4. Observation chambers: custom coverslip support (see Fig. 1C), coverslips, silicone grease, polyester mesh (500 μm), polyethylene tubing.
5. Fluid flow: polystyrene beads (0.5 μm in diameter, Sigma-Aldrich, St. Louis, MO 63178, US).

B. Solutions

1. Hanks' Balanced Salt Solution (Invitrogen Corp., Carlsbad, CA 92008, U.S.A.) supplemented with 25 mM Hepes, pH 7.4.
2. Dulbecco's Modified Eagle's Medium supplemented with 10% FBS, penicillin, and streptomycin.

C. Equipment

1. Vibratome (OTS-4000, Electron Microscopy Sciences)
2. Microscope (Olympus IX71 inverted microscope)
3. Objective (60×, NA 1.2, water immersion)
4. Camera (TM-6740, Pulnix, 640 × 480 pixels, 200 images per second, coupled with a frame grabber (DVR Express, IO Industries, Inc., London, Ontario N6H 5S1, Canada) linked to a computer hard-drive array)
5. Optional: zoom adaptor (Nikon)

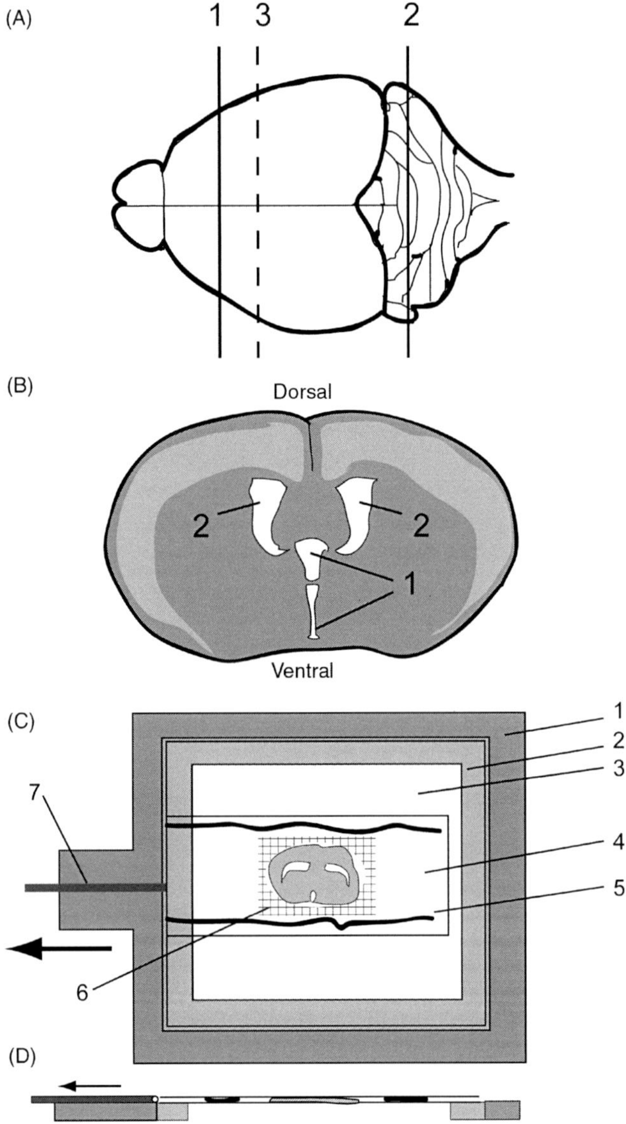

Fig. 1 Tissue preparation for *in vivo* imaging of ependymal cilia. (A) Dorsal view of a murine brain. Trim along lines 1 and 2 and glue section plane 2 of brain onto the specimen holder of the vibratome. (B) A coronal section cut approximately in the plane indicated with a dashed line (3) in panel A. The third (1) and the lateral (2) ventricles are marked. (C) Coverslip chamber consisting of a Plexiglas support (1) containing a milled groove (2) fitting a 45×50-mm coverslip (3). The section is placed between this and a second coverslip (4) using silicon grease (5) as spacers and to seal the sides of the chamber. Optional: A shim cut from polyester mesh (6) is placed around the section to keep it in place and prevent damage from pressure. A polyethylene tube (ID 0.045″, OD 0.062″) is attached to one side of the chamber for removal of fluid (7). For constant flow, a second tube can be attached to the opposite side of the chamber. To analyze cilia-generated fluid flow, beads are added to the edge of the coverslip opposite the outflow tube. Removal of fluid via the tube (we use a vacuum pump to accomplish this) then moves the beads to the section. (D) Side view of the observation chamber.

6. Digital image acquisition software (Video Savant V4, IO Industries)
   (The equipment we used is listed in parentheses).

III. Methods

A. Tissue Preparation

Inject mice intraperitoneally with a lethal dose of pentobarbital (0.5 mg/g body weight). Remove the skin from the head and open the skull from the base using scissors. Remove the brain by inserting a spatula below the brain from the back and wash brain in Hanks' Buffered Salt Solution (HBSS). Trim the brain using razor blades. For observation of cilia in the third ventricle in side view, trim the brain for coronal sectioning by cutting approximately in the middle between the olfactory bulbs and the Colliculus posterior (line 1 in Fig. 1A). To fasten the brain for vibratome sectioning, remove the cerebellum with a second cut parallel to the first one (Fig. 1A, line 2) and place the brain with this side down into a small drop of superglue on the specimen holder. Gently press down the brain with a spatula until the glue has polymerized. Process the brain accordingly for other views of the ventricular cilia.

B. Sectioning and Examination of Sections

Place the specimen holder into the vibratome reservoir filled with HBSS. Section 130-μm slices using a blade speed at a dial setting of ~3 and a blade advance setting of 0.2–0.5. Prior to observation of cilia at high magnification, it is useful to first locate the ciliated epithelium in sections (Fig. 1B) using an inverted microscope at low magnification. To do this, carefully transfer sections to a drop of HBSS in a glass-bottom culture dish (MatTek Corporation, Ashland, MA, USA) using a spatula. Transfer suitable sections to a coverslip (we use 45 × 50 mm, number 1) supported on a custom-built Plexiglas support (Fig. 1C). Use a syringe to apply two lines of silicon grease as spacers and carefully lower a second coverslip (we use 40 × 20 mm, number 1) onto the bottom coverslip to form a chamber; fill the chamber with HBSS (Fig. 1C and D). For prolonged observation, a polyester mesh shim should be placed around the section to minimize pressure on the tissue, and the buffer should be replaced regularly by removing buffer from one side and adding fresh buffer to the other side of the chamber. Usually, we try to analyze sections within 30–60 min after sacrificing the animal, but sections also can be stored in culture medium at 37°C. After 24 h the ciliated epithelium appeared intact and cilia beat vigorously but sections became sticky and more difficult to handle.

C. Imaging

Image acquisition will vary with each microscope. Good optics, Koehler illumination, and adjustment of differential interference contrast are required. An objective

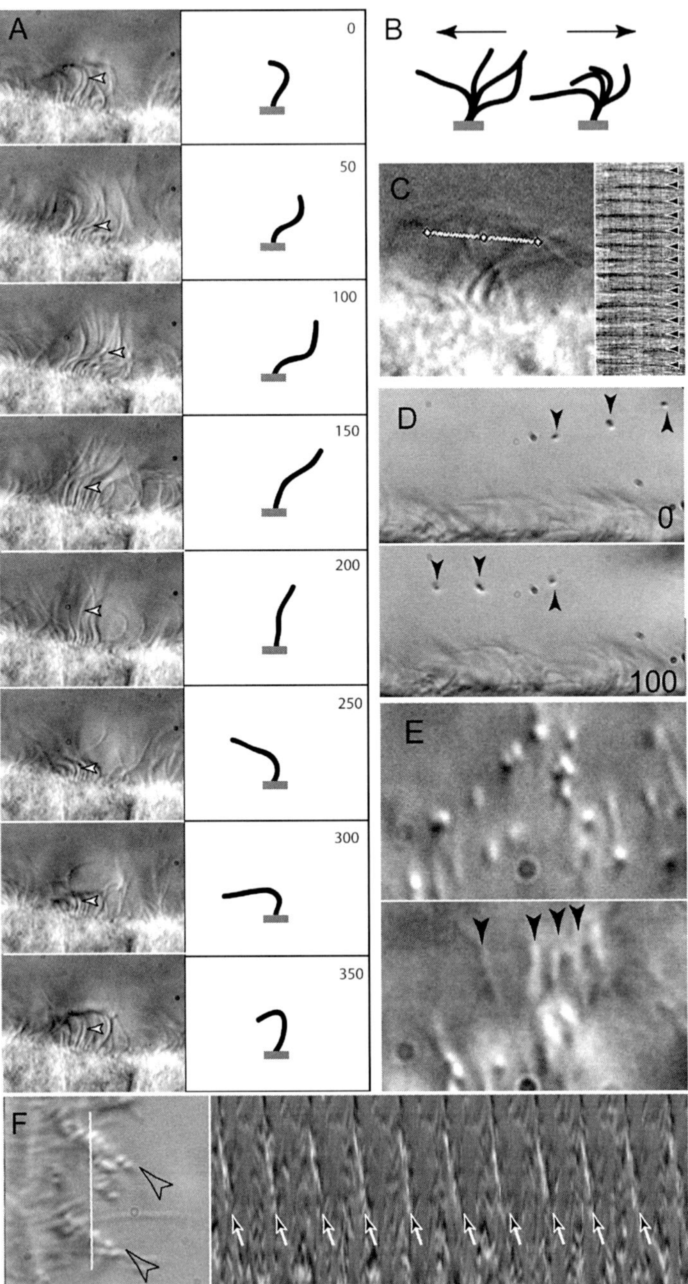

Fig. 2 (*Continued*)

with a relatively long working distance facilitates the examination of thicker tissue slices. In our hands, cilia are easier to observe at the surface of the section, but ciliary motility is usually better in the middle of the section where the tissue is less affected by the sectioning. Top views and front views of cilia can be obtained in brain sections cut horizontally approximately at a level connecting the top third of the olfactory bulbs to the middle of the cerebellum or by sagital sectioning close to the midline (Fissura longitudinalis cerebri). To visualize the fluid flow generated by the cilia, polystyrene beads can be added to one side of the chamber and moved near the section by removing fluid from the other side. Because the beads are rapidly captured by the cilia, start the recording as soon as the first beads reach the ciliated surface (Fig. 2D). Often, floating cell debris is sufficient to determine the velocity of the fluid flow. The software Video Savant provides the ability to record extended image sequences that are only limited by the size of the array hard drive. Image sequences can be analyzed within Video Savant using custom scripts or archived in a number of other file formats for analysis using other programs. Sample videos are available in the supplementary materials (http://www.elsevierdirect.com/companions/9780123749734) and from http://jcb.rupress.org/cgi/content/full/jcb.200710162/DC1 (Lechtreck *et al.*, 2008).

D. Data Analysis

1. Bending pattern: To determine the bending pattern, follow an individual cilium recorded in side view through consecutive images (Fig. 2A). We use a graphics tablet (Wacom Technology Corp., Vancouver, WA 98683, U.S.A.) and the Adobe Illustrator paintbrush tool (B), which somewhat smoothes the line, to track individual cilia through one beat cycle (Fig. 2B). Tracking of individual cilia recorded in top view will allow determination of whether the beating is planar (Fig. 2E). Alternatively, a running average can be generated using ImageJ; planar movements will generate a straight line, rotating cilia will generate circles.

Fig. 2 Motion analysis of ependymal cilia. (A) Bending pattern. A series of images from a recording of ependymal cilia and the corresponding tracings of an individual cilium (marked with arrowheads). The time in ms is indicated. (B) Various phases of the power stroke (left side) and recovery stroke (right side) of the cilium traced in part A. (C) CBF. Left side: Image from a movie showing the line used to generate a line scan. Right side: Line scan corresponding to 1 s (200 frames) of recording showing a regular pattern of diagonal lines (arrowheads) formed by the moving cilia. The CBF was estimated to be ~15 Hz. (D) Fluid flow. Two images spaced 100 ms apart; three polystyrene beads marked with arrowheads have moved about 12 µm. (E) Upper panel: Video image showing cilia from above (top view). Bottom panel: 8-frame image average of the same region showing parallel lines (arrowheads) indicating that the cilia beat in parallel planes. (F) Left panel: Image showing cilia in front view. Note the sequential positions in the beat cycles of neighboring cilia, which results in metachronal waves (arrowheads). The line used to generate the linescan shown in the right panel is indicated. Right panel: Line scan showing regular diagonal lines (arrows) indicating metachrony of the cilia. Part F of this figure is reprinted from © Lechtreck *et al.* (2008). Originally published in *J. Cell Biol.* doi: 10.1083/jcb.200710162.

2. CBF: A line scan is usually sufficient to determine the beat frequency. Top views of beating cilia are well suited for this analysis, but side or front views also can be used. Cilia passing through the line will generate a regular pattern on the scan (Fig. 2C and F). ImageJ or Scion Image with the appropriate plug-in ("multiplekymogram" for ImageJ) are examples of programs useful for generating line scans. Delmotte and Sanderson (Delmotte and Sanderson, 2006) describe an alternative method to determine the CBF based on digital videos.
3. Coordination: Top views will reveal if the direction of beating is similar for cilia in a tissue sample (Fig. 2E). Cilia moving with metachrony will generate diagonal lines on line scans of side and front views (Fig. 2F).
4. Video editing: In addition to the use of Video Savant software to create digital videos, images can be saved as individual files, for example, in Tiff format. These can be opened in Adobe Photoshop and rotated, cropped, adjusted, labeled, and saved using the "Actions" command, which ensures that all pictures of a stack are manipulated identically. QuickTime and other programs can be used to generate movies from the image sequences.

IV. Discussion

The above technique allows monitoring of ciliary motility in thick sections, which preserves tissue structure better than other techniques. In previous studies, tissue brushings obtained from mouse brain generated fluid flow at an average of 22 μm/s at room temperature (Ibanez-Tallon *et al.*, 2004); velocities of ~28 μm/s were recorded in the lateral and forth ventricles using prewarmed medium (Banizs *et al.*, 2005). By comparison, the fluid flow observed above the ciliated surface of the lateral and third ventricles of our vibratome sections had a velocity of 80–100 μm/s (Lechtreck *et al.*, 2008), indicating a better preservation of ciliary motility. We analyzed cilia at ambient temperature, which is probably why the CBF was below that reported in other studies (~18 Hz compared to ~40 Hz in rats) (Mönkkönen *et al.*, 2008). The CBF of airway cilia almost doubles when the temperature is increased by 8–10°C (Delmotte and Sanderson, 2006).

V. Summary

Vibratome thick sections of the brain allow analysis of ependymal cilia from wild-type and mutant animals. Depending on the direction from which the cilia are viewed, various parameters (CBF, bending pattern, beat plane, coordination, and fluid flow) can be easily analyzed. The technique is useful to determine the effect of certain mutations on the motility of ependymal cilia and for physiological studies of wild-type cilia.

Acknowledgments

This work was supported by National Institutes of Health grants HL071930 (M.J.S.) and GM30626 (G.B. W.) and by the Robert W. Booth Fund at the Greater Worcester Community Foundation.

References

Afzelius, B.A. (2004). Cilia-related diseases. *J. Pathol.* **204**, 470–477.

Banizs, B., Pike, M.M., Millican, C.L., Ferguson, W.B., Komlosi, P., Sheetz, J., Bell, P.D., Schwiebert, E.M., and Yoder, B.K. (2005). Dysfunctional cilia lead to altered ependyma and choroid plexus function, and result in the formation of hydrocephalus. *Development* **132**, 5329–5339.

Bloomfield, I.G., Johnston, I.H., and Bilston, L.E. (1998). Effects of proteins, blood cells and glucose on the viscosity of cerebrospinal fluid. *Pediatr. Neurosurg.* **28**, 246–251.

Chilvers, M.A., and O'Callaghan, C. (2000). Analysis of ciliary beat pattern and beat frequency using digital high speed imaging: Comparison with the photomultiplier and photodiode methods. *Thorax* **55**, 314–317.

Chilvers, M.A., Rutman, A., and O'Callaghan, C. (2003). Ciliary beat pattern is associated with specific ultrastructural defects in primary ciliary dyskinesia. *J. Allergy Clin. Immunol.* **112**, 518–524.

Clarke, J. (2006). Cell migration: Neurons go with the flow. *Curr. Biol.* **16**, R337–R339.

Delmotte, P., and Sanderson, M.J. (2006). Ciliary beat frequency is maintained at a maximal rate in the small airways of mouse lung slices. *Am. J. Respir. Cell Mol. Biol.* **35**, 110–117.

Gadelha, C., Wickstead, B., and Gull, K. (2007). Flagellar and ciliary beating in trypanosome motility. *Cell Motil. Cytoskeleton* **64**, 629–643.

Ibanez-Tallon, I., Pagenstecher, A., Fliegauf, M., Olbrich, H., Kispert, A., Ketelsen, U.P., North, A., Heintz, N., and Omran, H. (2004). Dysfunction of axonemal dynein heavy chain Mdnah5 inhibits ependymal flow and reveals a novel mechanism for hydrocephalus formation. *Hum. Mol. Genet.* **13**, 2133–2141.

Ikeda, T., Ikeda, K., Enomoto, M., Park, M.K., Hirono, M., and Kamiya, R. (2005). The mouse ortholog of EFHC1 implicated in juvenile myoclonic epilepsy is an axonemal protein widely conserved among organisms with motile cilia and flagella. *FEBS Lett.* **579**, 819–822.

Ishijima, S. (1995a). High-speed video microscopy of flagella and cilia. *Methods Cell Biol.* **47**, 239–243.

Ishijima, S. (1995b). Micromanipulation of sperm and other ciliated or flagellated single cells. *Methods Cell Biol.* **47**, 245–249.

Ishijima, S., and Witman, G.B. (1987). Flagellar movement of intact and demembranated, reactivated ram spermatozoa. *Cell Motil. Cytoskeleton* **8**, 375–391.

King, S.M. (2006). Axonemal protofilament ribbons, DM10 domains, and the link to juvenile myoclonic epilepsy. *Cell Motil. Cytoskeleton* **63**, 245–253.

Lechtreck, K.F., Delmotte, P., Robinson, M.L., Sanderson, M.J., and Witman, G.B. (2008). Mutations in Hydin impair ciliary motility in mice. *J. Cell Biol.* **180**, 633–643.

Mönkkönen, K.S., Hirst, R.A., Laitinen, J.T., and O'Callaghan, C. (2008). PACAP27 regulates ciliary function in primary cultures of rat brain ependymal cells. *Neuropeptides* **42**, 633–640.

Ruffer, U., and Nultsch, W. (1998). Flagellar coordination in *Chlamydomonas* cells held on micropipettes. *Cell Motil Cytoskeleton* **41**, 297–307.

Sanderson, M.J., and Dirksen, E.R. (1985). A versatile and quantitative computer-assisted photoelectronic technique used for the analysis of ciliary beat cycles. *Cell Motil.* **5**, 267–292.

Sanderson, M.J., and Dirksen, E.R. (1995). Quantification of ciliary beat frequency and metachrony by high-speed digital video. *Methods Cell Biol.* **47**, 289–297.

Sapiro, R., Kostetskii, I., Olds-Clarke, P., Gerton, G.L., Radice, G.L., and Strauss, J.F. III (2002). Male infertility, impaired sperm motility, and hydrocephalus in mice deficient in sperm-associated antigen 6. *Mol. Cell Biol.* **22**, 6298–6305.

Sawamoto, K., Wichterle, H., Gonzalez-Perez, O., Cholfin, J.A., Yamada, M., Spassky, N., Murcia, N.S., Garcia-Verdugo, J.M., Marin, O., Rubenstein, J.L., Tessier-Lavigne, M., Okano, H., *et al.* (2006). New neurons follow the flow of cerebrospinal fluid in the adult brain. *Science* **311**, 629–632.

Sutto, Z., Conner, G.E., and Salathe, M. (2004). Regulation of human airway ciliary beat frequency by intracellular pH. *J. Physiol.* **560**, 519–532.

Suzuki, T., Miyamoto, H., Nakahari, T., Inoue, I., Suemoto, T., Jiang, B., Hirota, Y., Itohara, S., Saido, T.C., Tsumoto, T., Sawamoto, K., Hensch, T.K., *et al.* (2009). Efhc1 deficiency causes spontaneous myoclonus and increased seizure susceptibility. *Hum. Mol. Genet.* **18**, 1099–1109.

Weibel, M., Pettmann, B., Artault, J.C., Sensenbrenner, M., and Labourdette, G. (1986). Primary culture of rat ependymal cells in serum-free defined medium. *Brain Res.* **390**, 199–209.

Wood, C.R., Hard, R., and Hennessey, T.M. (2007). Targeted gene disruption of dynein heavy chain 7 of *Tetrahymena thermophila* results in altered ciliary waveform and reduced swim speed. *J. Cell Sci.* **120**, 3075–3085.

Zhang, Z., Tang, W., Zhou, R., Shen, X., Wei, Z., Patel, A.M., Povlishock, J.T., Bennett, J., and Strauss, J.F., III (2007). Accelerated mortality from hydrocephalus and pneumonia in mice with a combined deficiency of SPAG6 and SPAG16L reveals a functional interrelationship between the two central apparatus proteins. *Cell Motil. Cytoskeleton* **64**, 360–376.

CHAPTER 14

Observation of Nodal Cilia Movement and Measurement of Nodal Flow

Yasushi Okada *and* Nobutaka Hirokawa

Department of Cell Biology and Anatomy, University of Tokyo, Graduate School of Medicine, 7-3-1 Hongo, Bunkyo-ku, Tokyo, Japan 113-0033

Abstract

Mammalian left–right determination is a good example of how multiple cell biological processes coordinate in the formation of a basic body plan, but until recently its mechanism was totally elusive. In the past 10 years, molecular genetic studies of kinesin and dynein motor proteins, live-cell imaging techniques, and theoretical studies

978-0-12-374973-4
DOI: 10.1016/S0091-679X(08)91014-1

of fluid mechanics revealed unexpected mechanisms of left–right determination. The leftward movement of fluid at the ventral node, called nodal flow, is the central process in symmetry breaking on the left–right axis. Nodal flow is autonomously generated by the rotation of posteriorly tilted cilia that are built by transport via the KIF3 motor on cells of the ventral node. Recent evidence suggests that nodal flow transports sheathed lipidic particles, called nodal vesicular parcels (NVPs), to the left edge of the node, which results in the activation of the noncanonical Hedgehog signaling pathway, an asymmetric elevation in intracellular Ca^{2+}, and changes in gene expression. This chapter reviews techniques for the observation of nodal cilia movement and nodal flow in living vertebrate embryos.

I. Introduction and Historical Background

Although the human body is apparently bilaterally symmetrical on the surface, the visceral organs are arranged asymmetrically in a stereotyped manner. The heart, spleen, and pancreas reside on the left side of the body, whereas the gall bladder and most of the liver are on the right side (Fig. 1A). In human, mouse, and other mammals, the embryo is cylindrically symmetrical when it implants itself into the wall of the uterus. The dorso-ventral (DV) axis is the first to be specified as the proximal–distal axis from the implantation site. Subsequently, the anterior–posterior (AP) axis is arbitrarily determined in the plane perpendicular to the DV axis (Alarcon and Marikawa, 2003; Beddington and Robertson, 1999). Left–right (LR) is thus the last axis to be determined and needs to be consistent with the preceding DV and AP axes. Since the chirality of the body is predetermined by chiral molecules, such as amino acids and nucleic acids, the laterality or orientation of the LR axis is established theoretically or potentially once the AP and DV axes are determined. The problem is how this potentially established laterality is materialized through developmental events. This mechanism is still totally unknown for invertebrates. However, recent studies of the mouse embryo clarified the LR determination mechanism in mammalian embryos (Hirokawa *et al.*, 2006).

More than 30 years ago, studies of a human genetic disease called Kartagener's syndrome suggested a link between ciliary motility and LR determination (Afzelius, 1976), but the mechanism was not known. Molecular biological studies identified several genes that are asymmetrically expressed in the LR orientation prior to LR asymmetric morphogenesis of the embryo (Capdevila *et al.*, 2000; Hamada, 2002; Harvey, 1998; Levin, 2005; Yost, 1999), but the upstream phenomena that cause asymmetrical expression of these genes remained enigmatic.

Many studies have suggested that the so-called node, a concave triangular region transiently formed during gastrulation at the ventral midline surface of early

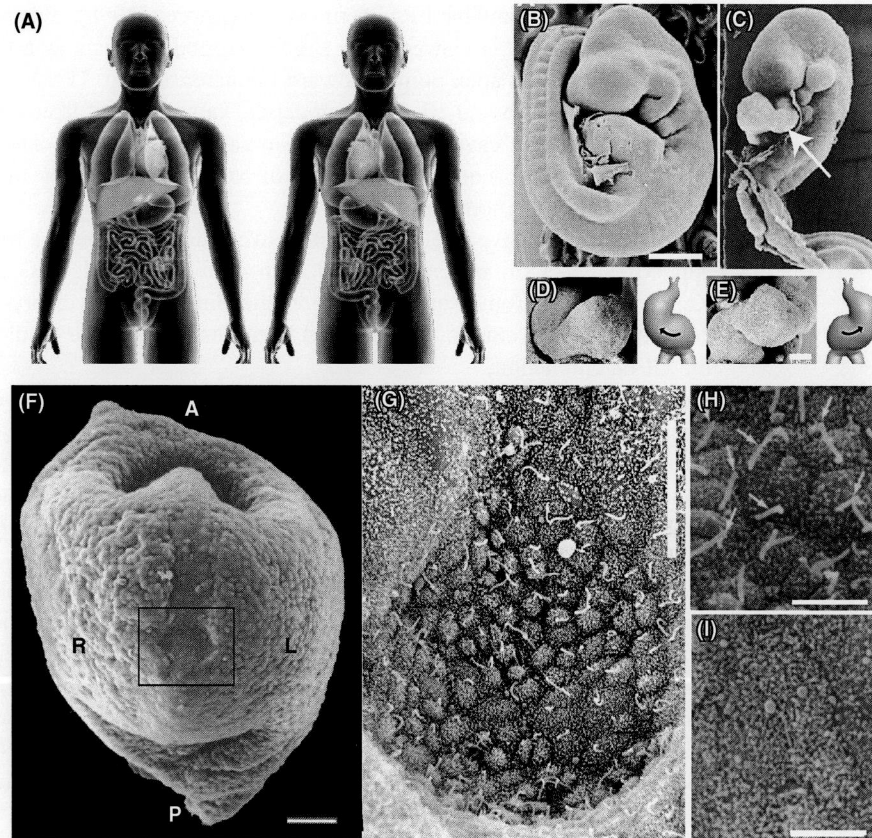

Fig. 1 (A) Left–right asymmetric arrangements of internal organs in the human body. Normal arrangement (*situs solitus*) (left). Most humans (>99%) have the heart on the left side and the liver on the right side. Mirrored arrangement (*situs inversus*) (right). Half of patients with Kartagener's syndrome have this arrangement, whereas the remaining patients are normal. Therefore, the left–right bilateral symmetry is randomly broken in this disease. (B–E) Scanning electron micrographs of wild-type (B, D) and *Kif3b*$^{-/-}$ (C, E) mouse embryos. (B, C) Full-length images. Wild-type embryos at this stage have already turned with a right-sided tail (B), whereas *Kif3b*$^{-/-}$ embryos remain unturned (C). In panel (C), the dilated pericardial sac has been removed, and the heart loop is inverted (*arrow*). (D, E) Higher magnification images and schematic representations of the heart loops showing a normal loop in the wild-type embryo (D) and an inverted loop in the mutant embryo (E). (F–I) Scanning electron micrographs of a mouse node. (F) Low-magnification view of a mouse embryo at 7.5 days post coitum. Reichert's membrane is removed, and the embryo is observed from the ventral side. The node is indicated by a black rectangle. The orientation is indicated in the panel as anterior (A), posterior (P), left (L), and right (R). Scale bar = 100 μm. (G) Higher magnification view of the mouse node. The orientation is the same as in panel (A). Scale bar = 20 μm. (H) Higher magnification view of the nodal cilia (*arrows*) and nodal pit cells. Scale bar = 5 μm. (I) Nodal pit cells of *Kif3b*$^{-/-}$ embryos. Nodal cilia are absent in these genetically manipulated embryos. Panels A was reproduced with permission from JT Biohistory Research Hall/TokyoCinema, (B–I) were modified from Nonaka *et al.* (1998), Okada *et al.* (2005), and Hirokawa *et al.* (2006).

embryos, is important for LR determination (Harvey, 1998). When the ventral side of a mouse embryo is viewed from above, the node appears as a roughly triangular depression with the apex pointed toward the anterior (Fig. 1F and G). It is typically 50–100 μm in width and 10–20 μm in depth. This nodal pit is covered by Reichert's membrane, and the cavity is filled with extraembryonic fluid. The ventral embryonic surface of the nodal pit consists of an epithelial sheet of a few hundred monociliated cells (nodal pit cells).

Nodal pit cells have one or sometimes two cilia that appear as rodlike protrusions approximately 5 μm in length and 0.3 μm in diameter (Fig. 1G and H). Because Kartagener's syndrome suggested a potential link between ciliary motility and LR determination, the cilia in the node had been postulated to be motile and responsible for LR determination.

However, the ultrastructure of these nodal cilia is similar to that of immotile primary cilia. The central pair that is important for the determination of the beat plane is missing, and nodal cilia have a 9 + 0 microtubule arrangement like other immotile primary cilia. Therefore, based on their ultrastructure and initial videomicroscopic observations, nodal monocilia were originally considered immotile (Bellomo *et al.*, 1996).

Through studies of the molecular motors of the kinesin superfamily, we serendipitously discovered that nodal flow is the key mechanism for LR determination. By generating knockout mice of *Kif3a* and *Kif3b*, we showed that these mammalian kinesin-2 proteins are essential motors for intraflagellar transport as reported by other groups in lower eukaryotes (for reviews, see Rosenbaum and Witman, 2002; Scholey, 2003). More importantly, this defective cilia phenotype accompanied defects in LR determination. Approximately 50% of KIF3A-deficient and KIF3B-deficient mice show reversed heart loops, whereas the rest are normal (Fig. 1C and E). Abnormal expression of *Lefty-2*, one of the earliest left-defining genes (Marszalek *et al.*, 1999; Nonaka *et al.*, 1998, Takeda *et al.*, 1999), accompanied the loss of the nodal cilia (Fig. 1I).

At the same time, a mutation in one isoform of axonemal dynein motor in mouse was reported to cause the randomization of the LR determination as occurs with the human Kartagener syndrome (Supp *et al.*, 1997). We, therefore, developed procedures for the video microscope observation of the node in living mouse embryos, which are described in the following sections. Surprisingly, we found that the monocilia are in fact vigorously rotating at approximately 600 rpm (10 Hz), and these rotating cilia generate leftward flow of fluid in the node cavity (nodal flow) (Nonaka *et al.*, 1998; Okada *et al.*, 1999; Takeda *et al.*, 1999). The directionality of this nodal flow was shown to be necessary and sufficient for the determination of the LR axis through studies of LR mutant mice (Okada *et al.*, 1999) and by culturing mouse embryos under artificial flow conditions (Nonaka *et al.*, 2002). Experimental and theoretical studies clarified how the direction of nodal flow is determined. High temporal resolution observation of nodal cilia movement (500 frames/s [fps]) demonstrated that the rotating axis of the nodal cilia is posteriorly tilted, and theories of fluid mechanics confirmed that this tilted rotation produces directional flow (Buceta *et al.*, 2005; Cartwright *et al.*, 2007; Nonaka *et al.*, 2005; Okada *et al.*, 2005; Smith *et al.*, 2007, 2008).

For these studies, the observation of nodal flow and nodal cilia in living embryos is essential. As suggested from the failures in earlier studies, it is crucial to maintain the embryos' health during the observation. At the same time, the microscope system needs enough resolution for the observation of submicron structures of nodal cilia in thick living embryos. In this chapter, we describe our microscope system for these studies as well as the preparation of the embryos.

II. Observation of Nodal Cilia Motility in Mouse Embryo

A. Overview

The diameter of the nodal cilia is below the optical diffraction limit, and the nodal cilia exist on the surface of embryo, which consists of at least two layers of cells. Hence, you need a high contrast image with thin optical sectioning capability. DIC optics are best suited for this purpose. With proper adjustment, currently available DIC optics have enough quality for the visualization of the nodal cilia. You can easily see rotating nodal cilia with your naked eye if the embryo is healthy and the optics are optimum. One of the most critical factors is the good preparation of the embryo. An advantage of DIC observation is that you can easily monitor the health of the embryo by the movement of the organelles in the nodal pit cells and other cells in the embryo. We, therefore, recommend starting from this experiment as practice for handling the embryos.

B. Microscope System

1. Optics

To enable the manipulation of embryos during observation, we use a fixed-stage-type upright microscope Olympus BX51WI (Fig. 2), but either a standard upright or an inverted microscope can be used. We have tried both Nikon and Zeiss upright microscopes, and they similarly worked well. Good DIC optics are required for high-contrast observation of nodal cilia. We use the universal condenser U-UCD8 with an oil immersion high NA condenser lens (U-TLO, Olympus, Tokyo, Japan) and a series of DIC prisms of different shear amount (DICTHC, DICTHR, and DICT, Olympus). A long shear prism (DICTHC) gives a high contrast image of small objects like cilia, so that the identification of cilia in the image analysis becomes easier, but the resolution is compromised. With thick samples like a whole fish embryo, the images are affected by a strong halo effect, which inhibits visualization of fine structures like cilia. Shorter shear prisms (DICTHR and DICT) improve resolution and enable thin optical sections of thick samples, but the contrast is compromised. It is, therefore, important to

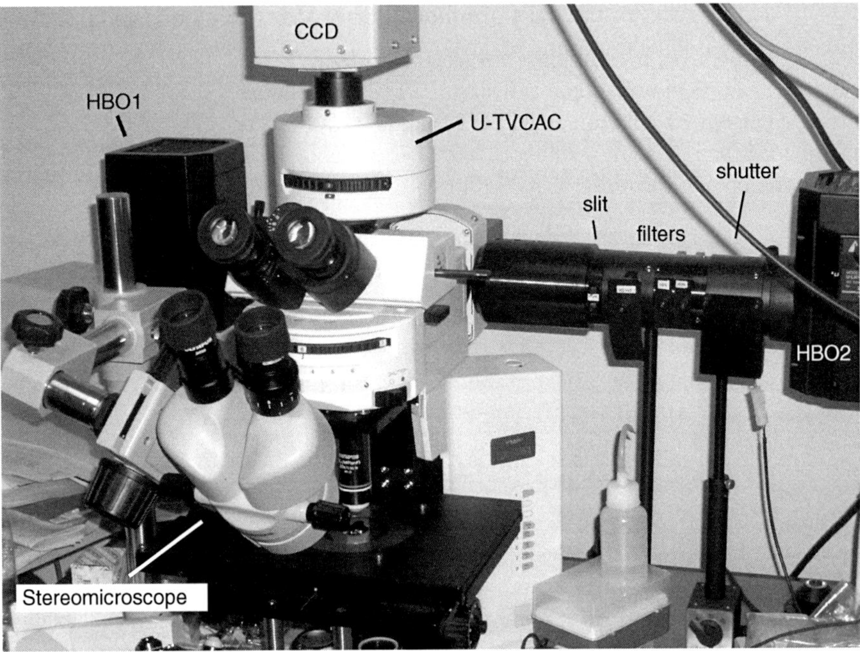

Fig. 2 Our microscope system. Fixed stage upright microscope is combined with a stereomicroscope for the dissection and manipulation of the embryo on the same microscope stage. The CCD camera is connected to the microscope with a C-mount video magnification unit U-TVCAC. A mercury arc lamp (HBO1) is used for the fluorescent excitation with the visible light. Another mercury arc lamp (HBO2) is used for the local photoactivation of caged-fluorescent proteins with an ultraviolet beam. An adjustable slit is placed at the field stop position, and filters and shutter are placed between the lamp house and the slit.

choose the best prism according to the thickness of the sample and the required contrast for detailed analysis. In most cases, a standard DIC prism (DICT) gives enough contrast for visual inspection and image analysis.

Choosing the right objective lens is also important for optimal results. Considering the size of the node, the field of a 100× objective is too narrow. We use a UPlanApo 60× water immersion lens (NA1.20, Olympus) to avoid spherical aberration. A water immersion 40× objective is another good choice. If you do not have a high NA water immersion lens, you can use oil immersion objectives (40× or 60×) or high NA dry objectives (e.g., 40× NA0.9), although the image quality is compromised with these choices.

2. Light Source and Camera

A stable and bright light source is necessary. We currently use a 250 W metal halide light source (PCS-UMX250, Fortissimo, Tokyo, Japan) with a glass light guide to enable high speed (1000 fps) recordings, but a standard halogen lamp

(12 V, 100 W) gives enough signal for video rate recordings (30 fps) or visual inspection. For video recording and image analysis, it is important to reduce the ripple of your power supply. Some standard microscope power supplies have such large ripples that the periodic changes of the light intensity remain in the final images. Before switching to the metal halide light source, we powered the halogen lamp with a highly stabilized DC power supply that we made from a used PC-AT power supply unit.

For video rate recordings, any CCD camera can be used, but fast frame rate is necessary. The nodal cilia rotate about 10 times/s. Thus, the frame rate should be significantly faster than 10 Hz. The cheapest choice is an analog CCD camera. The frame rate of the analog video is 30 Hz (in Japan and the United States, EIA format) or 25 Hz (in Europe, CCIR format), but each frame is composed of two fields that operate successively (interlaced scan). Thus, you can easily achieve a frame rate of 60 or 50 Hz by separating the fields (a standard function of most image analysis software). We used a 1/2″ interline CCD camera (XC75, Sony, Tokyo, Japan) for the initial studies, and we are now using a sensitive 1/2″ interline CCD camera Neptune 100 (Watec, Yamagata, Japan) for both DIC imaging of cilia and fluorescent imaging of flow. If your camera has the electronic shutter function, you can get sharp image suitable for image analysis by restricting the exposure time to 1–2 ms.

For faster recording than video rate, we have tried high-speed cameras from several different manufacturers, but the image quality with MOS imaging sensors was too poor for the imaging of cilia. Best results were obtained with a CCD-based high-speed monochrome camera, EktaPro HG Imager, model 2000 (Kodak, Rochester, New York, USA). Recently, we tried a high-speed EM-CCD camera Luca (Andor, Belfast, UK). Although the size of the field was restricted, good movies of motile cilia were obtained at >500 fps.

The projection magnification should be optimized according to the magnification of the objective and the size of the CCD detector. We use 1× magnification for the whole view of the node, and 2× or 4× for the observation of the nodal cilia.

The raw DIC video image should be preprocessed by introducing negative offset and contrast enhancement. We use a real-time video processor ARGUS-10 (Hamamatsu, Hamamatsu, Japan) for both preprocessing and the selective enhancement of the motile cilia (see Section E), and we record the image using a digital video recorder NV-DM1 (Panasonic, Osaka, Japan) for subsequent analyses. Alternatively, you can use a PC for recording and processing of the video signal.

C. Preparation of Medium

The failure of earlier studies to observe nodal cilia movement was most plausibly caused by unhealthy embryos due to poor culture conditions. In fact, our first experiment failed. We used Hank's balanced salt solution (HBSS) as the medium, and nodal cilia showed no movement. We noticed that organelles in most cells in the embryo showed only Brownian movement, which suggested to us that the embryo was dying or already dead. We, therefore changed the medium to one used for the *in vitro* culture of whole embryos.

1. Dissection and Preservation Medium, PB1

For the dissection of the embryos, we use modified Dulbecco's phosphate-buffered saline (D-PBS) PB1, which is reported to improve the preservation of embryos (Wood *et al.*, 1987). Prepare PB1 just before use by the addition of 1 mg/ml (final concentration) glucose, 0.33 mM sodium pyruvate, and 3 mg/ml bovine serum albumin to D-PBS. Sterilize by filtration, if necessary. We do not use antibiotics.

2. Observation Medium, DR50

We use DR50 medium to maintain embryos in the microscope chamber. This medium was developed for the *in vitro* culture of postimplantation embryos (Tam and Snow, 1980). DR50 medium is prepared before use by mixing equal volumes of Fresh D-MEM (Invitrogen, Frederick, MD, USA, without pyruvate, with glutamine and glucose 4000 mg/l) and immediately centrifuged rat serum (see below), followed by L-glutamine (final 2 mM) and sodium pyruvate (final 1 mM).

3. Preparation of Rat Serum

Since the hemolysis of serum affects embryogenesis, it is important to prepare immediately centrifuged serum with special care to avoid hemolysis.

1. Anesthetize the rat with ether, which can easily be removed from the serum.
2. Collect blood from abdominal aorta. About 15 ml of blood is collected from each rat.
3. Immediately centrifuge the blood at $2000 \times g$ for 5 min to separate serum.
4. Collect serum by decantation after squeezing the whitish fibrin clot.
5. Remove contaminating blood cells by re-centrifugation at $2000 \times g$ for 5 min.
6. Discard the serum, if any sign of hemolysis is found. The serum should appear amber in color without a red tint.
7. Heat inactivate the serum in a water bath at 56°C for 30 min. The lid should be loosened to allow the evaporation of remaining ether.
8. Aliquot and store the serum. The serum can be stored for years at −80°C.

D. Preparation of Mouse Embryo

The embryos are dissected as described in Hogan *et al.* (1986) from timed pregnant mice at 7.5–7.75 days post coitum. Nodal flow is a transient phenomenon during development and can only be observed in embryos from midneural plate stage to 3–4 somite stage (Okada *et al.*, 1999). Thus, the time window for the observation is very narrow and varies by the mouse strain or even by the mouse supplier.

1. Collection of the Decidua

The decidua are exposed by cutting the antimesometrial wall of the uterus with the tips of fine scissors. Collect deciduum with forceps into PB1 medium.

2. Collection of the Embryos

Under a stereomicroscope, the decidual tissue is then torn into two halves in PB1 medium by pulling apart the cleft at the mesometrial pole with fine forceps (Fig. 3A). One half of the deciduum retains the embryo. The remaining half of the deciduum can

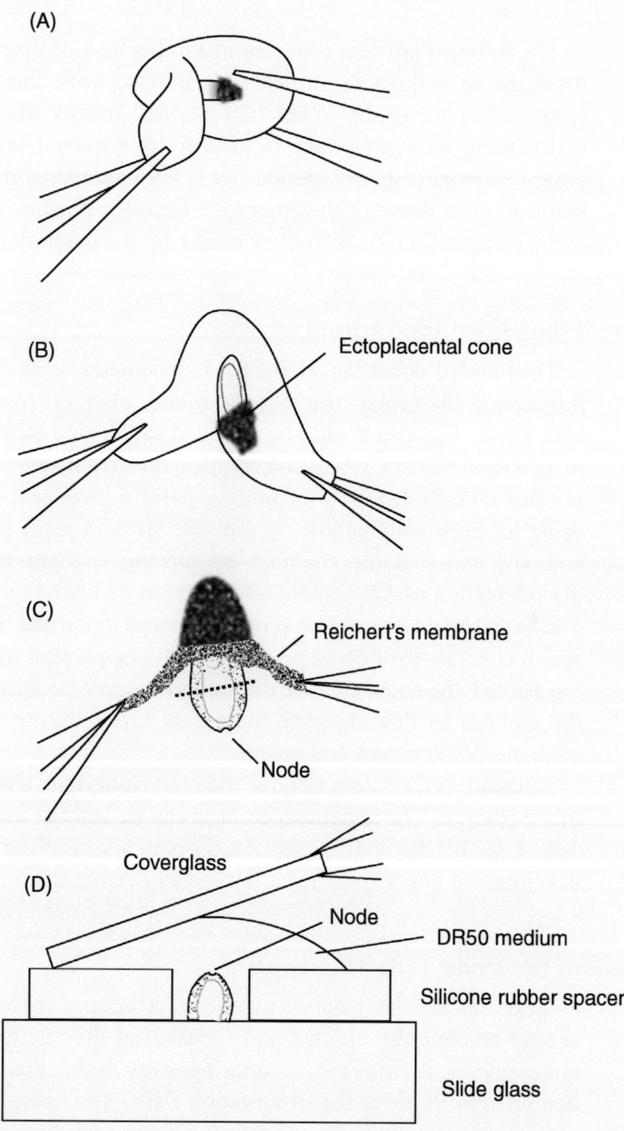

Fig. 3 Dissection and preparation of the mouse embryo. See main text for detail.

be removed similarly by peeling it off along the long axis of the embryo (Fig. 3B). Finally the embryo is detached from the remaining strip of deciduum with the tip of fine forceps. Alternatively, you can cut the whole deciduum into two halves along the long axis of the surface of the embryo with fine scissors and shell out the exposed embryo with the tips of fine forceps.

3. Dissection of the Embryo

The embryos are then collected in another dish of fresh PB1 medium and are further dissected to remove the Reichert membrane with fine forceps or tungsten needles (Fig. 3C). Finally, the lower half of the embryo (the ectoplacental cone and the surrounding extraembryonic ectoderm) is removed with tungsten needles or fine Noyes scissors (Fig. 3C, dotted line). The node should be located on the top of the hemispherical dome of the embryo. Check the position of the node and the developmental stage of the embryo by zooming up the magnification of the stereomicroscope.

4. Preparation of the Observation Chamber

The embryo dome thus dissected is mounted in a silicone chamber. A small hole is punched in the center of a silicone rubber plate (25 mm × 25 mm × 0.3 mm) using a cork borer. The size should be matched to the size of the embryo. This holed rubber plate is attached to a silane-coated glass slide (Matsunami, Osaka, Japan), and the hole is filled with fresh DR50 medium. Carefully transfer the embryo dome into this hole using an Eppendorf yellow tip (cut the tip with razor blade to match the size of the embryo) and seal the chamber by placing a cover glass on it (Fig. 3D). Check the orientation of the embryo and position of the node by using a stereomicroscope. The node should be on the top of the embryo dome and the ventral surface of the node, which is the layer of nodal pit cells, should be parallel to the cover glass. If the node is too slanted, the nodal cilia are difficult to observe because of halo effects. You can keep the embryo in this chamber for a few hours during which time the embryo will continue development and growth.

Alternatively, you can remove the extraembryonic tissues on the right and left sides of the embryo. The remaining embryo can be laid flat in the hole of the silicone rubber spacer, so that the node is positioned optimally for observation. The embryo continues development and growth for a few hours even in this highly invasive preparation.

E. Observation of the Nodal Cilia Movement

After setting the embryo on the microscope, the objective and condenser lens should be carefully aligned and focused on the ventral surface at the center of the node cavity. To minimize scattering stray light, the field stop should be opened just enough to cover the observation field. The aperture should be fully opened to maximize resolution. By adjusting the retardation bias, you should see the nodal cilia on the surface of the nodal pit cells. The nodal cilia should be vigorously rotating in the center of the node, while they are immotile or less motile near the edge of the node. If you can only see the immotile or nearly immotile cilia in the center of the node, check the healthiness of the nodal pit cells by observing the movement of organelles.

For recording and analysis, the DIC prism should be adjusted to introduce a retardation bias about $\lambda/9$ for best resolution after video enhancement. With this condition, the contrast of the raw image is too low for the naked eye. Adjust the light intensity so that the signals of the raw image cover the full dynamic range of the camera. Then set the gain and offset of the camera for best image contrast and brightness. By adjusting the focus, you will see the rotating nodal cilia on the video monitor. This raw image contains both rotating nodal cilia and stationary structures of nodal pit cells (Fig. 4A, Movie 1, http://www.elsevierdirect.com/companions/9780123749734). We routinely use the background subtraction function of an Argus-10 image processor (Hamamatsu) for selective enhancement of nodal cilia. The initial 30–60 frames are averaged and used as the "background" for the subsequent

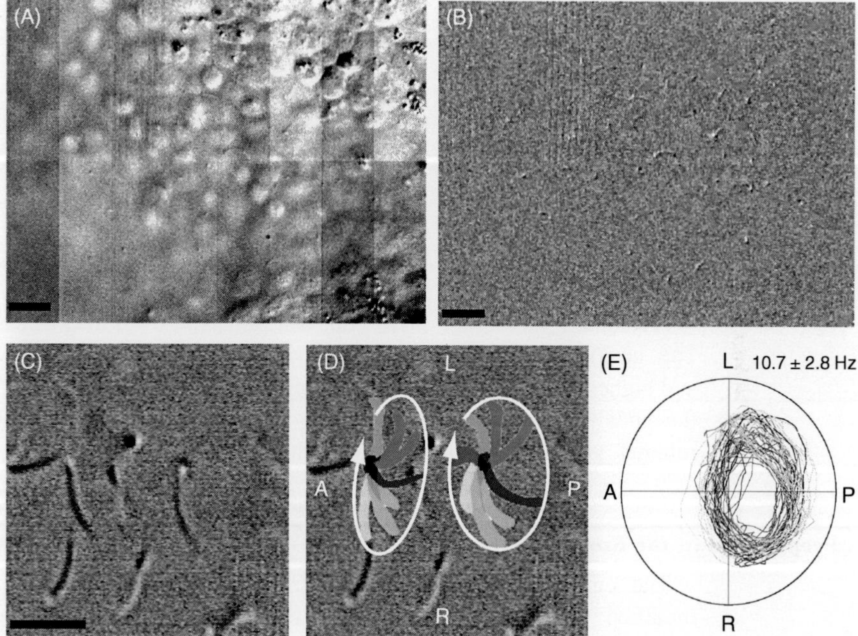

Fig. 4 Visualization of the nodal cilia and their analysis. (A) An example of the raw image taken with the high-speed CCD camera (500 fps). Immotile structures such as the surface of the nodal pit cells and the boundaries of the CCD panels obscure the nodal cilia. (B) Only motile cilia remain after subtracting the averaged image. Scale bar = 10 μm. (C) Nodal cilia are clearly visible after median filtering and adjustment of the contrast. Bar = 5 μm. (D) The positions of cilia are traced for each frame (10-ms interval in this image). (E) The trajectories of the cilia tips and their rotation rate are analyzed. The orientation is indicated in the panel as anterior (A), posterior (P), left (L), and right (R). Panels (D) and (E) are reproduced from Okada *et al.* (2005). (See Plate no. 16 in the Color Plate Section.)

real-time subtraction. This causes stationary structures along with fixed pattern mottle and shading to be subtracted as background causing the motile cilia to be selectively enhanced (Fig. 4B and C, Movies 2 and 3, http://www.elsevierdirect.com/companions/9780123749734). The position of each nodal cilium is recorded for each frame (Fig. 4D) and analyzed (Fig. 4E, Okada *et al.*, 2005).

III. Observation of Nodal Flow in Mouse Embryo

A. Overview

The nodal flow can be observed by authentic flow markers. If you carefully observe nodal ciliary movement with DIC optics, you will notice that small cellular fragments are flowing to the left side of the node (Movie 3, http://www.elsevierdirect.com/companions/9780123749734). These fragments, at least some of them, are the nodal vesicular parcels (NVPs), membrane-sheathed lipoprotein particles secreted from the nodal pit cells. They contain morphogens such as sonic hedgehog and retinoic acid, and play crucial roles in triggering the signaling cascade downstream to nodal flow (Tanaka *et al.*, 2005).

However, it is difficult to analyze the nodal flow from this authentic flow marker alone, due to the low number of particles and the low image contrast. The flow can be more clearly visualized by introducing flow markers. Fluorescent latex beads are most suitable for this purpose.

B. Microscope System

A standard fluorescent microscope can be used. Lower magnification objectives like 10× or 20× are suitable for the analysis of the global flow in the whole node cavity, and higher magnification objectives can be used for the analysis of the local flow near the nodal cilia. Since the nodal flow is much slower (~5 μm/s) than nodal cilia rotation, you can use any camera for the recording.

C. Preparation of the Embryo

The embryo is prepared as described in Section II.D. We use carboxylate-modified fluorescent latex beads (1.0 μm, yellow-green fluorescent, Invitrogen F-8823). The latex beads should be well dispersed by sonication and added to the DR50 medium just before mounting. It is not necessary to wash the beads before use, and we usually add the beads directly from the original stock. D-PBS or protein-free medium causes aggregation of the beads and should be avoided for the dilution or wash. Use pure water or DR50 for dilution. The concentration of the beads should be optimized according to the purpose of the experiment, but we recommend 1/100–1/1000 dilution of the original solution (2% solid) as the starting point. If the embryo is correctly prepared, you can easily see the unidirectional leftward nodal flow near the surface of the node (Fig. 5A–C, Movie 4, http://www.elsevierdirect.com/companions/9780123749734). Some of the beads might attach to the nodal cilia by nonspecific interactions (Fig. 5A, Movie 4, http://www.elsevierdirect.com/companions/9780123749734). You will also notice that the nodal flow is three dimensional (Fig. 5C–E, Movie 5, http://www.elsevierdirect.com/companions/

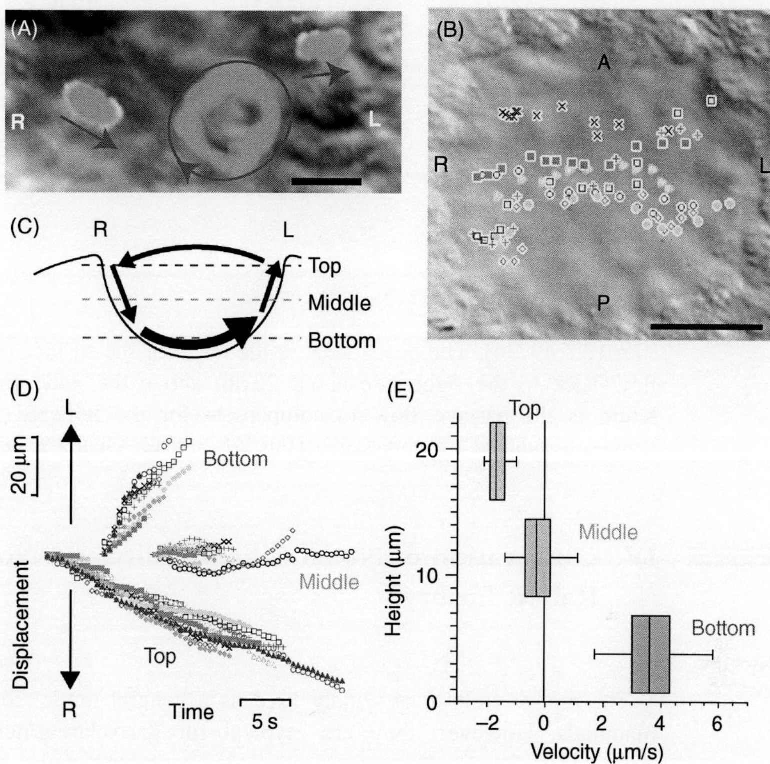

Fig. 5 Visualization of the nodal flow and its analysis. (A) Fluorescent latex beads flow to the left side near the bottom of the nodal pit. Some of the beads attach to the tip of the cilia and show rotatory movement. Fluorescent image (1-s exposure) is overlayed with DIC image. Bar = 5 µm. (B) The positions of beads that entered the node from the right edge traced for 4 s at 0.33-s intervals. Different symbols indicate different beads. Most beads go straight to the left edge of the node. Scale bar = 20 µm. (C) Three-dimensional profile of the nodal flow and its counter flow. Strong unidirectional leftward flow occurs at the bottom of the nodal pit (nodal flow), which is accompanied by the slower rightward return flow about 20 µm above. (D, E) Movement of beads near the surface of the nodal pit (bottom), about 10 µm above (middle) and about 20 µm above (top). Beads flow rapidly to the left in the bottom, show no directional movement in the middle, and flow slowly back to the right in the top. (F, G) Numerical simulation of the return flow. Fluid in the rectangular pit is propelled to the left by the constantly moving floor at the bottom. Then, the fluid near the bottom surface moves rapidly to the left and the fluid returns to the right side by the return flow in the region about 20 µm above. Panel (F) shows the flow line and (G) shows the velocity profile. Panels (A–E) are modified from Okada *et al.* (1999, 2005). (See Plate no. 17 in the Color Plate Section.)

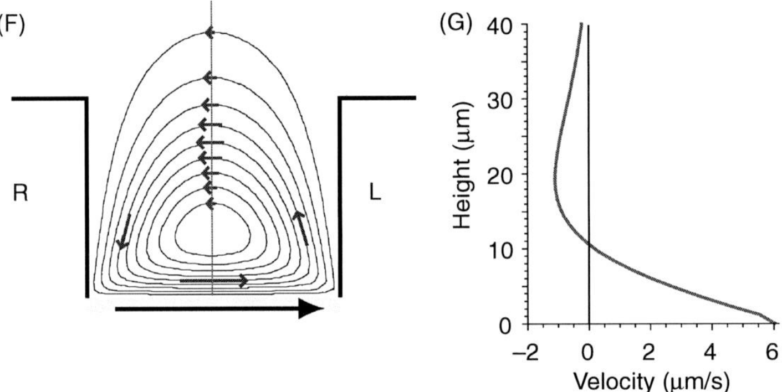

Fig. 5 (*Continued*)

9780123749734). The beads flow to the left near the surface of the node and slowly flow back to the right side about 20 μm above the node surface. This rightward return is the passive flow to compensate for the leftward nodal flow, which is actively produced by nodal cilia (Fig. 5F and G; Okada *et al.*, 2005).

IV. Observation of Nodal Cilia Motility and Nodal Flow in Rabbit Embryo

A. Overview

The mouse embryo is widely used as a general model for the development of mammals. However, they are atypical for gastrulating/neurulating mammalian embryos. Before embryonic turning, mouse and rat embryos form a cup-shaped "egg cylinder," while other mammalian embryos (including human) adopt a flat disc shape. Therefore, the rabbit, which develops via a flat blastodisc, is a better model to study archetypical mammalian development like that seen in humans. Furthermore, the flat embryonic disc of a rabbit embryo is more suitable for microscopic analysis and experimental manipulations. In the rabbit embryo, the ventral surface of the notochordal plate corresponds to the node of the mouse embryo (Fig. 6A–E), and the leftward nodal flow is produced in the groove of the notochordal plate by rotating cilia (Fig. 6F–H, Movies 6 and 7, http://www.elsevierdirect.com/companions/9780123749734; Okada *et al.*, 2005).

B. Microscope System

The same microscope system can be used as described in Sections II.B and III.B. We use an XLUMPlanFL 20× dipping objective (NA 0.95) with a WI-DICHRA prism for

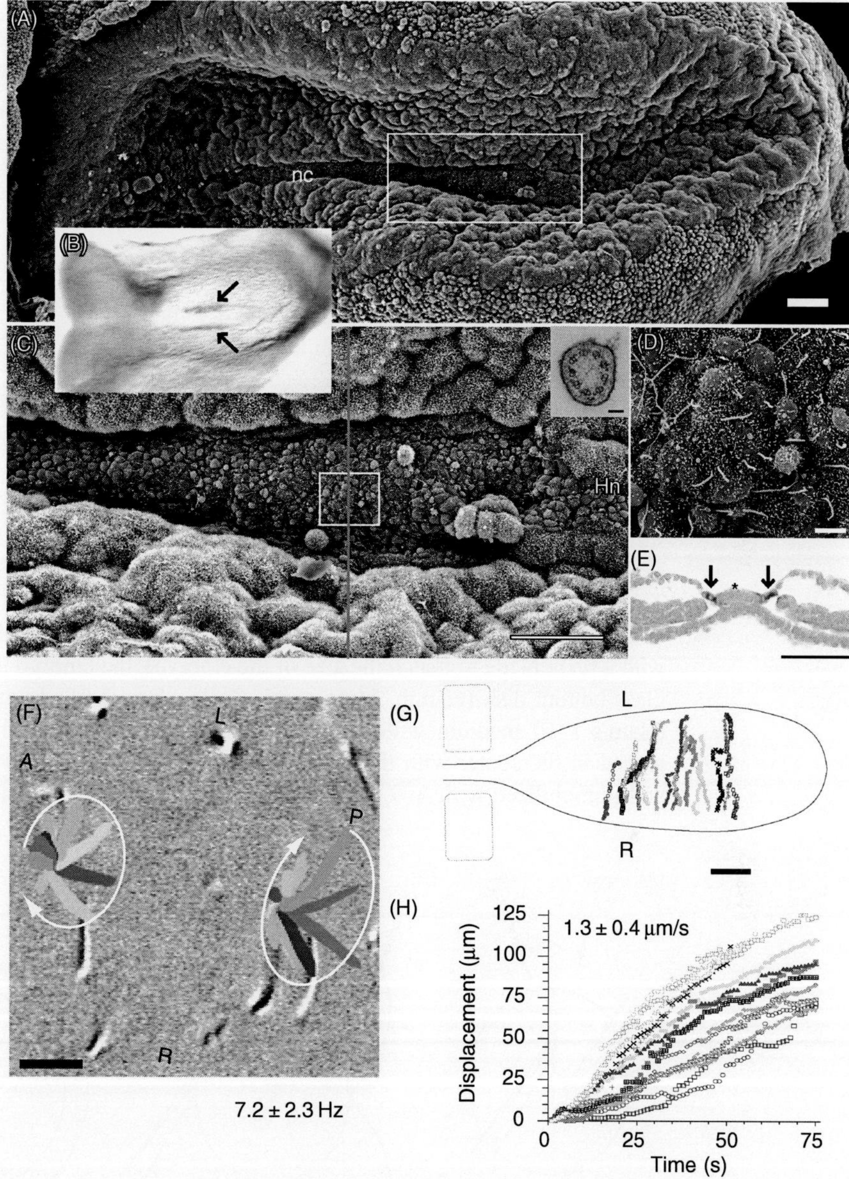

Fig. 6 The notochordal plate serves as the node in the rabbit embryo. (A–D) Ventral views of rabbit embryo with the anterior side (head) on the left and the posterior side (tail) on the right. (E) Transverse section of the notochordal plate, corresponding to the position of the line in (C). Panels (A), (C), and (D) show scanning electron microscope views. The white rectangle region is magnified in the next view. Notochord is indicated by "nc." Hn shows the position of Hensen's node. The inset of panel (C) shows a cross-section of the cilia. Panels (B) and (E) show the expression of *nodal*, a molecular marker for the node (arrows). (F) Posteriorly tilted rotation of the cilia. (G, H) Trajectories of the beads in the notochordal plate. Bars: (A) 100 μm; (C) 100 μm; (C) inset, 0.1 μm; (D) 5 μm; (E) 100 μm; (F) 5 μm; (G) 50 μm. Figure taken from Okada *et al.* (2005). (See Plate no. 18 in the Color Plate Section.)

analysis of the flow and video rate recording of rabbit nodal cilia rotation. For high-speed imaging, we use a UPlanApo 60× water immersion objective. The node (notochordal plate) of the rabbit embryo is much larger than the node of the mouse, and the nodal flow and the rotation of nodal cilia are proportionally slower (Fig. 6F–H). Therefore, you can use a lower magnification objective and slower frame rate camera.

C. Preparation of the Embryo

The embryos are dissected from timed pregnant rabbits at 8.0 days post coitum.

1. Remove the uterus intact by cutting across the cervix and the two utero-tubal junctions, and place in a dish containing HBSS (Invitrogen). Cut into individual swellings.
2. Fill a silicone rubber-coated dissection dish with fresh HBSS and pin down a single part of the uterus with the mesometrium side down (Fig. 7A).
3. Carefully open the bulging uterine tissue opposite to the implantation site, which will allow you to see the embryo disc. Detach the embryo disc together with the surrounding extraembryonic tissue by cutting the periphery of the disc with fine Noyes scissors (Fig. 7B).
4. The observation chamber described in Section II.D can be used for rabbit embryos, but we usually use a larger chamber. By using an Eppendorf blue tip (cut the tip with a razor blade to match the size of the embryo), the embryo is transferred into a glass bottom dish (60 mm, Matsunami, Osaka, Japan) filled with culture medium (Ham's F-10 medium with 10% rabbit serum, Invitrogen). The embryo disc is placed in the center with the ventral side up (down for the inverted microscope) and immobilized with two strips of cover glass (18 mm × 2 mm × 0.2 mm) put on

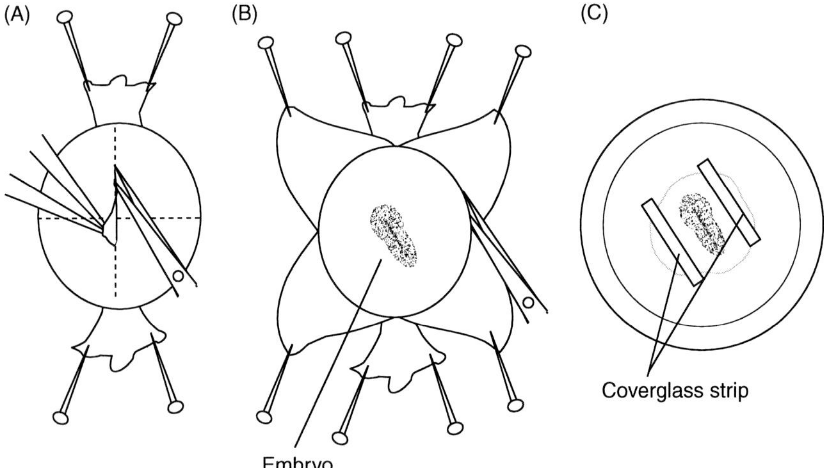

Fig. 7 Dissection and preparation of the rabbit embryo. See main text for detail.

the right and left extraembryonic region (Fig. 7C). With this chamber, you can manipulate the embryo during observation on the microscope stage.

5. You can add Fluorescent latex beads to the medium for the visualization of the flow as described in Section III.C, but we prefer to apply the bead solution directly to the node under the microscope.
6. The nodal flow in rabbit is also a transient phenomenon, and the developmental stage of the embryo is a critical factor for the successful imaging of nodal flow. The nodal flow is most prominent at the 2–4 somite stage. The nodal cilia of younger rabbit embryos (presomite stages) are shorter and immotile. They tend to grow from the center of the cells. They become longer and motile as development proceeds. The position of their root moves to the posterior. The vigorous rotation of posteriorly tilted nodal cilia and the unidirectional leftward nodal flow start around the emergence of the first somite pair, but the developmental stages of the nodal pit cells are not uniform along the AP axis of the notochordal plate.

V. Discussion

In this chapter, we describe the basic techniques for the observation of nodal flow and nodal cilia in living mouse and rabbit embryos. The same techniques can be applied to other vertebrates. Zebrafish and medaka fish have recently become popular model organisms for the development of vertebrates. Although they develop without implantation to the uterus (oviparity), their LR axis is dependent on the nodal flow system (Essner *et al.*, 2002; Kramer-Zucker *et al.*, 2005). Kupffer's vesicle (KV) corresponds to the node of mouse (Fig. 8A–D). The cilia in KV rotates at 40 Hz and produces leftward flow (Fig. 8E–G, Movies 8 and 9, http://www.elsevierdirect.com/companions/9780123749734; Okada *et al.*, 2005). Since the fish egg is transparent, the preparation is very simple. The dechorionated egg is mounted into a hole of a silicone rubber spacer slightly thinner than the egg (Fig. 8H). You can observe KV from any angle through the transparent embryo and egg (Fig. 8I, Movie 9, http://www.elsevierdirect.com/companions/9780123749734).

This technique is not limited to the simple observation of nodal flow and nodal cilia. Fluorescent microscopy techniques such as fluorescence decay after photoactivation (FDAP) can be used for the measurement of the dynamics of proteins in the node cavity (Movie 10, http://www.elsevierdirect.com/companions/9780123749734; Okada *et al.*, 2005). Application of fluorescent indicators enables functional imaging in the same preparations together with genetic manipulation of the embryo or pharmacological perturbations. For example, the application of the lipophilic dye DiI enabled the visualization of the dynamics of the surface lipid (Fig. 9A, Movie 11, http://www.elsevierdirect.com/companions/9780123749734). This study led to the identification of the NVP, which transports morphogens such as sonic hedgehog and retinoic acid by nodal flow (Fig. 9B; Tanaka *et al.*, 2005). Another example is the calcium indicator. Elevation of intracellular calcium was detected in the cells at the left edge of the node (Fig. 9C). It was first proposed that the nodal flow might stimulate the

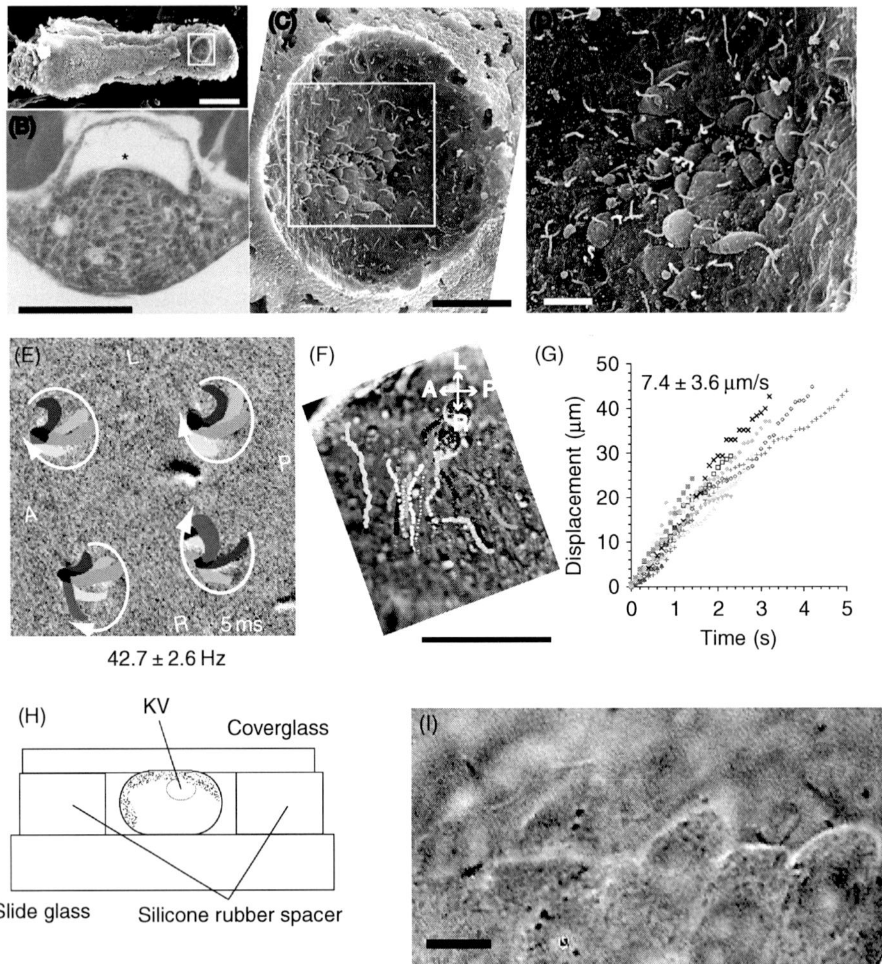

Fig. 8 Kupffer's vesicle (KV) serves as the node in the fish embryo. (A–D) KV of medaka fish embryo. Panels (A), (C), and (D) show ventral scanning electron micrograph views with the anterior side (head) on the left and the posterior side (tail) on the right. The white rectangle region is magnified in the next view. The apical surface of the monociliated cells is exposed by removing the membrane that covers KV. (B) Transverse section of KV. Asterisk shows the cavity that corresponds to the nodal pit of the mouse embryo. (E) Posteriorly tilted rotation of the cilia. (F, G) Trajectories of the authentic flow markers in KV. (H) Preparation of the fish embryo for the observation. (I) Side view of the nodal cilia in KV of the medaka fish. Bars: (A, B) 100 μm; (C) 100 μm; (D) 5 μm; (F) 50 μm; (I) 5 μm. Figure taken from Okada *et al.* (2005). (See Plate no. 19 in the Color Plate Section.)

cells directly via mechanical stimulation (McGrath *et al.*, 2003). However, pharmacological studies demonstrated that the FGF-signaling inhibitor SU5402 stops the calcium elevation without affecting the nodal flow itself and that exogenous SHH restored the calcium elevation (Tanaka *et al.*, 2005). The observation of NVPs showed

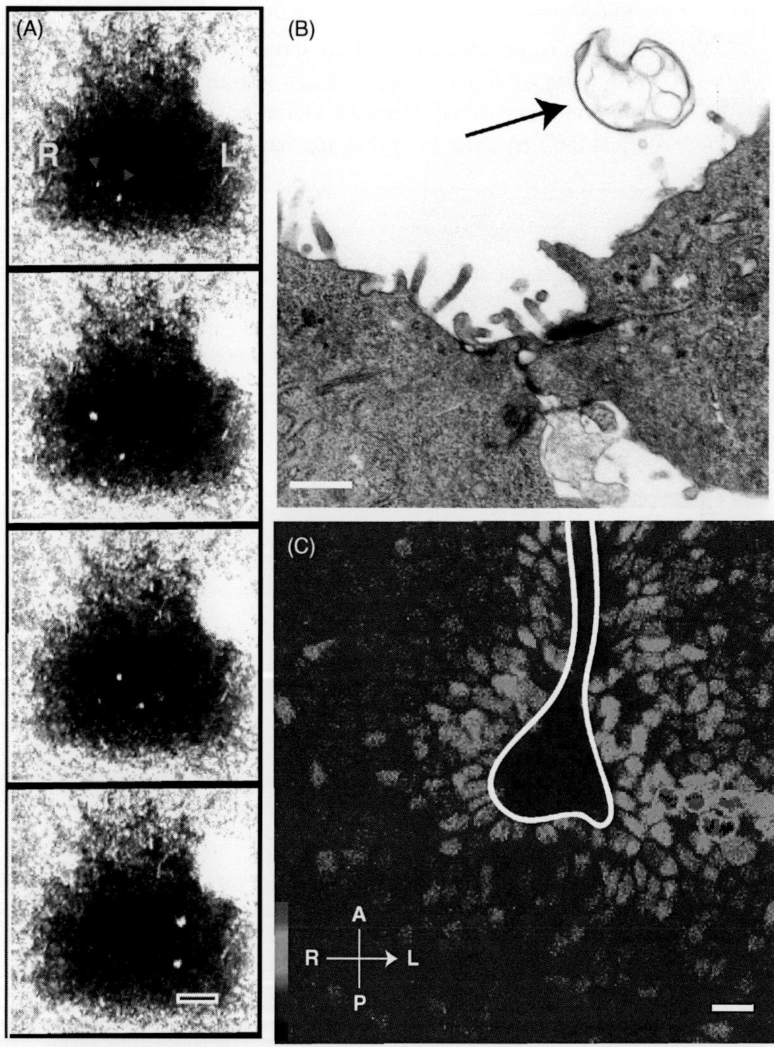

Fig. 9 Signaling by nodal vesicular parcel (NVP). (A) The flow of NVP is visualized by staining the membrane lipid with the lipophilic fluorescent dye DiI. Two-second interval image sequence is shown. Bar = 10 μm. (B) Transmission electron micrograph of NVP (arrow). Bar = 1 μm. (C) Leftward transport of morphogens such as sonic hedgehog triggers calcium elevation on the left side of the node. Bar = 20 μm. Figure taken from Tanaka *et al.* (2005). (See Plate no. 20 in the Color Plate Section.)

that the release of NVPs from the nodal pit cells is dependent on FGF signaling. These results suggest that the calcium response is not triggered directly by the mechanical stimulus of the flow, but by NVP-mediated morphogen transport to the left edge of the node (Hirokawa *et al.*, 2006; Tanaka *et al.*, 2005).

More importantly, the techniques described in this section can be applied to other processes of development. Many developmental processes have been described only from series of fixed embryos. High-resolution imaging of live embryos will clarify many yet unanswered enigmas. Defining the mechanism of LR determination would be just the first success of this approach.

Acknowledgments

We thank Masato Ohta, Kazuhiro Eto, and Patrick P. L. Tam for the techniques of manipulation of mouse embryos; Tomohiro Furukawa and Katsuyuki Abe for advise on the design of the microscope. We also thank Yosuke Tanaka, Sen Takeda, and other collaborators of our studies on nodal flow. This work was supported by the Ministry of Education, Culture, Sports, Science and Technology of Japan, Grant-in-Aid for Specially Promoted Research to N.H.

References

Afzelius, B.A. (1976). A human syndrome caused by immotile cilia. *Science* **193**, 317–319

Alarcon, V.B., and Marikawa, Y. (2003). Deviation of the blastocyst axis from the first cleavage plane does not affect the quality of mouse postimplantation development. *Biol. Reprod.* **69**, 1208–1212

Beddington, R.S.P., and Robertson, E.J. (1999). Axis development and early asymmetry of mammals. *Cell* **96**, 195–209.

Bellomo, D., Lander, A., Harragan, I., and Brown, N.A. (1996). Cell proliferation in mammalian gastrulation: The ventral node and notochord are relatively quiescent. *Dev. Dyn.* **205**, 471–485.

Buceta, J., Ibañes, M. Rasskin-Gutman, D., Okada, Y., Hirokawa, N., and Izpisúa-Belmonte, J.C. (2005). Nodal cilia dynamics and the specification of the left/right axis in early vertebrate embryo development. *Biophys. J.* **89**, 2199–2209.

Capdevila. J., Vogan, K.J., Tabin, C.J., and Izpisua-Belmonte, J.C. (2000). Mechanisms of left–right determination in vertebrates. *Cell* **101**, 9–21.

Cartwright, J.H., Piro, N., Piro, O., and Tuval, I. (2007). Embryonic nodal flow and the dynamics of nodal vesicular parcels. *J. R. Soc. Interface* **4**, 49–55.

Essner, J.J., Vogan, K.J., Wagner, M.K., Tabin, C.J., Yost, H.J., and Brueckner, M. (2002). Conserved function for embryonic nodal cilia. *Nature* **418**, 37–38.

Hamada, H. (2002). Establishment of vertebrate left–right asymmetry. *Nat. Rev. Genet.* **3**, 103–113.

Harvey, R.P. (1998). Links in the left/right axial pathway. *Cell* **94**, 273–276.

Hirokawa, N., Okada, Y., and Tanaka, Y. (2006). Nodal flow and the generation of left–right asymmetry. *Cell* **125**, 33–45.

Hogan B, Costantini, F, and Lacy, E. (1986). "Manipulating the Mouse Embryo." Cold Spring Harbor Laboratory, New York.

Kramer-Zucker, A.G., Olale, F., Haycraft, C.J., Yoder, B.K., Schier, A.F., and Drummond, I.A. (2005). Cilia-driven fluid flow in the zebrafish pronephros, brain and Kupffer's vesicle is required for normal organogenesis. *Development* **132**, 1907–1921.

Levin, M. (2005). Left–right asymmetry in embryonic development: A comprehensive review. *Mech. Dev.* **122**, 3–25.

Marszalek, J.R., Ruiz-Lozano, P., Roberts, E., Chien, K.R., and Goldstein, L.S.B. (1999). Situs inversus and embryonic ciliary morphogenesis defects in mouse mutants lacking the KIF3A subunit of kinesin-II. *Proc. Nat. Acad. Sci. USA* **96**, 5043–5048.

McGrath, J., Somlo, S., Makova, S., Tian, X., and Brueckner, M. (2003). Two populations of node monocilia initiate left–right asymmetry in the mouse. *Cell* **114**, 61–73.

Nonaka, S., Shiratori, H., Saijoh, Y., and Hamada, H. (2002). Determination of left–right patterning of the mouse embryo by artificial nodal flow. *Nature* **418**, 96–99.

Nonaka, S., Tanaka, Y., Okada, Y., Takeda, S., Harada, A., *et al.* (1998). Randomization of left–right asymmetry due to loss of nodal cilia generating leftward flow of extraembryonic fluid in mice lacking KIF3B motor protein. *Cell* **95**, 829–837.

Nonaka, S., Yoshiba, S., Watanabe, D., Ikeuchi, S., Goto, T., Marshall, W.F., and Hamada, H. (2005). De novo formation of left–right asymmetry by posterior tilt of nodal cilia. *PLoS Biol.* **3**, e268.

Okada, Y., Nonaka, S., Tanaka, Y., Saijoh, Y., Hamada, H., and Hirokawa, N. (1999). Abnormal nodal flow precedes situs inversus in iv and inv mutant mice. *Mol. Cell* **4**, 459–468.

Okada, Y., Takeda, S., Tanaka, Y., Izpisúa-Belmonte, J.C., and Hirokawa, N. (2005). Mechanism of nodal flow: A conserved symmetry breaking event in left–right axis determination. *Cell* **121**, 633–644.

Rosenbaum, J.L., and Witman, G.B. (2002). Intraflagellar transport. *Nat. Rev. Mol. Cell. Biol.* **3**, 813–825.

Scholey, J.M. (2003). Intraflagellar transport. *Annu. Rev. Cell. Dev. Biol.* **19**, 423–443.

Smith, D.J., Blake, J.R., and Gaffney, E.A. (2008). Fluid mechanics of nodal flow due to embryonic primary cilia. *J. R. Soc. Interface.* **5**, 567–573.

Smith, D.J., Gaffney, E.A., and Blake, J.R. (2007). Discrete cilia modeling with singularity distributions: Application to the embryonic node and the airway surface liquid. *Bull. Math. Biol.* **69**, 1477–1510.

Supp, D.M., Witte, D.P., Potter, S.S., and Brueckner, M. (1997). Mutation of an axonemal dynein affects left–right asymmetry in inversus viscerum mice. *Nature* **389**, 963–966.

Takeda, S., Yonekawa, Y., Tanaka, Y., Okada, Y., Nonaka, S., and Hirokawa, N. (1999). Left–right asymmetry and kinesin superfamily protein KIF3A: New insights in determination of laterality and mesoderm induction by $kif3A^{-/-}$ mice analysis. *J. Cell Biol.* **145**, 825–836.

Tam, P.P., and Snow, M.H. (1980). The *in vitro* culture of primitive-streak-stage mouse embryos, *J. Embryol. Exp. Morphol.* **59**, 131–143.

Tanaka, Y., Okada, Y., and Hirokawa, N. (2005). FGF-induced vesicular release of Sonic hedgehog and retinoic acid in leftward nodal flow is critical for left–right determination. *Nature* **435**, 172–177.

Wood, M.J., Whittingham, D.G., and Rall, W.F. (1987). The low temperature preservation of mouse oocytes and embryos. *In* "Mammalian Development. A Practical Approach" (M. Monk, ed.), pp. 255–280. IRL Press, Oxford, UK.

Yost, H.J. (1999). Diverse initiation in a conserved left–right pathway? *Curr. Opin. Genet. Dev.* **9**, 422–426.

CHAPTER 15

Modification of Mouse Nodal Flow by Applying Artificial Flow

Shigenori Nonaka

Laboratory for Spatiotemporal Regulations, National Institute for Basic Biology, Nishigonaka 38, Myodaiji, Okazaki 444-8585 Aichi, Japan

Abstract

In mammalian development, the earliest left–right (L–R) asymmetry is nodal flow, which is a cilia-driven leftward fluid flow on the ventral surface of the node. The importance of nodal flow for L–R determination was demonstrated by experiments to modify nodal flow by imposing artificial fluid flow. In this system, cultured mouse

978-0-12-374973-4
DOI: 10.1016/S0091-679X(08)91015-3

embryos developed reversed L–R asymmetry when their node cavity had rightward flow, and normal L–R asymmetry when their node had leftward flow. This chapter describes details of the culture system that can modify nodal flow.

I. Introduction

Throughout the last quarter of the last century, people wondered what relationship ciliary movement had to the left–right (L–R) asymmetry of the mammalian body plan. This question was evoked by the discovery of ciliary defects in patients with situs inversus due to Kartagener's syndrome (Afzelius, 1976), and by similar reports in other mammalian species (reviewed by Afzelius, 1995).

The importance of nodal cilia to breaking symmetry of the early embryo came from studies of mouse development. It was shown that the monocilia on the ventral surface of the node in gastrulating mouse embryos beat in a rotational manner and produced leftward fluid flow (nodal flow) (Nonaka *et al.*, 1998). Knockout analysis of the Kif3A and Kif3B genes showed that these kinesin-II motors are required to assemble node cilia. The absence of node cilia and the resultant disappearance of nodal flow in these mutants coincided with randomization of L–R asymmetry (Nonaka *et al.*, 1998; Takeda, 1999). Studies of the *inversus viscerum* (*iv*) mutant mouse also demonstrated the coincidence between nodal flow and L–R determination. This mouse, which exhibits L–R randomization, has a defect in a gene coding for an axonemal dynein motor (Supp, 1997). Node cilia were found to be immotile in *iv/iv* embryos (Okada, 1999).

These findings strongly suggested a critical role for nodal flow in determining L–R asymmetry; however, the possibility remained that flow was not important at all. For example, if the motors had two separate roles in the node cells, one in assembling cilia and another in establishing cellular L–R polarity in the cytoplasm that is actually important for future development, nodal flow would merely be a flag of the established polarity (Wagner and Yost, 2000). Genetic manipulations would affect both ciliary and cytoplasmic functions in the same cells making it hard to use genetics to test the hypothesis.

Since genetic studies did not fully answer the question, a physical approach was employed as an alternative. If leftward nodal flow really serves to establish L–R asymmetry, perturbations to nodal flow should alter L–R development. This idea was tested using a flow culture system in which embryos received pump-driven artificial flow on their surface. Application of artificial nodal flow was able to alter the L–R asymmetry of the developing embryos. Application of rightward flow to wild-type embryos with a fast enough flow rate to reverse the cilia-driven leftward flow, reversed L–R development of the embryos, whereas weaker rightward flow did not change L–R development. Applications of leftward and rightward flows to *iv/iv* embryos resulted in normal and reversed L–R development, respectively (Nonaka *et al.*, 2002).

This chapter describes the methods needed to culture mouse embryos and manipulate nodal flow.

II. Solutions

Dissection medium
10% fetal bovine serum (FBS)
90% Dulbecco's modified Eagle's medium (DMEM) buffered with 25 mM HEPES–NaOH (pH 7.2)

Culture medium
50% rat serum (See Chapter 14 by Okada and Hirokawa, this volume, for preparation.)
50% DMEM buffered with 44 mM $NaHCO_3$ (pH 7.2)
Penicillin (50 U/ml)
Streptomycin (50 mg/ml)

III. Experimental Setup

A. Overview

As shown in the schematic diagram (Fig. 1A), the system is composed of a peristaltic pump, two depulsators, and a flow chamber. The peristaltic pump generates pulsatile flow that will be evened out by the depulsators and provide constant flow of

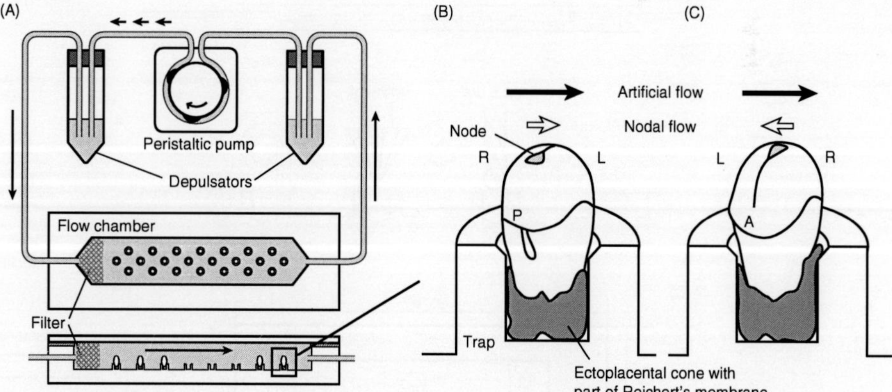

Fig. 1 (A) Diagram of flow culture system. Depulsators are inserted between the pump and flow chamber to flatten pulsative flow. (B, C) Embryos set in the traps. Part of ectoplacental cone and Reichert's membrane stuffed in the trap will stabilize the embryo in the trap and avoid it being dislodged by the artificial flow. Orientation of the embryos in the trap make the pump-driven artificial flow leftward (B) or rightward (C) with respect to the embryos.

culture medium to the flow chamber. A filter is inserted at the inlet of the chamber in order to avoid turbulence. The flow chamber has traps to hold embryos in a fixed orientation, that is, the embryos will receive unidirectional flow on their surface. By orienting the embryos in the traps, the system can apply leftward (Fig. 1B) or rightward (Fig. 1C) artificial flow to the embryos.

B. Flow Chamber

The flow chamber is made of a chamber body, a gasket, a lid, and a filter. Figure 2A–C provides dimensions. The chamber body is made of polymethyl methacrylate (PMMA) and 18-G stainless-steel pipes (Fig. 2A). These chambers can be made by a machine shop or can be built by yourself under a stereomicroscope.

Figure 2B shows dimensions of the trap and how to make it. The best size of the hole is 0.5-mm diameter for holding 7.5-day embryos. Scraping the edge of the hole slightly with a milling bit (PROXXON GmbH, Im Spanischen, Niersbach, Germany No. 28710) will help inserting embryos. Although the original setup had bulging of the traps from the chamber floor (left), this was found unnecessary.

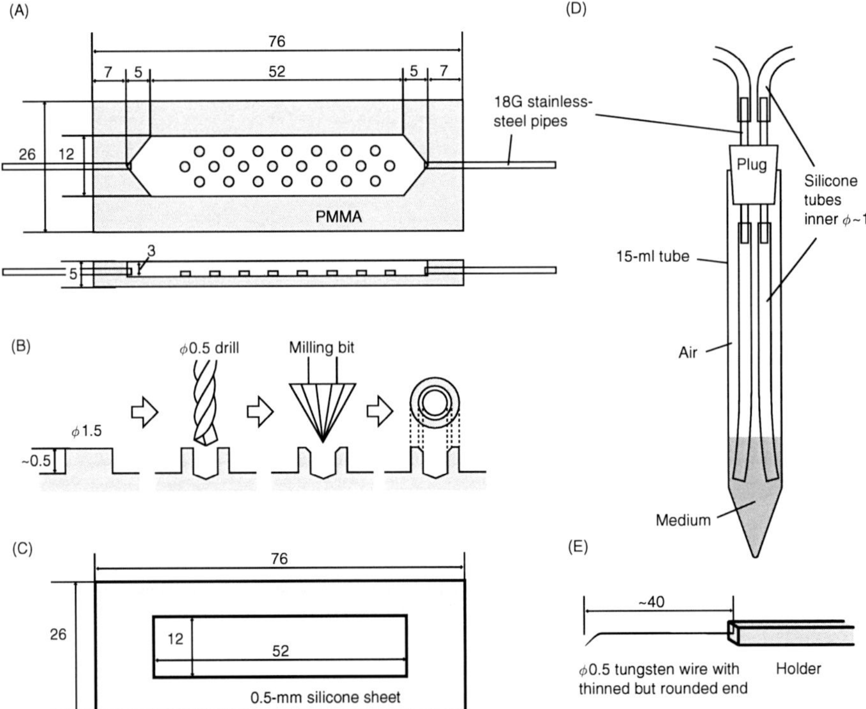

Fig. 2 (A)–(C) Dimensions of a flow chamber. The chamber body (A), the traps (B), and the gasket between the chamber body and the lid (C) are shown. (D, E) Design of a depulsator (D) and a Takopin (E), a tool for manipulating embryos in the flow chamber.

The gasket is made from 0.5-mm-thick silicone rubber sheet (Fig. 2C). The lid is a standard slide glass (76 mm × 26 mm), and the filter is wadding taken from a serological pipette.

C. Depulsator

Design of the depulsator is illustrated in Fig. 2D. Note that the lower end of the silicone tubes in the 15-ml tube is in the culture medium.

D. Takopin (Pusher Needle)

Aligning embryos in the trap holes is the most important and skill-requiring procedure in flow culture experiments. A specialized tungsten needle (illustrated in Fig. 2E) is useful for this manipulation: the tip is thinned to hook embryos, but rounded to avoid scratching them, and the last few millimeters of the tip is bent slightly (~30°) in order not to interfere with the objective of the stereomicroscope when the needle tip is inserted into the trap hole. The needle should be attached to a wooden chopstick or something similar to create a handle. I call this tool "Takopin" after a tool used in cooking the Japanese food Takoyaki (also known as Samurai ball).

IV. Methods

A. Preassembly of Culture System

1. Connect the silicone tubes to the peristaltic pump and depulsators. Do not connect the flow chamber at this point. Fill the system with culture medium.
2. Fill the chamber body with culture medium and carefully exclude bubbles. To remove bubble in the traps, blow them out with a 200-μl pipette or scrape out using a Takopin. Make sure that the stainless-steel pipes are also filled with medium.
3. Set the filter in the chamber and remove bubbles by pipetting.
4. Place the gasket on the chamber body and push to adhere.
5. Connect the flow chamber to the silicone tube.

B. Preparation of Embryos

Collect embryos from ~7.5-day pregnant mice (precise time will vary by strains and breeding conditions), as described in Chapter 14 by Okada and Hirokawa, this volume. I use 10% FBS DMEM–HEPES as the dissection medium instead of PB1 described in their protocol. This is not expected to make much difference. Leave the ectoplacental cone and some of Reichert's membrane attached to the embryo as these work as an adhesive to stabilize the embryos in the traps.

After collection, choose appropriate stages for the experiment. Only presomitic embryos can be used for the experiment to change L–R development, and embryos in somitegenesis are not sensitive to artificial flow (Nonaka *et al.*, 2002). I use late bud, early headfold, and late headfold stages. Staging is based on Downs' Criteria (Downs, 1993).

C. Trapping Embryos in the Chamber

Transfer all useful embryos into the chamber using a pipette with an end-cut tip. Use a Takopin to pick each embryo up by hooking the ectoplacental cone, tuck it into a trap, and rotate to the desired orientation (Fig. 3).

Add culture medium to the chamber so that its surface comes above the chamber's top. Then place the lid onto the gasket being careful not to bring bubbles into the chamber. Wipe up spilled medium and clamp the chamber with clips as shown in Fig. 4.

D. Running the System

Turn on the pump and adjust the flow rate as desired. Note that flow rate means the average flow rate in the chamber, which is not the same as the velocity of the flow on the surface of the embryos. The flow rate should be determined by measuring the volume of flow in a certain time and dividing it by cross-sectional area of the flow chamber (42 mm^2, including thickness of the gasket). Empirically, I chose 110 and 5.7 μm/s as "fast flow" and "slow flow" conditions. The former effectively reverses the situs of wild-type embryos when the artificial flow is imposed rightward, while the latter does not change the situs (Nonaka *et al.*, 2002). Place the whole system into an incubator at 37°C with 5% CO_2 and culture for 14 h.

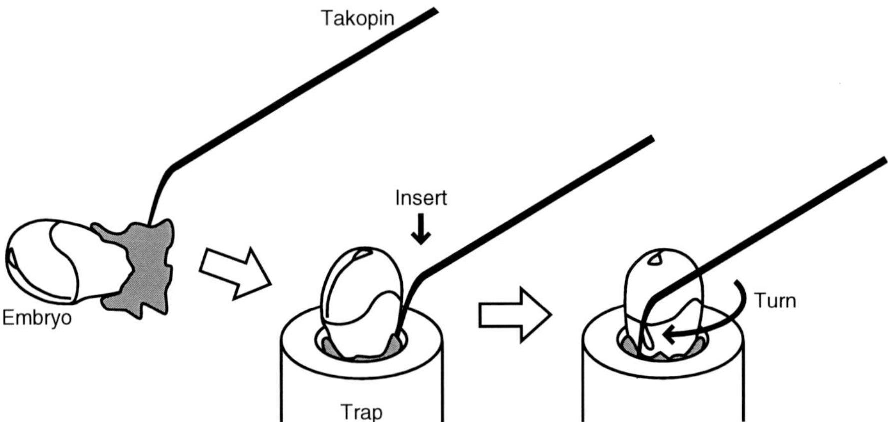

Fig. 3 Illustration of how to set the embryo in the chamber and orient them using a Takopin. Untrapping the embryo is the opposite procedure.

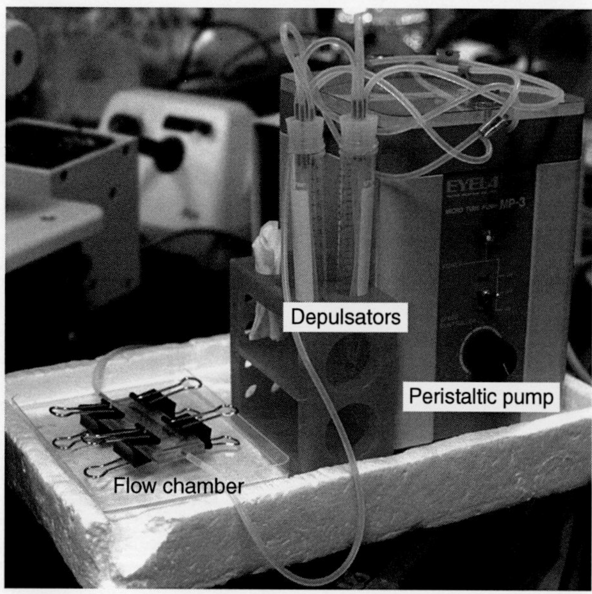

Fig. 4 Picture of the flow culture setup. It is convenient to place the setup on a tray for moving in and out of the incubator.

E. Untrapping

Turn off the pump, open the lid, and collect the grown embryos. However this process is not easy. The chamber parts are strongly adhered and simply opening the lid will harm the embryos by lowering the surface level of the culture medium. Instead, clamp off the silicone tube at the outlet of the flow chamber and run the pump for a few seconds to apply positive pressure within the chamber. Slip the tip of a pair of forceps between the lid and the gasket (or between the gasket and the chamber body) and slide the forceps carefully to extend the dislodged area (Fig. 5A). Once enough of the lid has been dislodged with the forceps, carefully lift the lid away with your fingers (Fig. 5B). Remove the embryos from the traps using the Takopin, and transfer them with a pipette to medium in a 35-mm dish.

Check the collected embryos by microscopy. Remove collapsed and underdeveloped ones for further experiments. Embryos below the four-somite stage are discarded.

F. Rotation Culture

While L–R symmetry has been broken at this stage, the embryos are morphologically symmetric: Heart looping starts at the eight-somite stage and axial turning occurs much later, around 9.5 days. To see these events, the embryos need to be cultured for an additional 32 h in a conventional rotation culture. This is longer than would be required *in vivo* because growth of embryos in culture is slightly delayed.

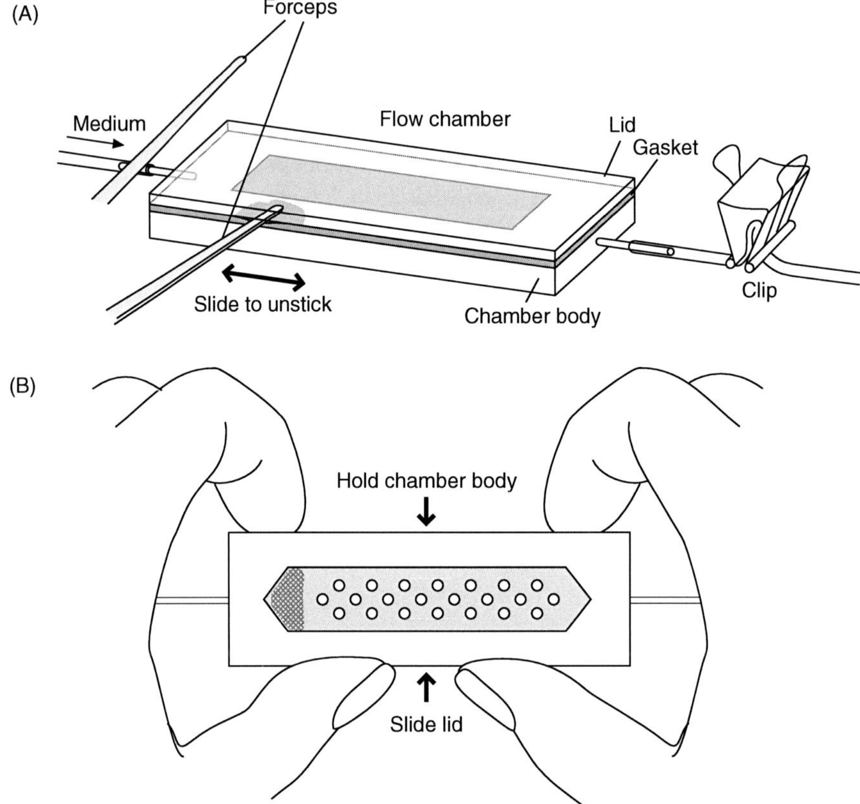

Fig. 5 Disassembly of the flow chamber after flow culture. (A) Apply a positive pressure to the chamber by sending medium into the chamber while the outlet is closed with a clip, unstick the lid from the gasket by inserting the tip of a forceps (B), and slide the lid horizontally against the chamber body to open the chamber.

The protocol for mouse rotation culture is based on Nagy (2002), but is slightly simplified. The embryos and culture medium are transferred to a 50-ml tube with loosened lid, placed on a rotator in a CO_2 incubator (Fig. 6A) and cultured for 32 h. The volume of culture medium should be more than 0.5 ml per embryo and at least 2 ml in total.

G. Typing L–R Asymmetry

After the rotation culture, the embryo is covered with a ballooned yolk sac (Fig. 6B, left). The direction of axial rotation is easily recognizable from the position of the tail, which is right-sided in normal development (Kaufman, 1992). The sidedness of the tail should be determined before breaking the yolk sac because sometimes the tail goes to the opposite side after removal of the physical constraint.

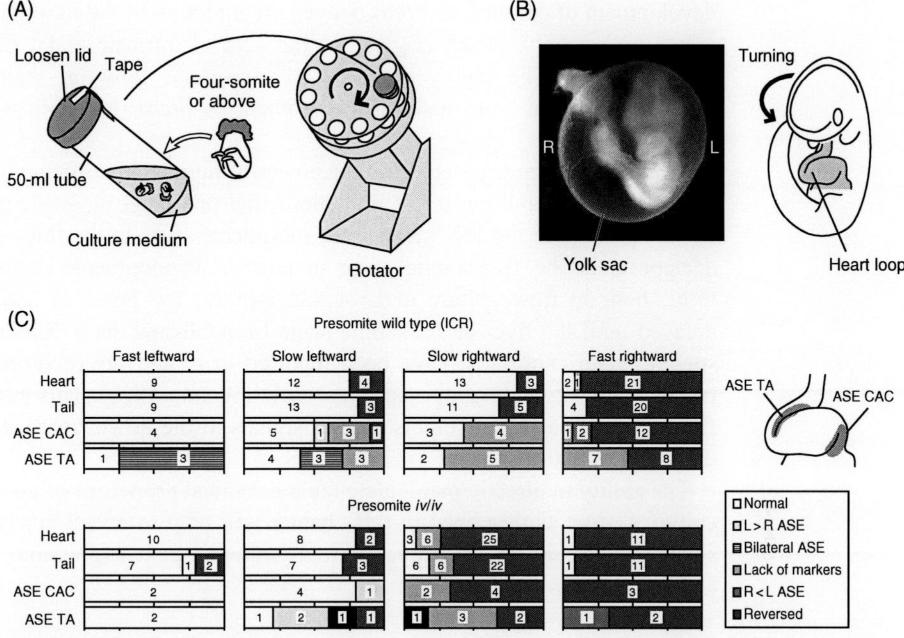

Fig. 6 (A) Rotation culture. Flow-cultured embryos are further cultured to the 9.5-day stage in a 50-ml tube with medium rotated in a CO_2 incubator. (B) Left: front view of a 9.5-day embryo with intact yolk sac and right-sided tail after normal axial turning. Right: illustration showing heart looping and axial turning of normal development. (C) The results of flow culture experiments. Numbers in the graphs indicate the sample numbers. Fast rightward flow efficiently reversed L–R development of both wild-type and *iv/iv* embryos, whereas slow rightward flow only reversed *iv/iv* embryos.

After scoring the position of the tail, tear open the yolk sac so that the heart can be clearly seen. Normal heart looping is called a dextral loop (D-loop; Fig. 6B, right) because it develops from a rightward shift of a straight heart tube at the midline.

To see both morphological asymmetry and asymmetric gene expression, a transgenic line carrying *Pitx2* ASE (A left side–specific enhancer)-*lacZ* (Shiratori, 2001) is useful. This gene is expressed on the left side of the common atrium chamber (CAC) and truncus arteriosus (TA) in normal 9.5-day embryos (Fig. 6B, right). Left-specific expression of nodal or lefty are no longer detectable at this stage.

V. Discussion

The results of L–R typing are shown in Fig. 6C. By a combination of experiments with wild-type and *iv/iv* mutant embryos, artificial flow directing leftward and rightward, and fast (110 μm/s) or slow (5.7 μm/s) flow conditions, this system demonstrated the critical role of nodal flow for L–R determination (Nonaka *et al.*, 2002). L–R development of cultured embryos obeyed the direction of the fluid flow on the surface of the node cavity, which is given as the sum of intrinsic nodal flow and imposed artificial flow. The embryos receive the artificial flow on their entire surface, which is different from natural development *in utero*, but it does not affect L–R development.

While mouse embryo culture is a powerful approach to manipulate the developing mammalian embryo, it has limitations that one must be aware of. Expression of *nodal* and *lefty* at the left lateral plate mesoderm starts at the three-somite stage and disappears at the five-somite stage in normal development. Under culture conditions, both in flow culture and rotation culture, the onset of *nodal* expression is delayed until the five- or six-somite stage (unpublished data). Similar delays of left-specific gene expression have been reported in *iv* and *inv* (inversion of embryonic turning) mutants with abnormal nodal flow (Okada, 1999). This suggests that nodal flow in these conditions inputs weaker signals to the downstream signaling pathway (Nakamura, 2006).

The ability to directly manipulate the mechanical properties of nodal flow in a flow culture system as described in this chapter will help to reveal unanswered questions such as how fluid flow is converted to asymmetric gene expression.

Acknowledgments

I thank my collaborators Hidetaka Shiratori, Yukio Saijoh, Hiroshi Hamada, and other members of the Hamada laboratory for technical help and valuable comments. I also thank Itsushi Minoura for helpful discussion of hydrodynamics. This work was supported by CREST (Core Research for Evolutional Science and Technology) of the Japan Science and Technology Corporation and by a fellowship from the Japan Society for the Promotion of Science for Japanese Junior Scientists to S. N.

References

Afzelius, B.A. (1976). A human syndrome caused by immotile cilia. *Science* **193**, 317–319.

Afzelius, B.A. (1995). Situs inversus and ciliary abnormalities. What is the connection? *Int. J. Dev. Biol.* **39**, 839–844.

Downs, K.M. and Davies, T. (1993). Staging of gastrulating mouse embryos by morphological landmarks in the dissecting microscope. *Development* **118**, 1255–1266.

Kaufman, M.H. (1992). "The Atlas of the Mouse Development." Academic Press, London.

Nagy, A., Gertsenstein, M., Vintersten, K., and Behringer, R. (2002). "Manipulating the Mouse Embryo: A Laboratory Manual." Cold Spring Harbor Laboratory Press, New York.

Nakamura, T., Mine, N., Nakaguchi, E., Mochizuki, A., Yamamoto, M., Yashiro, K., Meno, C., and Hamada, H. (2006). Generation of robust left-right asymmetry in the mouse embryo requires a self-enhancement and lateral-inhibition system. *Dev. Cell* **11**, 495–504.

Nonaka, S., Shiratori, H., Saijoh, Y., and Hamada, H. (2002). Determination of left-right patterning of the mouse embryo by artificial nodal flow. *Nature* **418**, 96–99.

Nonaka, S., Tanaka, Y., Okada, Y., Takeda, S., Harada, A., Kanai, Y., Kido, M., and Hirokawa, N. (1998). Randomization of left-right asymmetry due to loss of nodal cilia generating leftward flow of extraembryonic fluid in mice lacking KIF3B motor protein. *Cell* **95**, 829–837.

Okada, Y., Nonaka, S., Tanaka, Y., Saijoh, Y., Hamada, H., and Hirokawa, N. (1999). Abnormal nodal flow precedes situs inversus in iv and inv mice. *Mol. Cell* **4**, 459–468.

Shiratori, H., Sakuma, R., Watanabe, M., Hashiguchi, H., Mochida, K., Sakai, Y., Nishino, J., Saijoh, Y., Whitman, M., and Hamada, H. (2001). Two-step regulation of left-right asymmetric expression of Pitx2: Initiation by nodal signaling and maintenance by Nkx2. *Mol. Cell* **7**, 137–149.

Supp, D.M., Witte, D.P., Potter, S.S., and Brueckner, M. (1997). Mutation of an axonemal dynein affects left-right asymmetry in inversus viscerum mice. *Nature* **389**, 963–966.

Takeda, S., Yonekawa, Y., Tanaka, Y., Okada, Y., Nonaka, S., and Hirokawa, N. (1999). Left-right asymmetry and kinesin superfamily protein KIF3A: New insights in determination of laterality and mesoderm induction by kif3A-/- mice analysis. *J. Cell Biol.* **145**, 825–836.

Wagner, M.K., and Yost, H.J. (2000). Left-right development: The roles of nodal cilia. *Curr. Biol.* **10**, R149–151.

CHAPTER 16

Measuring Cilium-Induced Ca^{2+} Increases in Cultured Renal Epithelia

Helle A. Praetorius
Department of Physiology and Biophysics, Aarhus University, 8000 Aarhus C, Denmark

Abstract

The primary cilium is a sensory organelle in many mammalian cells. Functional analysis has established the cilium to be instrumental in the detection of mechanical and environmental changes. The cilium detects these stimuli and activates various signal-transduction pathways including alteration of intracellular Ca^{2+} concentration ($[Ca^{2+}]_i$). This chapter describes methods of measuring primary cilium-dependent $[Ca^{2+}]_i$ signaling.

978-0-12-374973-4
DOI: 10.1016/S0091-679X(08)91016-5

I. Introduction

In recent years there has been steady progress in the identification of interesting proteins that reside in primary cilia. Some of these proteins are involved in the growth and maintenance of the structure whereas others are involved in signal-transduction processes. With the discovery that primary cilia are intimately involved in development of vertebrates and in tissue homeostasis, the interest in evaluating potential cell biological and physiological functions of primary cilia has become increasingly relevant. Much of our knowledge of the primary cilium as a sensory organelle derives from the groundbreaking work with *Caenorhabditis elegans.* In this organism, mechanosensation, chemotaxis, thermotaxis, and male mating behavior have been used as readouts of ciliary function. It was inspiration from work performed with *C. elegans* (Barr and Sternberg, 1999; Barr *et al.*, 2001; Liedtke *et al.*, 2003) that led to the identification of TRP ion channels (TRPP2 and TRPV4) in primary cilia of mammals.

The identification of readouts of primary cilia function in mammalian cells is not as straight forward as in *C. elegans*. A number of laboratories have established that signals using intracellular Ca^{2+} ($[Ca^{2+}]_i$) as a second messenger can be initiated in primary cilia. As Ca^{2+} is a common second messenger for multiple agonist and mechanical stimuli, $[Ca^{2+}]_i$ measurements can be used to monitor many types of signal-transduction pathways. These approaches have been implemented by many groups investigating ciliary function (Gradilone *et al.*, 2007; Kotsis *et al.*, 2007; Kottgen *et al.*, 2008; Masyuk *et al.*, 2006; Praetorius and Spring, 2001, 2003a).

Genetically encoded Ca^{2+} sensors open new doors for intracellular Ca^{2+} measurements (Palmer and Tsien, 2006). Targeted expression of Ca^{2+}-sensitive probes has made it possible to make reliable *in vivo* $[Ca^{2+}]_i$ measurements with single-cell resolution (Wallace *et al.*, 2008). In this context, the progress of *in vivo* imaging of intrarenal signaling by the use of multiphoton microscopy (Kang *et al.*, 2006; Peti-Peterdi, 2006; Peti-Peterdi *et al.*, 2009) is likely to allow us to obtain an integrated understanding of the physiological relevance of primary cilia. These indicators also have great potential for traditional live-cell imaging as they can be used to scrutinize $[Ca^{2+}]_i$ signaling in cellular subdomains (Lee *et al.*, 2006; Mao *et al.*, 2008; Petreanu *et al.*, 2009). For example, genetically encoded Ca^{2+} sensors targeted to the cilium may allow one to detect $[Ca^{2+}]_i$ signaling in a single cilium. This is difficult with classic Ca^{2+}-sensitive dyes as it has been hard to detect the ciliary signal over the background cytosolic fluorescence. Due to this constraint, ciliary Ca^{2+} changes have only been detected in a limited number of studies (Leinders-Zufall *et al.*, 1997).

II. Methods

A. Perfusion Chambers—Open or Closed?

Perfusion chambers are available commercially or can be custom built to fit your specific needs—the important point is that the design should be chosen according to the intended use. Both open and closed perfusion chambers can be used and an

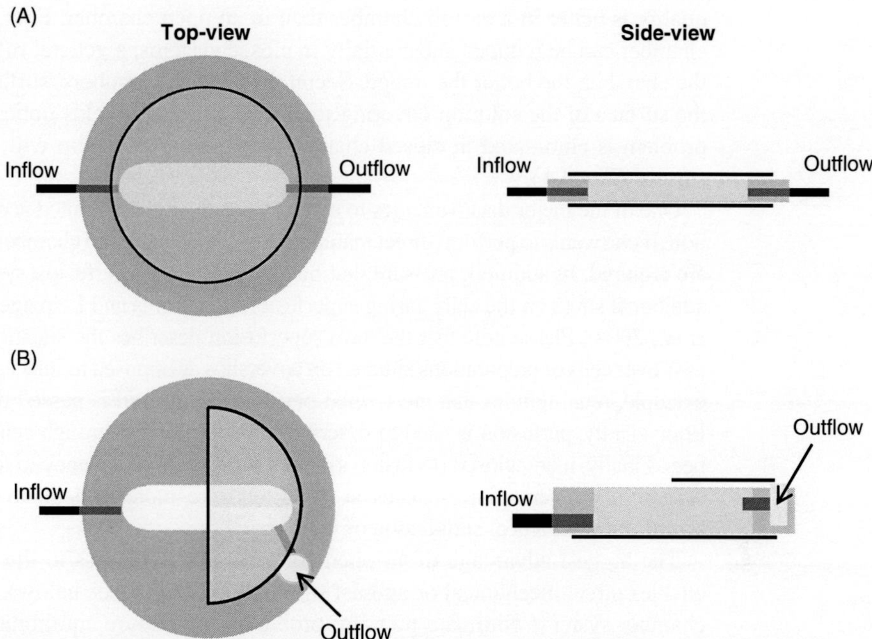

Fig. 1 Examples of perfusion chambers. (A) shows a closed chamber from the top and from the side and (B) shows the same projections for an open chamber.

example of a design is shown in Fig. 1. The critical issue is to attain laminar flow over the cells in the field of view. This means that the flow chamber should have a slit shape with a rather broad inflow. The inflow should be horizontal and as close as possible to the same plane as the coverslip containing the preparation. Some commercially available chambers have the inflow perpendicular to the flow direction, which will always result in turbulent flow and are not suitable for this application.

The outflow should be unhindered and again as close to the plane of the coverslip as possible. In open chambers, a system where the outflow spills into a small well is preferable. This allows one to put a suction device into the well without disturbing the outflow from the chamber. If a suction device is placed directly in the outflow path, it often causes oscillatory flow in the chamber. The oscillatory flow can be observed as a rising and falling of the chamber surface.

With regard to the choice of open or closed perfusion systems, one should be aware of the advantages and disadvantages of each type. The advantage to a closed system is that it is easier to maintain a steady environment with regard to pH, CO_2/HCO_3^- and osmolality. It is also much easier to avoid turbulence and get a good flow profile in a closed chamber. There are a number of closed chamber systems on the market with very nice flow profiles (RC-21BRFS and RC-31, Warner Instruments, Hamden, CT). These can be complete perfusion systems with integrated microscope slides or chambers designed to hold cells grown on coverslips. Another important point is that image quality is better in a closed chamber than in an open chamber. First, the height of the chamber can be reduced substantially in closed systems; a general rule is the shallower the chamber, the better the image. Secondly, in open chambers, surface tension curves the surface of the solution covering the preparation and yields optical distortion. This problem is eliminated in closed chambers as the top coverslip will level the solution surface (Fig. 1A).

One of the major disadvantages to a closed chamber is the limited access to the preparation. If one wants to perform direct manipulation of cilia in closed chambers, optical tweezers are required. In addition, pressure that builds up in closed perfusion systems can produce additional stress on the cells during superfusion (Praetorius and Leipziger, 2009; Praetorius *et al.*, 2004). Please note that the term superfusion describes the situation where solutions pass over cells or preparations situated on coverslips as opposed to intact preparations of, for example, renal tubules that are termed perfused when fluid is passed through the tubule. Equivalently, perfusion is used to describe solution passing through cell incubation chambers. Finally, many closed perfusion systems suffer from a tendency to form bubbles in the system, a problem that escalates at 37°C, and are more difficult to assemble without significant mechanical stimulation of the preparation.

The biggest advantage of an open chamber is easy access to the preparation. This enables direct mechanical or agonist stimulation of cilia (see below). In addition, open chamber systems eliminate pressure problems as pressure equilibrates as soon as the solution enters the open bath. The same is true for air bubbles. However, bubbles in the system should be avoided as they create uneven flow.

The drawbacks of open perfusion systems are plentiful. As mentioned above, the image quality in transmitted light suffers in an open chamber system as a result of the curvature of the solution covering the cells; this is a significant problem when imaging cilia. Also the aperture of the chamber leads to evaporation and consequently increasing osmolality. In addition, the fact that pressure is released when the solution enters the chamber necessitates that vacuum be applied to the outflow to remove the solution from the chamber. The introduction of suction can cause oscillatory flow within the chamber. This can be avoided by letting the solution from the perfusion chamber spill into a small well from which the excess solution is vacuumed off (see Fig. 1B).

Many of the problems with an open perfusion chamber can be minimized by the use of a semi-open chamber (see Fig. 1B). This type of chamber can be manufactured by modification to either open or closed chamber systems. A simple way to do this is to place a cut cover glass over the aperture of an open chamber. The image quality instantly improves by flattening the superfusate. The semi-open chamber also reduces evaporation but still allows for equilibration of pressure and access to the preparation. For tests of the perfusion chambers please see *Flow or agonist stimulation of cilia*.

B. Temperature Control

As in any kind of preparation of living cells or tissues, control of temperature is crucial in live-cell imaging. There are many ways to control the temperature of the preparation. For high-resolution imaging, one must consider maintaining temperature

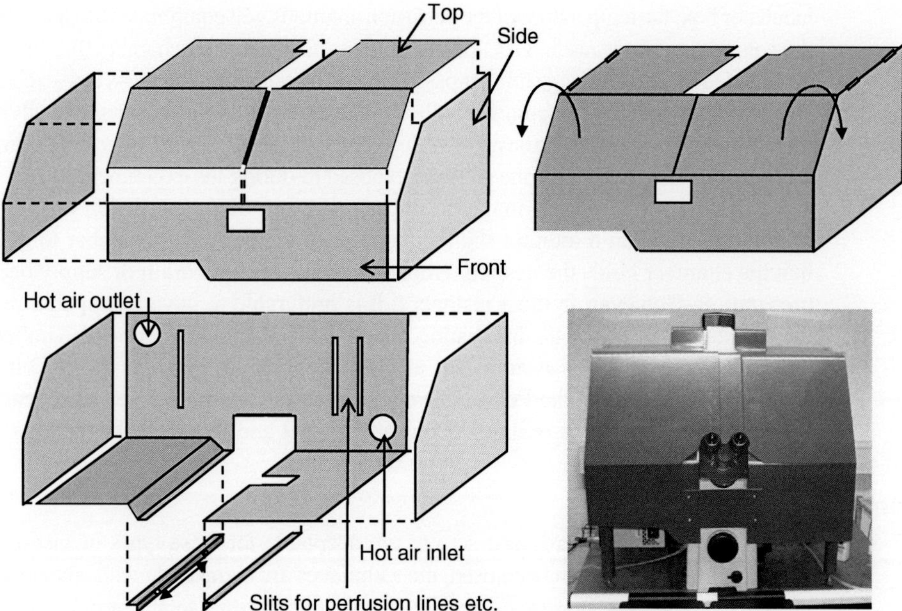

Fig. 2 Example of a microscope incubator box. The box shown here was originally designed to fit a Nikon 2000ES, but can be altered to fit any microscope; note openings in the back of the box. The circular holes allow entrance of hot air provided by a temperature regulation unit placed outside the box and slits allows entrance of micromanipulator holders, perfusion lines, temperature controllers, etc.

of: (1) the environment including the microscope, (2) the perfusate, and (3) the perfusion chamber.

When imaging in high resolution, the contact between the coverslip and the objective lens will serve as a heat sink and lower the temperature of the cells one wants to image. This problem can be overcome by different approaches. One can buy or customize a heated box to surround the entire microscope including the headpiece and objectives (Fig. 2). The temperature in this box is regulated by blowing hot air in the bottom and letting it flow out at the top, with the amount of heat added regulated by a thermostat. Commercially available objective heaters or heating the preparation by a hot air stream aimed at the preparation and headpiece of the microscope are alternatives to enclosing the microscope. The latter is a low-cost solution that is suitable for some applications. However, one should be aware that it often creates temperature gradients and causes excessive evaporation in open chamber systems. Furthermore, an uneven temperature distribution along the axis of the objective decreases the quality of the optics and thereby hinders imaging of structures close to the diffraction-limited resolution, such as primary cilia.

The perfusate must be brought up to temperature before it enters the observation chamber. If one is using a temperature-controlled box to maintain the microscope temperature, it can also serve as heater for the perfusion lines. By passing the perfusion lines through the incubator box, the temperature of the perfusion solutions will equilibrate with the chamber. If the solution does not reach 37°C before entering the perfusion chamber, the lines can be extended to increase passage time or passed over a metal plate to increase heat exchange. As an alternative, one can use commercially available perfusion line heaters. It is advisable to monitor temperature at the inflow or outflow of the chamber to ensure accurate sampling of the solution temperature that the cells are exposed to during the experiment.

Lastly, it is important to maintain the temperature of the perfusion chamber. At a minimum, one should monitor the temperature of the perfusion chamber to make sure that the chamber holds the desired temperature and does not drain or supply heat to the preparation. However, in most instances, it is preferable to have a temperature adjustment system attached to the incubation chamber. This could either be a chamber heater or, better still, a Peltier system that allows for both heating and cooling. It can be advantageous to adjust the Peltier current so that the chamber never quite reaches the desired temperature; this ensures a constant current and less heat fluctuations.

C. Microscope Set Up

A large variety of microscope set ups are adaptable for these types of measurements. While wide field has often been used, the techniques are compatible with advanced microscope technologies to increase the resolution such as confocal-, total internal reflection- and stimulated emission depletion microscopy. However, I will only discuss the minimal equipment that is needed for flow or agonist stimulation and direct manipulation or local agonist addition to the primary cilium. More information on live-cell imaging and microscopy can be found at: http://micro.magnet.fsu.edu/primer/ and in Inoue and Spring (1997).

1. Inverted microscope equipped for fluorescence detection. The microscope can be either wide field or confocal. Upright microscopes can alternatively be used for closed perfusion systems.
2. High-magnification objective lenses (60× or 100×, with a high NA—over 1) for oil immersion. Water objectives are problematic when working at 37°C because of inevitable evaporation of the interface medium. For further information about the choice of interface media between objective lens and specimen please see Hell *et al.* (1993).
3. Contrast enhancing optics for transmitted light. Phase contrast or differential interference contrast (DIC) optics is obligatory to visualize primary cilia directly in transmitted light, as the cilia in the y–x plane is at the edge of resolution.
4. Detection device. A charge-coupled device camera is the minimum needed unless you are interested only in fluorescence changes. In this case, a photomultiplier attached to the microscope or a cuvette-based photomultiplier system could be used. For more information on improving resolution of the fluorescence signal please see Section II.D below.
5. Standard set up for fluorescence microscopy. Minimum requirements include an XBO lamp and appropriate excitation filters for the probe one wants to use. Ratiometric fluorescence measurements will also require a filter wheel. However, monochromators or laser systems are common and allow digital control of the intensities and wavelength.

In normal fluorescence microscopy it is usually not possible to combine simultaneous epi-fluorescence and transmitted light. However, in experiments requiring direct manipulation or isolated application to the primary cilia, it is a huge advantage to be able to visualize cilia simultaneously with fluorescence measurements (Fig. 3). The simplest and cheapest solution is to add a shutter to the transmitted light path. The shutter can be

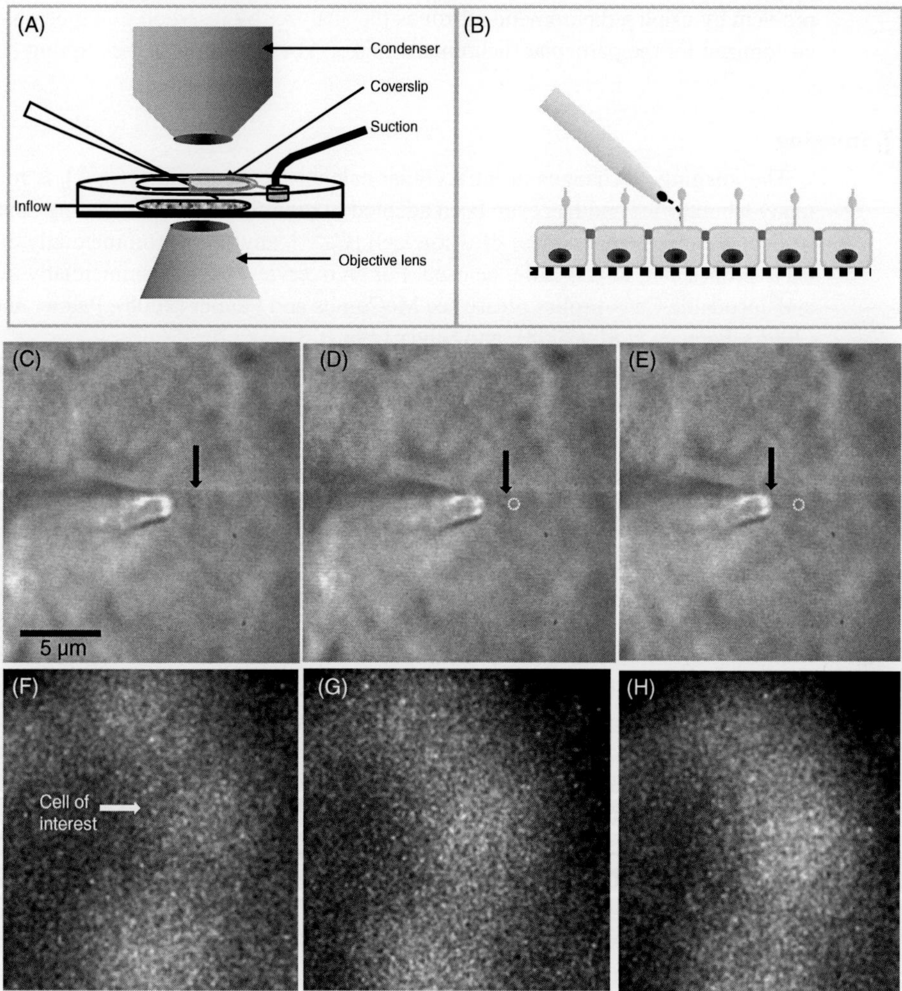

Fig. 3 Micromanipulations of a single cilium. (A) shows the microscope set up with a semi-open perfusion chamber in place. (B) shows a schematic drawing of the bending of a single cilium with a micropipette. (C, D, E) show DIC images obtained with a 100× objective (NA 1.4) and an extra 1.5× enhancement. (C) image of the cilium immediately before it is bent toward the micropipette by application of negative pressure. The arrow marks the position of the cilium. (D) and (E) are 1 and 4 s after bending, respectively. The arrow marks the position of the cilium; a dotted circle marks the original position of the cilium. (F, G, H) are the corresponding images of the fluorescence intensity obtained with fluo-4 loaded cells.

synchronized via the acquisition software to allow consecutive sampling of fluorescence and transmitted images. If one wants to combine epi-fluorescence measurements with DIC, one should be aware that the polarization of light in DIC reduces emission light considerably and forces one to use higher excitation intensities. This problem is caused by the analyzer, which is a polarization filter inserted below the lower Wallaston prism perpendicular to the direction of the polarized transmitted light. One can overcome this problem by using a dichromatic mirror as the analyzer by insertion of a green pass filter customized for the particular dichromatic mirror. For details please see Spring (1990).

D. $[Ca^{2+}]_i$ Imaging

The imaging of changes in intracellular calcium concentration $[Ca^{2+}]_i$ is routine in many laboratories and has even been adapted to high-throughput screening using plate readers. For the measurement of whole-cell $[Ca^{2+}]_i$ any of the commercially available Ca^{2+} probes can, in principle, be used. For an overview of the commercially available and encodable Ca^{2+} probes please see McCombs and Palmer (2008), Palmer and Tsien (2006), Paredes *et al.* (2008), and Silver (1998).

For experiments using fluorescent probes loaded as acetoxymethyl (AM) esters, the properties of a given probe should be considered in light of the particular application. If one is interested in subtle Ca^{2+} entries it is important to choose a fluorescence probe with a large dynamic range (e.g., Fluo-4). If the imaging set up precludes simultaneous use of transmitted light, baseline fluorescence can be used for orientation. In this situation, it might be easier to use a probe with higher baseline fluorescence such as calcium green. In general it is recommended to use ratiometric measurements for $[Ca^{2+}]_i$ and if one wants to compare baseline $[Ca^{2+}]_i$ levels between various cell types and culture conditions or calibrate the $[Ca^{2+}]_i$ signal, ratiometric probes are obligatory (e.g., fura-2). For considerations regarding ratio imaging, please consult (Dunn and Maxfield, 2003; O'Connor and Silver, 2007; Silver, 1998).

Special attention should be paid to the intracellular distribution of AM-loaded Ca^{2+} probes (Thomas *et al.*, 2000). Ideally, the fluorescence probe should be confined to the cytoplasm. This is seldom the case and so the fluorescence signal will report the total $[Ca^{2+}]_i$ response in the cytosol and in the organelles that are loaded with the probe. Therefore, it is important to inspect the preparation at high magnification to determine the probe's intracellular distribution. For example, fluo-3 and fluo-4 have a striking tendency to accumulate in the nucleus (Thomas *et al.*, 2000). This is of course an advantage if one wants to measure intranuclear Ca^{2+}. If it is a measurement for the cytosolic Ca^{2+} concentration that is needed, this is a significant draw back and one would have to verify that the Ca^{2+} signals measured in the nucleus are identical to the ones measured in the cytosol before using whole cell measurements for a given application. The vesicular and ER compartments cause the most difficulty, because these compartments often have very low luminal pH. As many of the Ca^{2+} probes are sensitive to large pH changes, vesicular distribution can create a false high intracellular baseline concentration. Furthermore, vesicles may release their contents to the exterior upon agonist stimulation and this may influence the results.

Photo- and temperature activation of the preparation is a serious concern especially if one wants to obtain a stable baseline signal from the cells in question during "no-flow" conditions. Problems with photo-activation can be avoided by using a fluorescence probe with a high dynamic range for easy detection of low light signals. With these probes, the light can be dampened by neutral density filters or the exposure time can be reduced. If high resolution is not a major requirement, binning of the camera signal is an easy way to reduce the demand for a high degree of light exposure. Finally, one should try to avoid UV-excitation probes when possible, as the short wavelengths needed to excite these probes are generally more damaging to cells than longer wavelengths. Temperature activation can be reduced by a constant low flow through the perfusion chamber.

Long-term loading—high probe concentration. High concentrations of Ca^{2+} probes should be avoided as Ca^{2+} probes also function as Ca^{2+} buffers and can change the parameter one wants to assess. Most probes are loaded as AM esters that allow the probes to cross the cell membrane. In the cytoplasm, the ester is removed rendering them unable to permeate the membrane This allows the probes to concentrate in the cell, and thus intracellular dye concentration may exceed the extracellular concentration. Formaldehyde production is another concern of high probe concentration as the AM group is hydrolyzed to formaldehyde and acetic acid within the cell. This means that it is crucial to optimize detection systems for low light level fluorescence rather that increasing the dye loading of the preparation. This is especially important for UV probes as the low excitation wavelength (below 360 nm) requires good quartz optics through the entire excitation pathway to assure reasonable intensities at the preparation.

Dye leak. Many cell types are capable of removing fluorescence probes from the cytosol. In most cases, this dye leak can be prevented by drugs like probenecid that inhibit various channels and anion exchangers. However, one should be careful that drug addition for methodological reasons does not interfere with the pathway under study.

E. Micromanipulation of a Single Primary Cilium

To manipulate a single cilium, one needs a precise micromanipulator with a minimum step size of less than a micron. Commercially available micromanipulators with step sizes of 0.2 μM are adequate. If one is using a temperature-controlled box to enclose the microscope and perfusion chamber, the micromanipulator and pipette holder is placed inside the box and the controller kept outside the box. In this way, one can keep the box closed during the entire experiment. Patch clamp pipettes with an electrical resistance of 1–2 MΩ are suitable for manipulating individual cilia. These micropipettes are fabricated on a dual stage puller and the tip of the pipette is rounded by heat-polishing. Before the experiment, the micropipette is filled with the normal external solution. As in patch clamping, the micropipette is attached to the pipette holder and connected to tubing that allows the regulation of pressure in the pipette. For these applications, it is necessary to use a semi-open perfusion bath. During the experiments, it is best to keep the cells under constant perfusion at a very low flow rate that does not excite the cells or move the cilia.

When the cells are ready, the tip of the micropipette is positioned under the half coverslip in a semi-open perfusion chamber. This can be done either with the gross-manipulation controls of the micromanipulator or by hand (Fig. 3A). The box is closed and the micropipette is positioned close to a primary cilium using the fine controls of the micromanipulator. When the pipette is in place, the whole set up is left undisturbed for about 30 min under constant quiet flow of perfusion solution as the preparation is warmed to 37°C. Each experiment starts by recording the baseline signal. The primary cilium can then be moved toward the pipette by applying slight negative pressure to the pipette. With the equipment described here, one can watch the cilium move toward the pipette (Fig. 3), while recording $[Ca^{2+}]_i$ from the cells to which the cilium is attached and the adjacent cells. It is important to perform time controls for this type of experiment, that is, observing the cells under the exact same conditions, with the micropipette in place—without manipulating the cilium.

F. Local Application to the Shaft of Cilium

Micromanipulator-controlled pipettes can also be used for local application of agonists or other activators to primary cilia. One strategy is to slowly engulf the primary cilium in a glass pipette. A wide bore patch pipette with a resistance of 1–2 MΩ is preferable for this approach. Cells should be observed with high magnification (100× with a 1.5× extra magnification if the microscope has this feature). The trick is to approach the tip of the primary cilium with the pipette under a slight negative pressure and then to move the pipette downward and toward the cell membrane while slowly absorbing the cilium into the pipette interior.

One major problem with this approach is that the cilium will be gradually stimulated by the solution in the patch pipette as the pipette is positioned for measurements. It is possible to establish an internal perfusion system within the pipette although this presents a risk of flushing the cilium back out again. The easier solution is to simply dip the pipette tip in the external solution followed by backfilling of the patch pipette with a stimulation solution containing a fluorescent probe with an excitation wavelength different from the measuring probe. By monitoring the fluorescence of the stimulation solution, one can determine when it reaches the cilium. To avoid stimulation of the plasma membrane by exhaust from the stimulation pipette, it is important to keep a constant low nonstimulatory flow in the incubation chamber. By having the aforementioned reporter probe in the stimulation pipette one can also ensure that the plasma membrane is not exposed to a significant amount of the stimulation solution.

G. Flow or Agonist Stimulation of Primary Cilia

Flow or agonist stimulation to a sheet of cells can be used to substantiate data from single cilia experiments. This method can also be used for screening purposes or on its own with the appropriate control experiments. For these experiments, it is critical that the perfusion chamber and experimental set up actually provide changes in laminar flow and laminar flow only. There are several steps that need to be considered in setting up a reliable perfusion system.

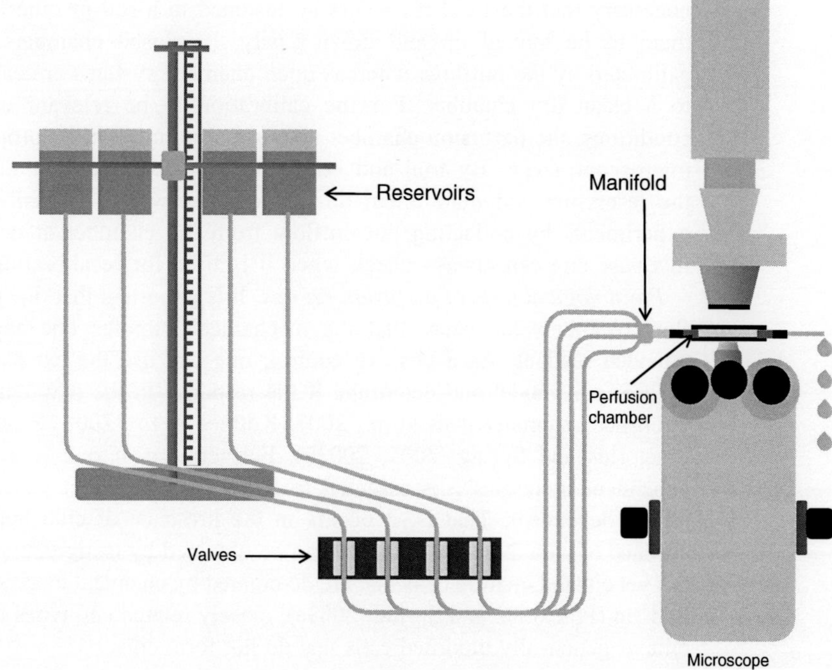

Fig. 4 Example of a perfusion system. The figure shows the component needed to set up a perfusion system for live-cell imaging. The valves are optional and increase the usability of the system.

Perfusion systems (Fig. 4). It is possible to set up a perfusion system with ordinary laboratory supplies or the perfusion systems can be purchased fully assembled. If the system is made in the laboratory, it is important that it have multiple reservoirs to allow solution changes during the experiments. The reservoirs should be open (preferable with a lose lid) so the height of the solutions can be adjusted during the experiments. It is very important when adding agonist/antagonist that the column height of the agonist/antagonist solutions are adjusted to the same height as the basal solution or one will cause a change in flow rate at the same time as the cells are being chemically stimulated. It is a good idea to choose large containers for the superfusate so that the flow rate diminishes very little during the experiment. In addition, one should standardize the inflow lines so that they are all made of the same tubing and have exactly the same diameter and length. Finally, for fast changes between the different solutions, the inflow lines are collected right before the perfusion chamber inflow with a manifold located that allows reasonably fast change between the different solutions. It is important to rinse the perfusion system after use and replace the tubing on a regular basis.

Calibrating the flow rate. Flow rates should be calibrated regularly. Gravity-controlled flow rates are adjusted by changing the elevation of the inflow fluid reservoir with respect to the outflow. To be able to adjust the flow rates, it is necessary that the fluid reservoirs be fastened to a rod or other device that allows them to be moved up and down easily. In closed chambers, the flow rate is calibrated by the outflow, whereas open chamber systems are calibrated via inflow to a clean dry chamber. For the calibration to be relevant under experimental conditions, the perfusion chamber should be mounted in its proper position on the microscope stage. By trial and error, one can determine the desired positions of the reservoirs and mark them on the rods. Flow rate is easily measured during experiments by collecting the outflow from the chamber at defined intervals. In this way one can always check when it is time for recalibration.

Physiological tests of perfusion set ups. It is important that one tests new perfusion flow chambers to ensure that the mechanical stimulus one applies produces the intended readout. As a positive control, one can use the extensively characterized MDCK cell model and determine if the response in the new chamber is similar to published reports (Kotsis *et al.*, 2007; Kottgen *et al.*, 2008; Praetorius *et al.*, 2004; Praetorius and Spring, 2001, 2003b). Furthermore, if one is investigating cilium-dependent properties it is essential to know that the effect you observe actually is cilium dependent. That is, it occurs in the presence of cilia but not absent in the absence of cilia. Non-ciliated cells can be obtained by using young cultures that have not yet ciliated, mature cells that are de-ciliated by chemical treatments such as chloral hydrate (Praetorius and Spring, 2003a), closely related cell types that do not develop cilia or genetically modified cells that do not form cilia.

III. Materials

The details below are based on studies of ciliary stimulation in MDCK cells but these methods can serve as a reference for other cell types.

A. Cell Culture

1. *Madin-Darby canine kidney (MDCK) cells*, type 1 (American Type Culture Collection, Rockville, MD, USA) are the cell type in which the cilium-dependent flow response was first described (Praetorius and Spring, 2001). This cell line was derived from the distal segments of a dog kidney and has a cell composition similar to the cortical collecting duct, that is, it is a culture of both intercalated and principle cells.
2. MDCK cells were purchased from ATCC at generation 53, expanded and frozen for long-term storage in liquid nitrogen at passages 54–56. After thawing, the cells were allowed to grow to confluency and passaged at least twice before they were used in experiments. After 10–12 passages, the cells were discarded.
3. Cell culture medium: Dulbecco's Modified Eagle Medium (DMEM, 21063-029 Invitrogen) supplemented with 10% fetal bovine serum. Addition of penicillin and streptomycin to the cell culture medium does not affect the results.

4. Trypsin-EDTA (0.5 g/l, 0.2 g/l, 25300-054, Invitrogen) is used to dissociate the cells. Confluent cells in 25 cm^2 cell culture flasks are split into 2 new cell culture flasks and 8–12 petri dishes (3-cm diameter) containing coverslips for experiments.
5. The cells are fed every other day, and typically reach confluency after 3–4 days. The cells do not develop primary cilia until the entire culture has reached confluency. Seven days after passage, the cilia are usually about 4–5 μm long. The cells can be used for experiments between days 7–11 after passage.

B. Live-Cell Imaging

Flow chamber, temperature control (microscope box, perfusion heating, and chamber heater) and microscope/imaging set up are described in detail above. For examples from the literature, please see: (Gradilone *et al.*, 2007; Kotsis *et al.*, 2007; Kottgen *et al.*, 2008; Malone *et al.*, 2007; Masyuk *et al.*, 2006; Praetorius and Spring, 2001, 2003a).

C. Experimental Solutions

A suitable bath solution may have the following composition, in mM: $[Na^+]$ 138, $[K^+]$ 5.3, $[Ca^{2+}]$ 1.8, $[Mg^{2+}]$ 0.8, $[Cl^-]$ 126.9, $[SO_4^{2-}]$ 0.8, Hepes 14, glucose 5.6, probenecid 5, pH 7.4 (37°C, 300 mOsmol/l). It is important to check the pH of all solutions, including those containing drugs, at 37°C immediately before initiating the experiments.

IV. Summary/Conclusion

With the understanding that primary cilia detect the extracellular environment and signal the cell body came the need to develop tools for studying this organelle. In this report, I describe methods for monitoring Ca^{2+} changes in live cells that result from stimulating cilia. These assays can be used to study a variety of inputs and will be widely useful in understanding the functional importance of newly discovered ciliary proteins.

References

Barr, M.M., DeModena, J., Braun, D., Nguyen, C.Q., Hall, D.H., and Sternberg, P.W. (2001). The Caenorhabditis elegans autosomal dominant polycystic kidney disease gene homologs lov-1 and pkd-2 act in the same pathway. *Curr. Biol.* **11**, 1341–1346.

Barr, M.M., and Sternberg, P.W. (1999). A polycystic kidney-disease gene homologue required for male mating behaviour in C. elegans. *Nature* **401**, 386–389.

Dunn, K., and Maxfield, F.R. (2003). Ratio imaging instrumentation. *Methods Cell Biol.* **72**, 389–413.

Gradilone, S.A., Masyuk, A.I., Splinter, P.L., Banales, J.M., Huang, B.Q., Tietz, P.S., Masyuk, T.V., and LaRusso, N.F. (2007). Cholangiocyte cilia express TRPV4 and detect changes in luminal tonicity inducing bicarbonate secretion. *Proc. Natl. Acad. Sci. USA* **104**, 19138–19143.

Hell, S., Reiner, G., Cremer, C., and Stelzer, E.H.K. (1993). Aberrations in confocal fluorescence microscopy induced by mismatches in refractive index. *J. Microsc.* **169**, 391–405.

Inoue, S., and Spring, K.R. (1997). "Video Microscopy, the Fundamentals." Plenum Press, New York.

Kang, J.J., Toma, I., Sipos, A., McCulloch, F., and Peti-Peterdi, J. (2006). Quantitative imaging of basic functions in renal (patho)physiology. *Am. J. Physiol. Renal. Physiol.* **291**, F495–F502.

Kotsis, F., Nitschke, R., Boehlke, C., Bashkurov, M., Walz, G., and Kuehn, E.W. (2007). Ciliary calcium signaling is modulated by kidney injury molecule-1 (Kim1). *Pflugers Arch.* **453**, 819–829.

Kottgen, M., Buchholz, B., Garcia-Gonzalez, M.A., Kotsis, F., Fu, X., Doerken, M., Boehlke, C., Steffl, D., Tauber, R., Wegierski, T., Nitschke, R., Suzuki, M., et al., (2008). TRPP2 and TRPV4 form a polymodal sensory channel complex. *J. Cell Biol.* **182**, 437–447.

Lee, M.Y., Song, H., Nakai, J., Ohkura, M., Kotlikoff, M.I., Kinsey, S.P., Golovina, V.A., and Blaustein, M.P. (2006). Local subplasma membrane Ca2+ signals detected by a tethered Ca^{2+} sensor. *Proc. Natl. Acad. Sci. USA* **103**, 13232–13237.

Leinders-Zufall, T., Rand, M.N., Shepherd, G.M., Greer, C.A., and Zufall, F. (1997). Calcium entry through cyclic nucleotide-gated channels in individual cilia of olfactory receptor cells: Spatiotemporal dynamics. *J. Neurosci.* **17**, 4136–4148.

Liedtke, W., Tobin, D.M., Bargmann, C.I., and Friedman, J.M. (2003). Mammalian TRPV4 (VR-OAC) directs behavioral responses to osmotic and mechanical stimuli in Caenorhabditis elegans. *Proc. Natl. Acad. Sci. USA* **100**(Suppl. 2), 14531–14536.

Malone, A.M., Anderson, C.T., Tummala, P., Kwon, R.Y., Johnston, T.R., Stearns, T., and Jacobs, C.R. (2007). Primary cilia mediate mechanosensing in bone cells by a calcium-independent mechanism. *Proc. Natl. Acad. Sci. USA* **104**, 13325–13330.

Mao, T., O'Connor, D.H., Scheuss, V., Nakai, J., and Svoboda, K. (2008). Characterization and subcellular targeting of GCaMP-type genetically-encoded calcium indicators. *PLoS. ONE* **3**, e1796.

Masyuk, A.I., Masyuk, T.V., Splinter, P.L., Huang, B.Q., Stroope, A.J., and LaRusso, N.F. (2006). Cholangiocyte cilia detect changes in luminal fluid flow and transmit them into intracellular Ca^{2+} and cAMP signaling. *Gastroenterology* **131**, 911–920.

McCombs, J.E., and Palmer, A.E. (2008). Measuring calcium dynamics in living cells with genetically encodable calcium indicators. *Methods* **46**, 152–159.

O'Connor, N., and Silver, R.B. (2007). Ratio imaging: Practical considerations for measuring intracellular Ca^{2+} and pH in living cells. *Methods Cell Biol.* **81**, 415–433.

Palmer, A.E., and Tsien, R.Y. (2006). Measuring calcium signaling using genetically targetable fluorescent indicators. *Nat. Protoc.* **1**, 1057–1065.

Paredes, R.M., Etzler, J.C., Watts, L.T., Zheng, W., and Lechleiter, J.D. (2008). Chemical calcium indicators. *Methods* **46**, 143–151.

Peti-Peterdi, J. (2006). Calcium wave of tubuloglomerular feedback. *Am. J. Physiol. Renal. Physiol.* **291**, F473–F480.

Peti-Peterdi, J., Toma, I., Sipos, A., and Vargas, S.L. (2009). Multiphoton imaging of renal regulatory mechanisms. *Physiology (Bethesda)* **24**, 88–96.

Petreanu, L., Mao, T., Sternson, S.M., and Svoboda, K. (2009). The subcellular organization of neocortical excitatory connections. *Nature* **457**, 1142–1145.

Praetorius, H.A., Frokiaer, J., and Leipziger, J. (2004). Transepithelial pressure pulses induce nucleotide release in polarized MDCK cells. *Am. J. Physiol. Renal. Physiol.* **288**, F133–F141.

Praetorius, H.A., and Leipziger, J. (2009). Released nucleotides amplify the cilium-dependent, flow-induced $[Ca^{2+}]_i$ response in MDCK cells. *Acta Physiol.* e-published ahead of print.

Praetorius, H.A., and Spring, K.R. (2001). Bending the MDCK cell primary cilium increases intracellular calcium. *J. Membr. Biol.* **184**, 71–79.

Praetorius, H.A., and Spring, K.R. (2003a). Removal of the MDCK cell primary cilium abolishes flow sensing. *J. Membr. Biol.* **191**, 69–76.

Praetorius, H.A., and Spring, K.R. (2003b). The renal cell primary cilium functions as a flow sensor. *Curr. Opin. Nephrol. Hypertens.* **12**, 517–520.

Silver, R.B. (1998). Ratio imaging: Practical considerations for measuring intracellular calcium and pH in living tissue. *Methods Cell Biol.* **56**, 237–251.

Spring, K.R. (1990). Quantitative imaging at low light levels: Differential Interference Contrast and flourescence microscopy without significant light loss. *In* "Optical Microsocpy for Biology" (B. Herman, and K. Jacobson, eds.), pp. 513–522. Wiley-Liss, New York.

Thomas, D., Tovey, S.C., Collins, T.J., Bootman, M.D., Berridge, M.J., and Lipp, P. (2000). A comparison of fluorescent Ca^{2+} indicator properties and their use in measuring elementary and global Ca^{2+} signals. *Cell Calcium* **28**, 213–223.

Wallace, D.J., Borgloh, S.M., Astori, S., Yang, Y., Bausen, M., Kugler, S., Palmer, A.E., Tsien, R.Y., Sprengel, R., Kerr, J.N., Denk, W., and Hasan, M.T. (2008). Single-spike detection in vitro and in vivo with a genetic Ca^{2+} sensor. *Nat. Methods* **5**, 797–804.

INDEX

O

P

T

U

V

W

X

Z

VOLUMES IN SERIES

Founding Series Editor
DAVID M. PRESCOTT

Volume 1 (1964)
Methods in Cell Physiology
Edited by David M. Prescott

Volume 2 (1966)
Methods in Cell Physiology
Edited by David M. Prescott

Volume 3 (1968)
Methods in Cell Physiology
Edited by David M. Prescott

Volume 4 (1970)
Methods in Cell Physiology
Edited by David M. Prescott

Volume 5 (1972)
Methods in Cell Physiology
Edited by David M. Prescott

Volume 6 (1973)
Methods in Cell Physiology
Edited by David M. Prescott

Volume 7 (1973)
Methods in Cell Biology
Edited by David M. Prescott

Volume 8 (1974)
Methods in Cell Biology
Edited by David M. Prescott

Volume 9 (1975)
Methods in Cell Biology
Edited by David M. Prescott

Volume 10 (1975)
Methods in Cell Biology
Edited by David M. Prescott

Volume 11 (1975)
Yeast Cells
Edited by David M. Prescott

Volume 12 (1975)
Yeast Cells
Edited by David M. Prescott

Volume 13 (1976)
Methods in Cell Biology
Edited by David M. Prescott

Volume 14 (1976)
Methods in Cell Biology
Edited by David M. Prescott

Volume 15 (1977)
Methods in Cell Biology
Edited by David M. Prescott

Volume 16 (1977)
Chromatin and Chromosomal Protein Research I
Edited by Gary Stein, Janet Stein, and Lewis J. Kleinsmith

Volume 17 (1978)
Chromatin and Chromosomal Protein Research II
Edited by Gary Stein, Janet Stein, and Lewis J. Kleinsmith

Volume 18 (1978)
Chromatin and Chromosomal Protein Research III
Edited by Gary Stein, Janet Stein, and Lewis J. Kleinsmith

Volume 19 (1978)
Chromatin and Chromosomal Protein Research IV
Edited by Gary Stein, Janet Stein, and Lewis J. Kleinsmith

Volume 20 (1978)
Methods in Cell Biology
Edited by David M. Prescott

Advisory Board Chairman
KEITH R. PORTER

Volume 21A (1980)
Normal Human Tissue and Cell Culture, Part A: Respiratory, Cardiovascular, and Integumentary Systems
Edited by Curtis C. Harris, Benjamin F. Trump, and Gary D. Stoner

Volume 21B (1980)
Normal Human Tissue and Cell Culture, Part B: Endocrine, Urogenital, and Gastrointestinal Systems
Edited by Curtis C. Harris, Benjamin F. Trump, and Gray D. Stoner

Volume 22 (1981)
Three-Dimensional Ultrastructure in Biology
Edited by James N. Turner

Volume 23 (1981)
Basic Mechanisms of Cellular Secretion
Edited by Arthur R. Hand and Constance Oliver

Volume 24 (1982)
The Cytoskeleton, Part A: Cytoskeletal Proteins, Isolation and Characterization
Edited by Leslie Wilson

Volume 25 (1982)
The Cytoskeleton, Part B: Biological Systems and *In Vitro* Models
Edited by Leslie Wilson

Volume 26 (1982)
Prenatal Diagnosis: Cell Biological Approaches
Edited by Samuel A. Latt and Gretchen J. Darlington

Series Editor
LESLIE WILSON

Volume 27 (1986)
Echinoderm Gametes and Embryos
Edited by Thomas E. Schroeder

Volume 28 (1987)
***Dictyostelium discoideum:* Molecular Approaches to Cell Biology**
Edited by James A. Spudich

Volume 29 (1989)
Fluorescence Microscopy of Living Cells in Culture, Part A: Fluorescent Analogs, Labeling Cells, and Basic Microscopy
Edited by Yu-Li Wang and D. Lansing Taylor

Volume 30 (1989)
Fluorescence Microscopy of Living Cells in Culture, Part B: Quantitative Fluorescence Microscopy—Imaging and Spectroscopy
Edited by D. Lansing Taylor and Yu-Li Wang

Volume 31 (1989)
Vesicular Transport, Part A
Edited by Alan M. Tartakoff

Volume 32 (1989)
Vesicular Transport, Part B
Edited by Alan M. Tartakoff

Volume 33 (1990)
Flow Cytometry
Edited by Zbigniew Darzynkiewicz and Harry A. Crissman

Volume 34 (1991)
Vectorial Transport of Proteins into and across Membranes
Edited by Alan M. Tartakoff

Selected from Volumes 31, 32, and 34 (1991)
Laboratory Methods for Vesicular and Vectorial Transport
Edited by Alan M. Tartakoff

Volume 35 (1991)
Functional Organization of the Nucleus: A Laboratory Guide
Edited by Barbara A. Hamkalo and Sarah C. R. Elgin

Volume 36 (1991)
***Xenopus laevis:* Practical Uses in Cell and Molecular Biology**
Edited by Brian K. Kay and H. Benjamin Peng

Series Editors
LESLIE WILSON AND PAUL MATSUDAIRA

Volume 37 (1993)
Antibodies in Cell Biology
Edited by David J. Asai

Volume 38 (1993)
Cell Biological Applications of Confocal Microscopy
Edited by Brian Matsumoto

Volume 39 (1993)
Motility Assays for Motor Proteins
Edited by Jonathan M. Scholey

Volume 40 (1994)
A Practical Guide to the Study of Calcium in Living Cells
Edited by Richard Nuccitelli

Volume 41 (1994)
Flow Cytometry, Second Edition, Part A
Edited by Zbigniew Darzynkiewicz, J. Paul Robinson, and Harry A. Crissman

Volume 42 (1994)
Flow Cytometry, Second Edition, Part B
Edited by Zbigniew Darzynkiewicz, J. Paul Robinson, and Harry A. Crissman

Volume 43 (1994)
Protein Expression in Animal Cells
Edited by Michael G. Roth

Volume 44 (1994)
***Drosophila melanogaster:* Practical Uses in Cell and Molecular Biology**
Edited by Lawrence S. B. Goldstein and Eric A. Fyrberg

Volume 45 (1994)
Microbes as Tools for Cell Biology
Edited by David G. Russell

Volume 46 (1995)
Cell Death
Edited by Lawrence M. Schwartz and Barbara A. Osborne

Volume 47 (1995)
Cilia and Flagella
Edited by William Dentler and George Witman

Volume 48 (1995)
***Caenorhabditis elegans:* Modern Biological Analysis of an Organism**
Edited by Henry F. Epstein and Diane C. Shakes

Volume 49 (1995)
Methods in Plant Cell Biology, Part A
Edited by David W. Galbraith, Hans J. Bohnert, and Don P. Bourque

Volume 50 (1995)
Methods in Plant Cell Biology, Part B
Edited by David W. Galbraith, Don P. Bourque, and Hans J. Bohnert

Volume 51 (1996)
Methods in Avian Embryology
Edited by Marianne Bronner-Fraser

Volume 52 (1997)
Methods in Muscle Biology
Edited by Charles P. Emerson, Jr. and H. Lee Sweeney

Volume 53 (1997)
Nuclear Structure and Function
Edited by Miguel Berrios

Volume 54 (1997)
Cumulative Index

Volume 55 (1997)
Laser Tweezers in Cell Biology
Edited by Michael P. Sheetz

Volume 56 (1998)
Video Microscopy
Edited by Greenfield Sluder and David E. Wolf

Volume 57 (1998)
Animal Cell Culture Methods
Edited by Jennie P. Mather and David Barnes

Volume 58 (1998)
Green Fluorescent Protein
Edited by Kevin F. Sullivan and Steve A. Kay

Volume 59 (1998)
The Zebrafish: Biology
Edited by H. William Detrich III, Monte Westerfield, and Leonard I. Zon

Volume 74 (2004)
Development of Sea Urchins, Ascidians, and Other Invertebrate Deuterostomes: Experimental Approaches
Edited by Charles A. Ettensohn, Gary M. Wessel, and Gregory A. Wray

Volume 75 (2004)
Cytometry, 4th Edition: New Developments
Edited by Zbigniew Darzynkiewicz, Mario Roederer, and Hans Tanke

Volume 76 (2004)
The Zebrafish: Cellular and Developmental Biology
Edited by H. William Detrich, III, Monte Westerfield, and Leonard I. Zon

Volume 77 (2004)
The Zebrafish: Genetics, Genomics, and Informatics
Edited by William H. Detrich, III, Monte Westerfield, and Leonard I. Zon

Volume 78 (2004)
Intermediate Filament Cytoskeleton
Edited by M. Bishr Omary and Pierre A. Coulombe

Volume 79 (2007)
Cellular Electron Microscopy
Edited by J. Richard McIntosh

Volume 80 (2007)
Mitochondria, 2nd Edition
Edited by Liza A. Pon and Eric A. Schon

Volume 81 (2007)
Digital Microscopy, 3rd Edition
Edited by Greenfield Sluder and David E. Wolf

Volume 82 (2007)
Laser Manipulation of Cells and Tissues
Edited by Michael W. Berns and Karl Otto Greulich

Volume 83 (2007)
Cell Mechanics
Edited by Yu-Li Wang and Dennis E. Discher

Volume 84 (2007)
Biophysical Tools for Biologists, Volume One: *In Vitro* Techniques
Edited by John J. Correia and H. William Detrich, III

Volume 85 (2008)
Fluorescent Proteins
Edited by Kevin F. Sullivan

Volume 86 (2008)
Stem Cell Culture
Edited by Dr. Jennie P. Mather

Volume 87 (2008)
Avian Embryology, 2nd Edition
Edited by Dr. Marianne Bronner-Fraser

Volume 88 (2008)
Introduction to Electron Microscopy for Biologists
Edited by Prof. Terence D. Allen

Volume 89 (2008)
Biophysical Tools for Biologists, Volume Two: *In Vivo* Techniques
Edited by Dr. John J. Correia and Dr. H. William Detrich, III

Volume 90 (2008)
Methods in Nano Cell Biology
Edited by Bhanu P. Jena

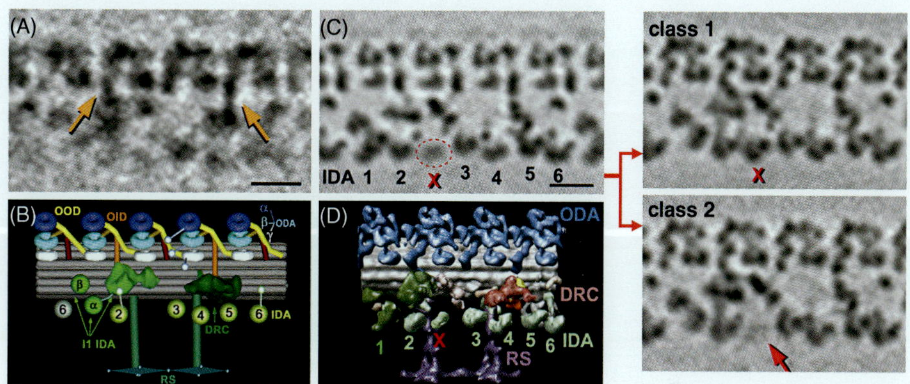

Plate 1 **(Figure 3 on page 10 of this volume)**

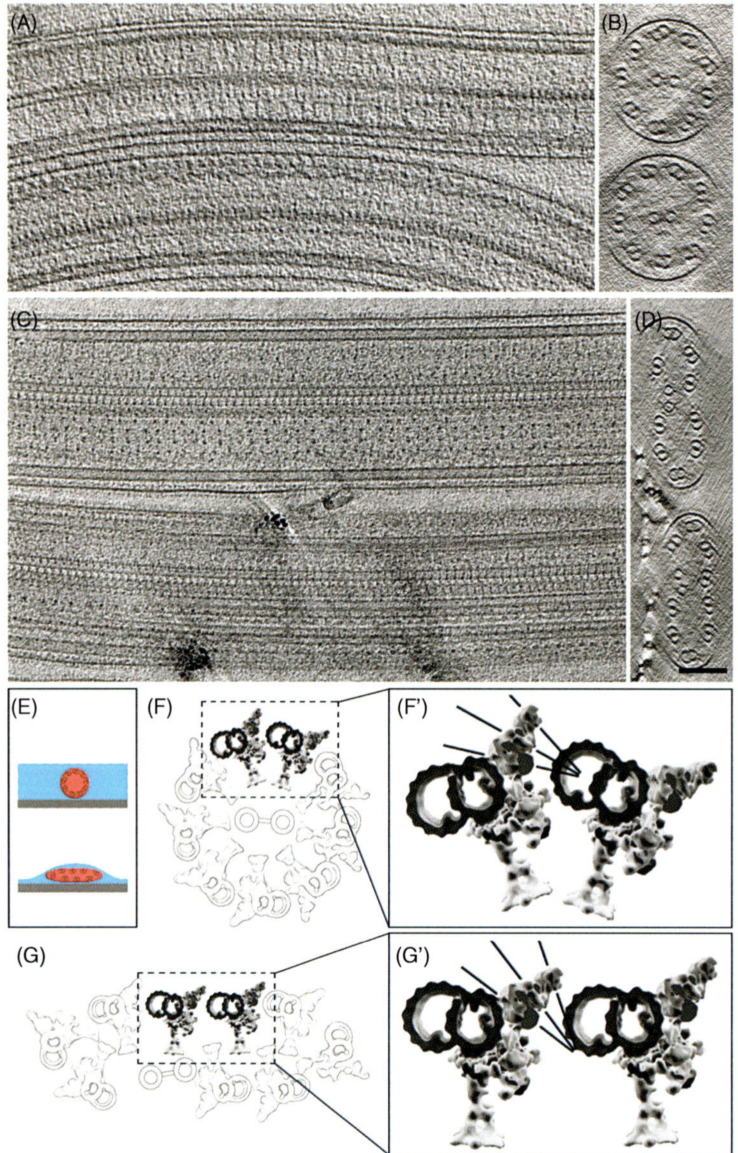

Plate 2 (Figure 6 on page 28 of this volume)

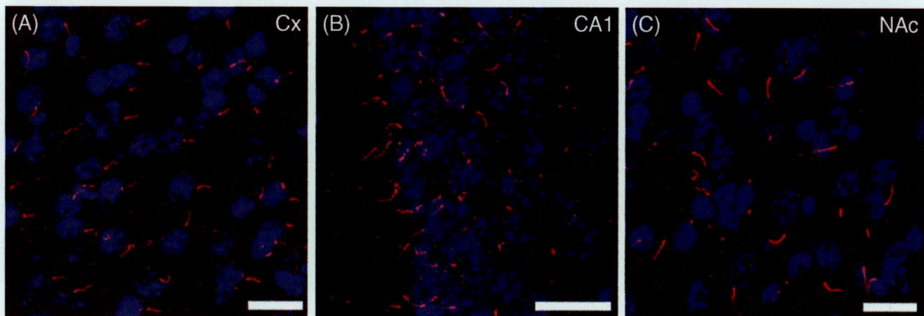

Plate 3 (Figure 2 on page 117 of this volume)

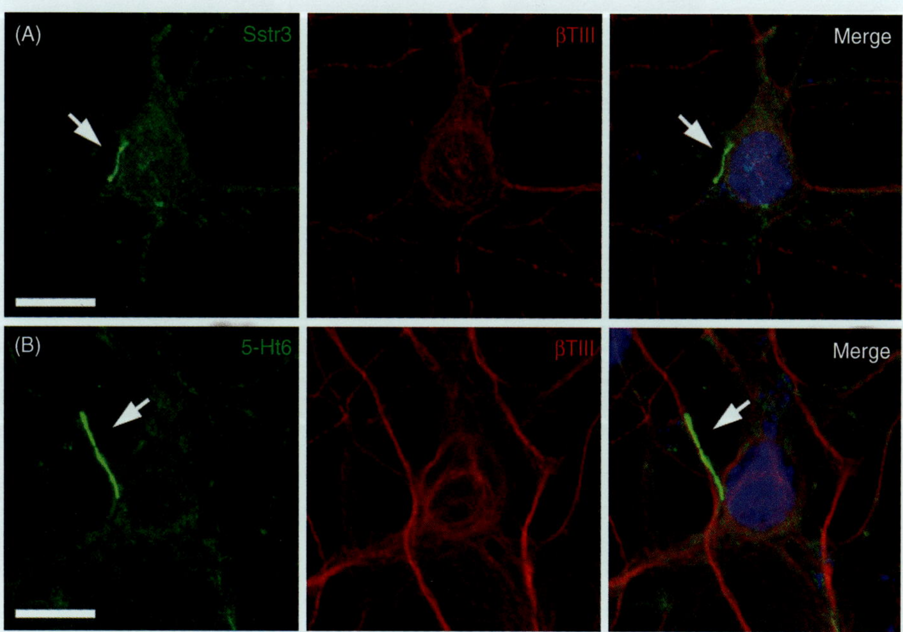

Plate 4 (Figure 3 on page 120 of this volume)

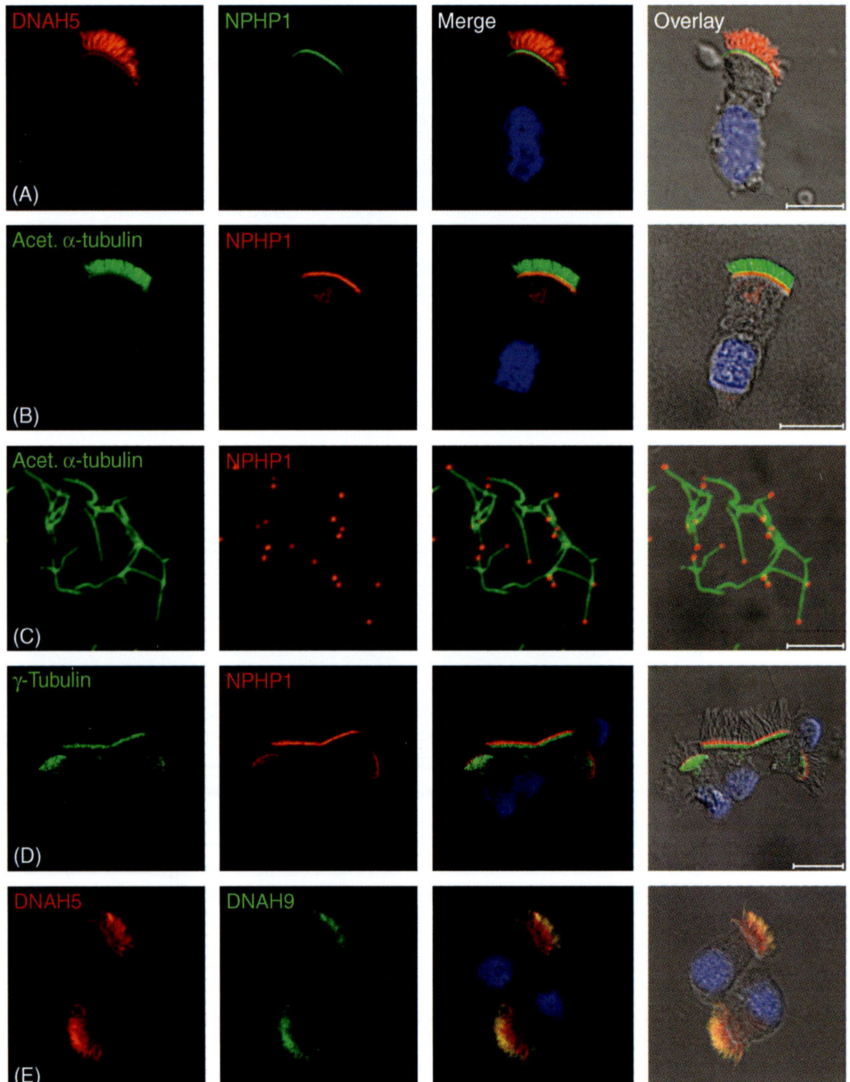

Plate 5 (Figure 2 on page 125 of this volume)

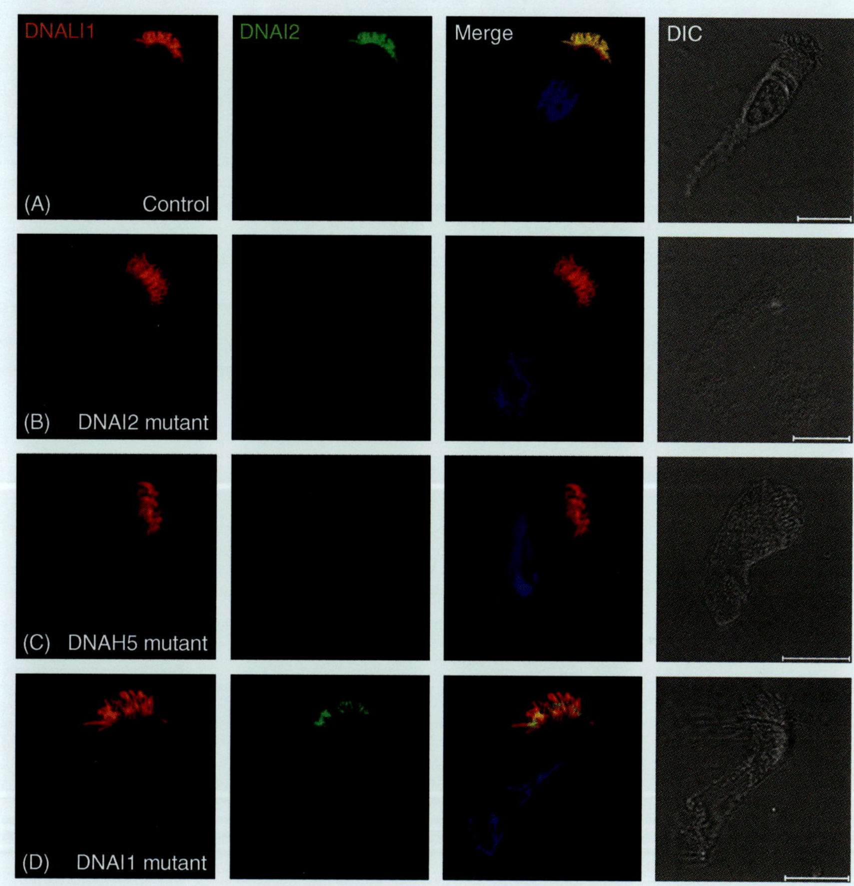

Plate 6 **(Figure 3 on page 127 of this volume)**

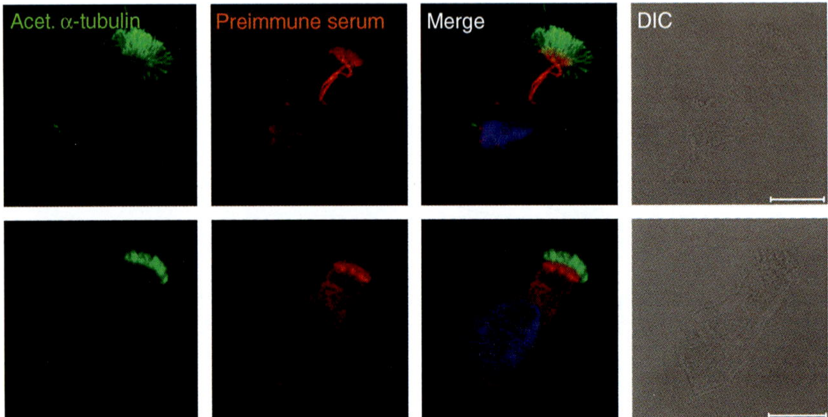

Plate 7 (Figure 4 on page 128 of this volume)

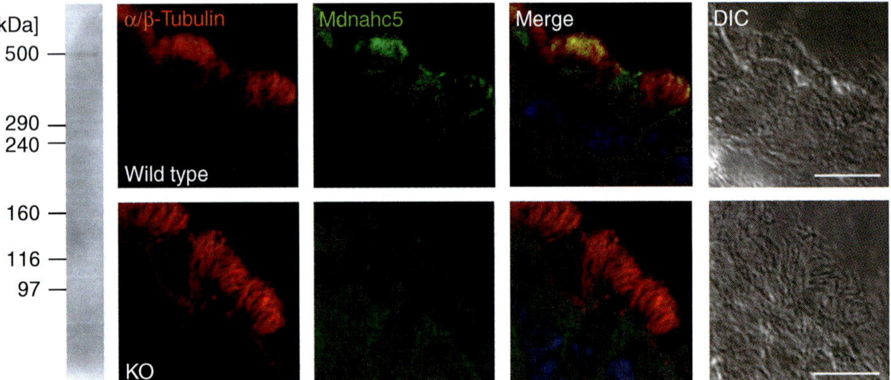

Plate 8 (Figure 5 on page 132 of this volume)

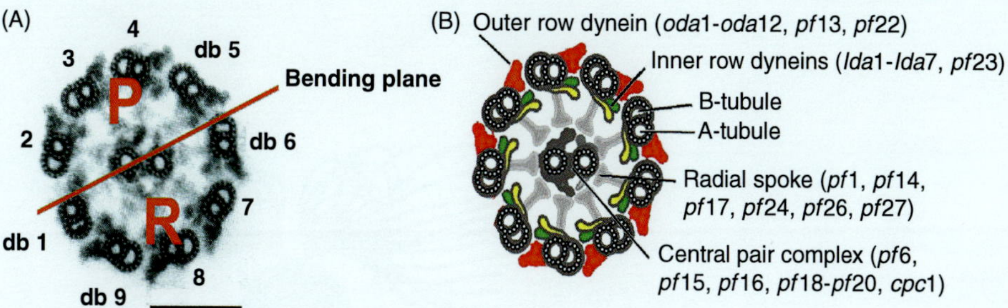

Plate 9 (Figure 2 on page 175 of this volume)

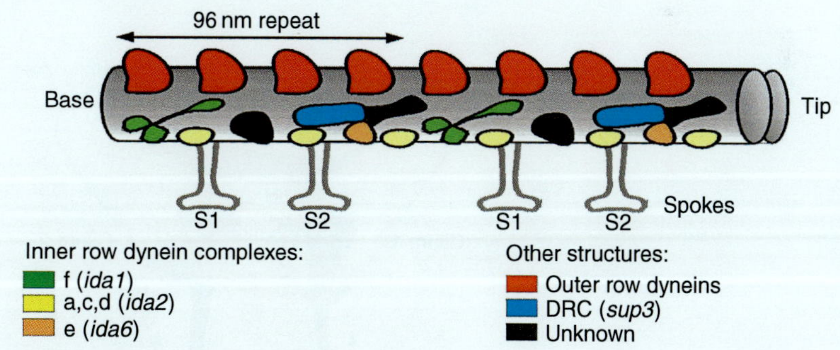

Plate 10 (Figure 3 on page 175 of this volume)

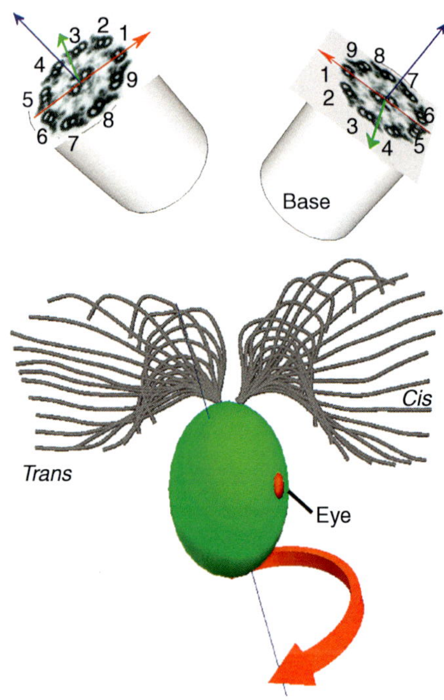

Plate 11 **(Figure 5 on page 177 of this volume)**

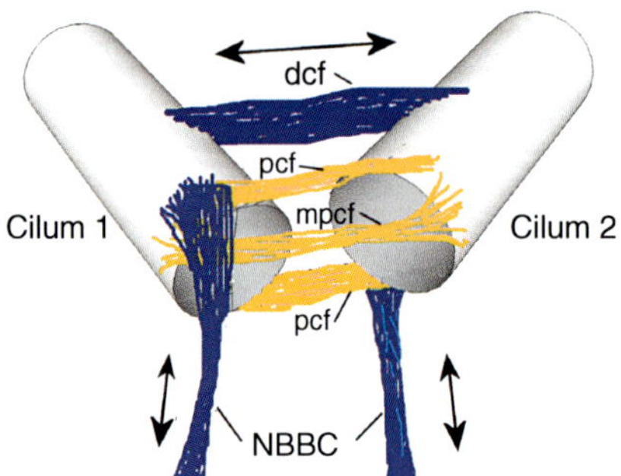

Plate 12 **(Figure 6 on page 178 of this volume)**

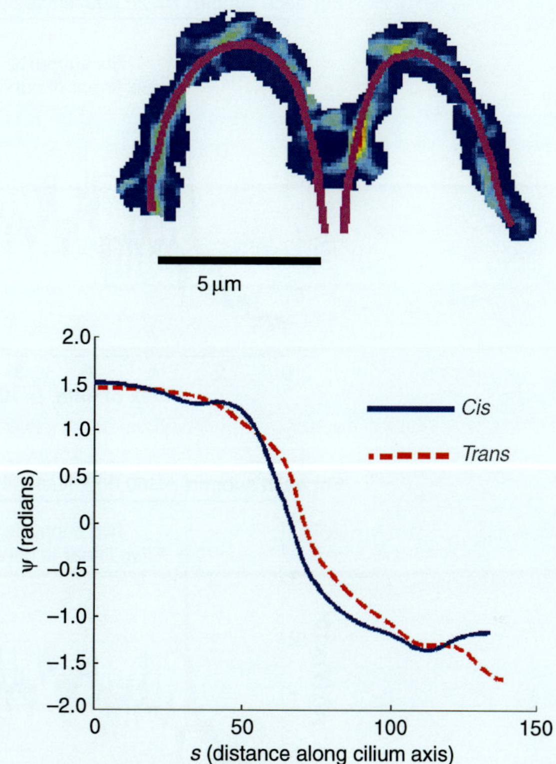

Plate 13 **(Figure 11 on page 193 of this volume)**

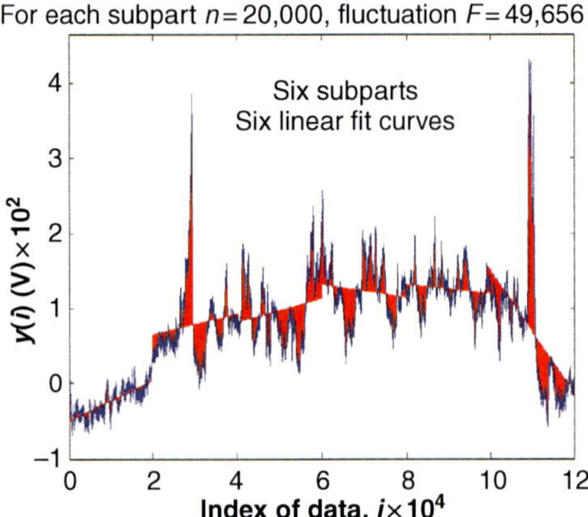

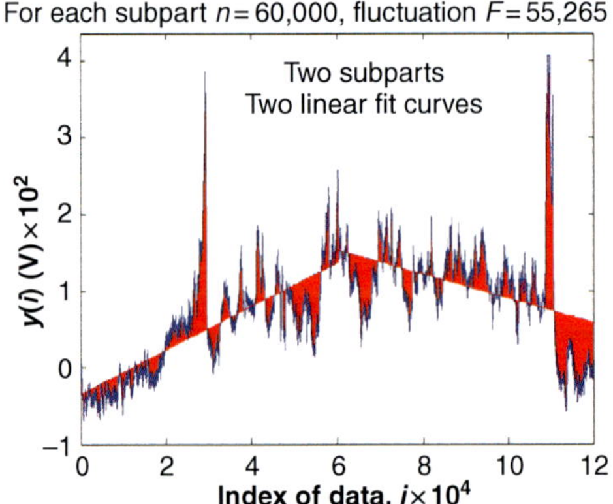

Plate 14 **(Figure 16 on page 202 of this volume)**

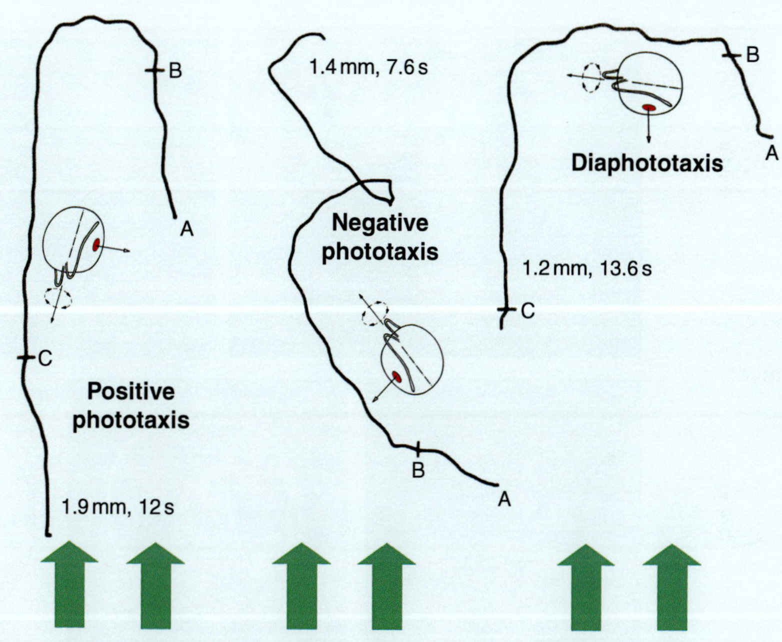

Plate 15 (Figure 22 on page 213 of this volume)

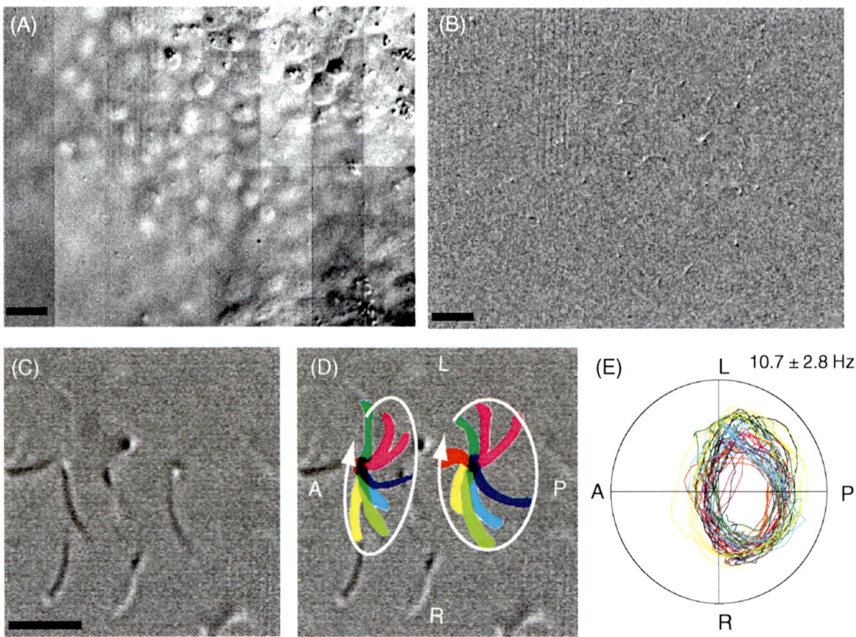

Plate 16 **(Figure 4 on page 275 of this volume)**

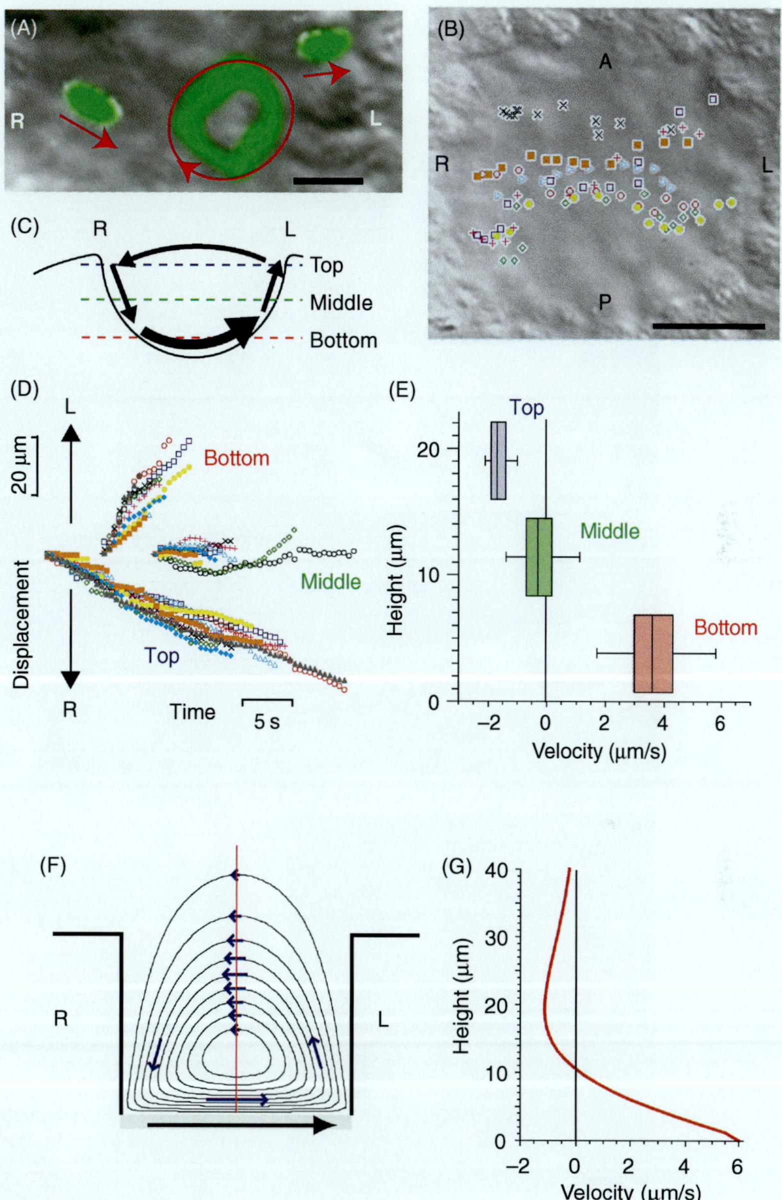

Plate 17 (Figure 5 on page 277 of this volume)

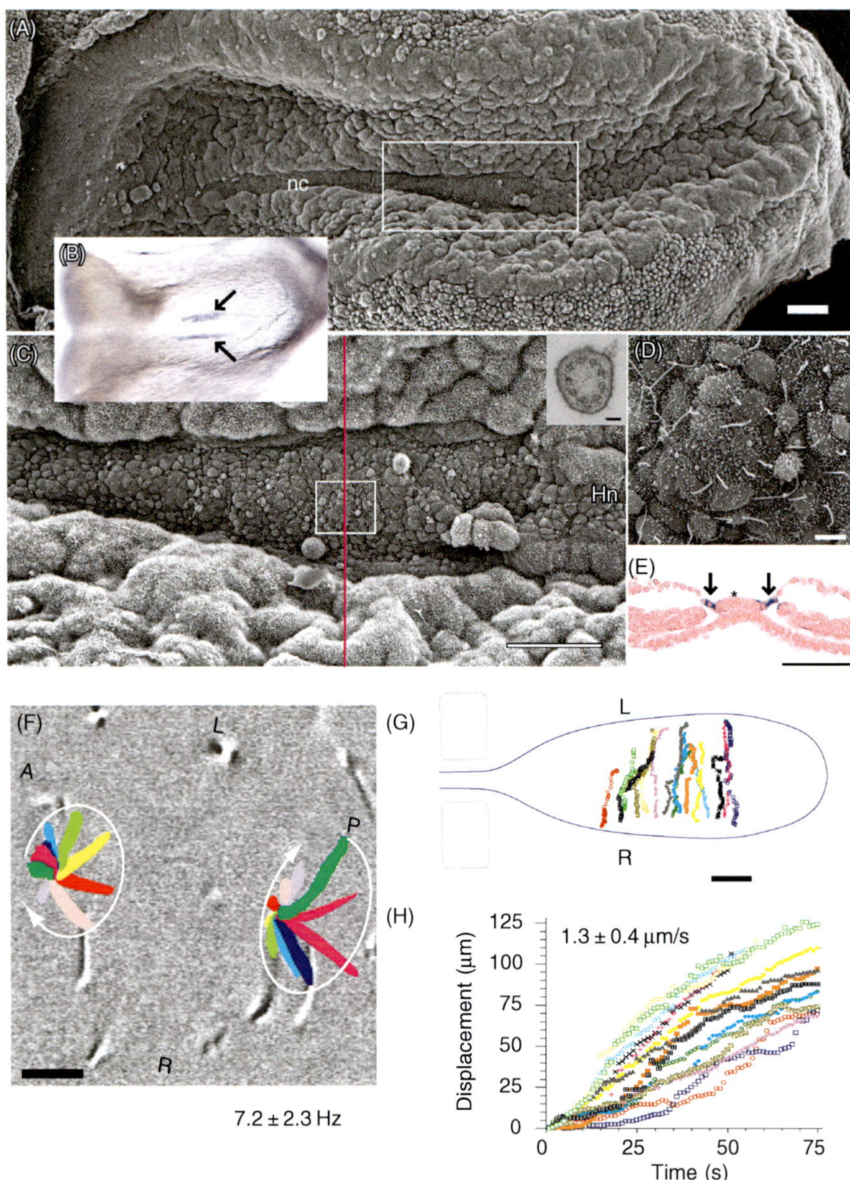

Plate 18 **(Figure 6 on page 279 of this volume)**

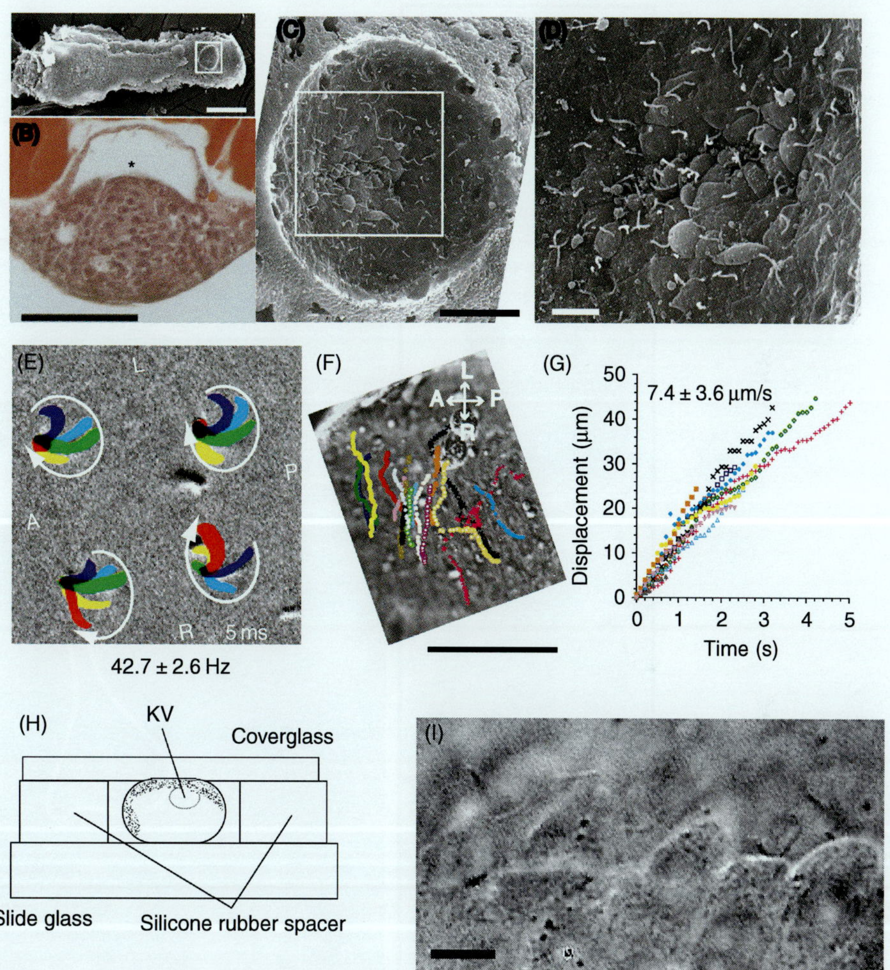

Plate 19 (Figure 8 on page 282 of this volume)

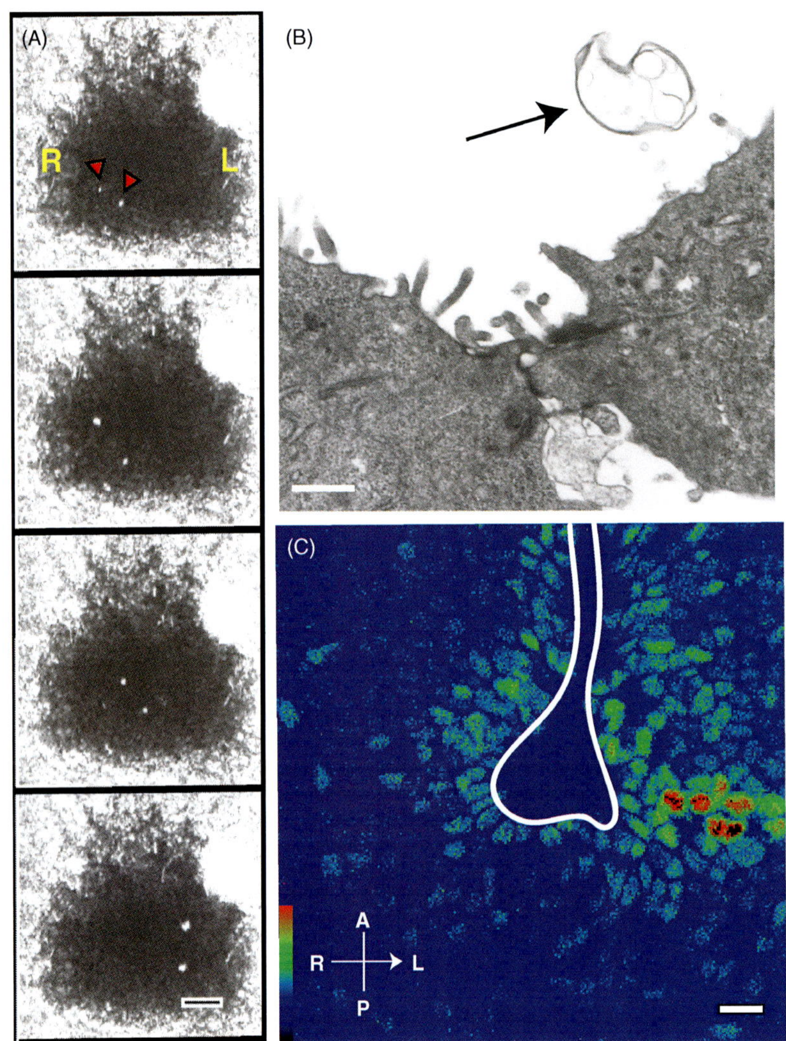

Plate 20 **(Figure 9 on page 283 of this volume)**

Edwards Brothers Malloy
Thorofare, NJ USA
January 24, 2015